Lecture Notes in Computer Science 12592

More information about this subseries at http://www.springer.com/series/7412

Esther Puyol Anton · Mihaela Pop ·
Maxime Sermesant · Victor Campello ·
Alain Lalande · Karim Lekadir ·
Avan Suinesiaputra · Oscar Camara ·
Alistair Young (Eds.)

Statistical Atlases and Computational Models of the Heart

M&Ms and EMIDEC Challenges

11th International Workshop, STACOM 2020
Held in Conjunction with MICCAI 2020
Lima, Peru, October 4, 2020
Revised Selected Papers

 Springer

Editors
Esther Puyol Anton
King's College
London, UK

Mihaela Pop
University of Toronto
Toronto, ON, Canada

Maxime Sermesant
Inria
Sophia Antipolis, France

Victor Campello
Universitat de Barcelona
Barcelona, Spain

Alain Lalande
Université de Bourgogne
Dijon, France

Karim Lekadir 🄳
Universitat de Barcelona
Barcelona, Spain

Avan Suinesiaputra
University of Leeds
Leeds, UK

Oscar Camara
Universitat Pompeu Fabra
Barcelona, Spain

Alistair Young
King's College
London, UK

ISSN 0302-9743   ISSN 1611-3349   (electronic)
Lecture Notes in Computer Science
ISBN 978-3-030-68106-7   ISBN 978-3-030-68107-4   (eBook)
https://doi.org/10.1007/978-3-030-68107-4

LNCS Sublibrary: SL6 – Image Processing, Computer Vision, Pattern Recognition, and Graphics

This Springer imprint is published by the registered company Springer Nature Switzerland AG
The registered company address is: Gewerbestrasse 11, 6330 Cham, Switzerland

Preface

Integrative models of cardiac function are important for understanding disease, evaluating treatment and planning intervention. In recent years, there has been considerable progress in cardiac image analysis techniques, cardiac atlases and computational models, which can integrate data from large-scale databases of heart shape, function and physiology. However, significant clinical translation of these tools is constrained by the lack of complete and rigorous technical and clinical validation, as well as benchmarking of the developed tools. To rectify this, common and available ground-truth data capturing generic knowledge on the healthy and pathological heart is required. Several efforts are now established to provide Web accessible structural and functional atlases of the normal and pathological heart for clinical, research and educational purposes. We believe that these approaches will only be effectively developed through collaboration across the full research scope of the cardiac imaging and modelling communities.

The 11th edition of the *Statistical Atlases and Computational Modelling of the Heart* workshop, STACOM 2020 (http://stacom2020.cardiacatlas.org), was held in conjunction with the MICCAI 2020 international conference (held virtually), following the ten previous editions: STACOM 2010 (Beijing, China), STACOM 2011 (Toronto, Canada), STACOM 2012 (Nice, France), STACOM 2013 (Nagoya, Japan), STACOM 2014 (Boston, USA), STACOM 2015 (Munich, Germany), STACOM 2016 (Athens, Greece), STACOM 2017 (Quebec City, Canada), STACOM 2018 (Granada, Spain) and STACOM 2019 (Shenzhen, China). STACOM 2020 provided a forum to discuss the latest developments in various areas of computational imaging and modelling of the heart, application of machine learning to cardiac image analysis, electro-mechanical modelling of the heart and statistical cardiac atlases.

The topics of the 11th edition of the STACOM workshop included: cardiac imaging and image processing, machine learning applied to cardiac imaging and image analysis, atlas construction, artificial intelligence, statistical modelling of cardiac function across different patient populations, cardiac computational physiology, model customization, atlas-based functional analysis, ontological schemata for data and results, and integrated functional and structural analyses, as well as the pre-clinical and clinical applicability of these methods. Besides regular contributed papers, additional efforts of the STACOM 2020 workshop were also focused on three challenges: 1) Multi-centre, multi-vendor and multi-disease cardiac image segmentation challenge (M&Ms); 2) Automatic evaluation of myocardial infarction from delayed-enhancement cardiac MRI challenge (EMIDEC); and 3) Myocardial pathology segmentation combining multi-sequence CMR challenge (MyopPS).

Along with regular papers, selected papers from two of these challenges (i.e., M&Ms and EMIDEC) are included in the current proceedings volume. The M&Ms and EMIDEC challenges are described in detail below.

The multi-centre, multi-vendor, multi-disease segmentation challenge (M&Ms) was proposed to promote the development of generalizable deep learning models that can maintain good segmentation accuracy on unseen vendors or imaging protocols in the cardiac MRI domain. A total of 14 participants submitted various solutions by implementing and combining methodologies such as data augmentation, domain adaptation and generative models. The initial results show the promise of the proposed solutions but also the need for more research to enhance future generalizability of deep learning in cardiac image segmentation. The M&Ms benchmarking dataset, totaling 375 cardiac MRI studies from four different scanners and six clinical centers, has now been made open-access on the challenge website to enable further research and benchmarking on multi-scanner and multi-centre cardiac image analysis.

More information about the M&Ms challenge can be found here:
https://www.ub.edu/mnms/

Automatic evaluation of myocardial infarction from delayed-enhancement cardiac MRI challenge (EMIDEC). MRI, in short-axis orientation, performed several minutes after the injection of a contrast agent (delayed enhancement MRI or DE-MRI), is a method of choice to evaluate the extent of a myocardial infarction. For the EMIDEC challenge, the available database consists of 150 exams (all from different patients). The database contains 50 cases with normal MRI after injection of a contrast agent and 100 cases with myocardial infarction (and then with a hyperenhanced area on DE-MRI). 100 cases were randomly chosen to perform the training and 50 to perform the testing. Along with MRI, clinical characteristics are provided to distinguish normal and pathological cases. The two main objectives of the EMIDEC challenge are first to classify normal and pathological cases from the clinical information with or without DE-MRI, and second to automatically detect the different relevant areas (the myocardial contours, the infarcted area and the permanent microvascular obstruction area (no-reflow area)) from a series of short-axis DE-MRI covering the left ventricle. Thirteen teams took part in the challenge: two of them entered both contests, 7 teams only entered the segmentation contest and the 4 remaining ones concentrated on the classification contest. Jun Ma won the classification contest with an accuracy of 92% (the best result was from Shi et al. with an accuracy of 94%, but they have not been ranked because their co-authors come from the challenge organization team). Yichi Zhang won the segmentation contest, with a global Dice index of 0.88 for the myocardium, 0.71 for the myocardial infarction and 0.79 for the no-reflow area.

More information about the EMIDEC challenge can be found here:
http://emidec.com/

In total, 43 regular and challenge papers were revised, selected and invited for publication in this LNCS Proceedings volume.

We hope that the results obtained from the challenges along with the regular paper contributions will act to accelerate progress in the important areas of heart function and structure analysis. We would like to thank all organizers, reviewers, authors and sponsors for their time, efforts, contributions and support in making STACOM 2020 a successful event.

October 2020

Esther Puyol Anton
Mihaela Pop
Maxime Sermesant
Victor Campello
Alain Lalande
Karim Lekadir
Avan Suinesiaputra
Oscar Camara
Alistair Young

Organization

Chairs and Organizers

Chairs

Mihaela Pop
Maxime Sermesant
Esther Puyol Anton
Alain Lalande
Oscar Camara
Karim Lekadir
Victor Campello
Avan Suinesiaputra
Alistair Young

Organizers

STACOM

Esther Puyol Anton	King's College London, UK
Mihaela Pop	Sunnybrook Research Institute, University of Toronto, Canada
Maxime Sermesant	Inria, Epione group, France
Avan Suinesiaputra	University of Leeds, UK
Oscar Camara	Universitat Pompeu Fabra, Spain
Alistair Young	King's College London, UK

M&Ms Challenge

Victor Campello	University of Barcelona, Spain
Karim Lekadir	University of Barcelona, Spain
José F. Rodríguez Palomares	Vall d'Hebron Hospital, Barcelona, Spain
Andrea Guala	Vall d'Hebron Hospital, Barcelona, Spain
Lucia La Mura	Vall d'Hebron Hospital, Barcelona, Spain
Mahir Karakas	Universitätsklinikum Hamburg-Eppendorf, Germany
Ersin Çavuş	Universitätsklinikum Hamburg-Eppendorf, Germany
Matthias G. Friedrich	McGill University Health Centre, Montreal, Canada
Julie Lebel	McGill University Health Centre, Montreal, Canada
Ria Garg	McGill University Health Centre, Montreal, Canada
Filipe Henriques	McGill University Health Centre, Montreal, Canada
Martín Descalzo	Hospital de la Santa Creu i Sant Pau, Barcelona, Spain
David Viladés	Hospital de la Santa Creu i Sant Pau, Barcelona, Spain

EMIDEC Challenge

Alain Lalande	Université de Bourgogne, Dijon, France
Fabrice Meriaudeau	ImViA Laboratory, Dijon, France
Dominique Ginhac	ImViA Laboratory, Dijon, France
Abdul Qayyum	ImViA Laboratory, Dijon, France
Khawla Brahim	ImViA Laboratory, Dijon, France
Thibaut Pommier	University Hospital of Dijon, Dijon, France
Raphaël Couturier	FEMTO-ST Institute, Belfort, France
Michel Salomon	FEMTO-ST Institute, Belfort, France
Gilles Perrot	FEMTO-ST Institute, Belfort, France
Zhihao Chen	FEMTO-ST Institute, Belfort, France

Additional Reviewers

Nicolas Cedilnik
Teodora Chitiboi
Eric Lluch Alvarez
Viorel Mihalef
Matthew Ng.

OCS - Springer Conference Submission/Publication System

Mihaela Pop	Sunnybrook Research Institute, University of Toronto, Canada
Maxime Sermesant	Inria, Epione group, France

Webmaster

Avan Suinesiaputra	University of Leeds, UK

Workshop Website

stacom2020.cardiacatlas.org

Contents

Multi-centre, Multi-vendor, Multi-disease Cardiac Image Segmentation Challenge (M&Ms)

**Automatic Evaluation of Myocardial Infarction from
Delayed-Enhancement Cardiac MRI Challenge (EMIDEC)**

Regular Papers

A Persistent Homology-Based Topological Loss Function for Multi-class CNN Segmentation of Cardiac MRI

Nick Byrne[1,2]([✉]), James R. Clough[2], Giovanni Montana[3],
and Andrew P. King[2]

[1] Medical Physics, Guy's and St. Thomas' NHS Foundation Trust, London, UK
[2] School of Biomedical Engineering and Imaging Sciences, King's College London,
London, UK
nicholas.byrne@kcl.ac.uk
[3] Warwick Manufacturing Group, University of Warwick, Coventry, UK

Abstract. With respect to spatial overlap, CNN-based segmentation of
short axis cardiovascular magnetic resonance (CMR) images has achieved
a level of performance consistent with inter observer variation. However, conventional training procedures frequently depend on pixel-wise
loss functions, limiting optimisation with respect to extended or global
features. As a result, inferred segmentations can lack spatial coherence, including spurious connected components or holes. Such results
are implausible, violating the anticipated topology of image segments,
which is frequently known a priori. Addressing this challenge, published
work has employed persistent homology, constructing topological loss
functions for the evaluation of image segments against an explicit prior.
Building a richer description of segmentation topology by considering
all possible labels and label pairs, we extend these losses to the task of
multi-class segmentation. These topological priors allow us to resolve all
topological errors in a subset of 150 examples from the ACDC short axis
CMR training data set, without sacrificing overlap performance.

Keywords: Image segmentation · CNN · Topology · MRI

1 Introduction

Medical image segmentation is a prerequisite of many pipelines dedicated to the
analysis of clinical data. In cardiac magnetic resonance (CMR) imaging, segmentation of short axis cine images into myocardial, and left and right ventricular

N. Byrne—Funded by a National Institute for Health Research (NIHR), Doctoral
Research Fellowship for this research project. This report presents independent research
funded by the NIHR. The views expressed are those of the author(s) and not necessarily those of the NHS, the NIHR or the Department of Health and Social Care. The
authors have no conflicts of interest to disclose.
G. Montana and A.P. King—Joint last authors..

E. Puyol Anton et al. (Eds.): STACOM 2020, LNCS 12592, pp. 3–13, 2021.
https://doi.org/10.1007/978-3-030-68107-4_1

components permits quantitative assessment of important clinical indices derived from ventricular volumes [17]. Motivated by the burden of manual segmentation, researchers have sought automated solutions to this task.

To this end, deep learning, and in particular convolutional neural networks (CNNs), have brought significant progress [4]. One key to their success has been the design of specialised architectures dedicated to image segmentation. Theoretically, these permit the learning of image features related to extended spatial context, such as anatomical morphology.

Whilst considerable effort has gone into investigating methods for the extraction of multi-scale image features, less attention has been paid to their role in network optimisation [7]. Instead, CNNs have often been trained using pixel-wise loss functions based on cross-entropy (CE) or the Dice Similarity Coefficient (DSC). Though having favourable numerical properties, these are insensitive to higher order image features such as topology. In the absence of this information, CNN optimisation can result in predicted segmentations with unrealistic properties such as spurious connected components or holes [15]. These errors can appear nonsensical, violating the most basic features of image segments. If small, the presence of such errors may not preclude a high spatial overlap with the ground truth and have little consequence for certain clinical indices. However, a wider array of downstream applications such as biophysical modelling or 3D printing, demand a high fidelity representation of such features.

In short axis CMR, prior knowledge dictates that the right ventricular cavity is bound to the left ventricular myocardium, which in turn fully surrounds the left ventricular blood pool. In contrast to pixel-wise objectives, this is a global description of segmentation coherence. However, whilst these constraints are simple to express qualitatively, the opaque nature of CNNs has made it difficult to explicitly exploit such prior information in model optimisation.

1.1   Related Work

At least in the context of large, homogeneous training datasets, CNN-based short axis segmentation has achieved a level of performance consistent with inter observer variation [4]. However, in studies of cardiovascular disease (for which datasets typically contain fewer subjects who exhibit greater morphological variation), a deficit remains. This gap is in part characterised by the anatomically implausible errors described.

To address these modes of failure, previous works have sought to leverage prior information. The combination of deep learning with atlas-based segmentation [7] and active contour refinement [1] have both been investigated. Whilst these extensions improve performance, their capacity to represent pathological variation in image features is limited. Accordingly, others have injected prior information directly into CNN optimisation, developing a supervisory signal from a learned, latent distribution of plausible anatomy [13]. Their implicit embedding, however, hinders an understanding of the extent to which such priors are related to morphology or topology as claimed. Bridging this gap, Painchaud et al.

augmented the latent space via a rejection sampling procedure, maintaining only those cases satisfying sixteen criteria related to anatomical plausibility [15].

Eleven of Painchaud et al.'s criteria concern anticipated anatomical topology. Structured losses have previously been designed to capture aspects of segmentation topology including hierarchical class containment [2] and adjacency [9,11]. More recently, CNN-based segmentation has benefited from the global, exhaustive and robust description of topology provided by persistent homology (PH). PH admits the construction of topological loss functions which, in contrast to those built on a latent representation of anatomical shape, allow evaluation against an explicit topological prior. Applications have included segmenting the tree-like topology of the murine neurovasculature [10]; and the toroidal topology of the myocardium in short axis CMR [5,6].

1.2 Contributions

To the best of our knowledge, no PH-based loss function has been proposed for the task of multi-class segmentation. Compared with the binary case, extension to this setting considers a richer set of topological priors, including hierarchical class containment and adjacency. short axis CMR segmentation is a useful test bed for this task, not only for its clinical significance, but also for its economic representation of such priors in a well-defined, four-class problem. In this context, our contributions are as follows:

1. We propose a novel topological loss function, based on PH, for the task of multi-class image segmentation.
2. We use the novel loss function for CNN-based segmentation of short axis CMR images from the ACDC dataset.
3. We demonstrate significant improvement in segmentation topology without degradation in spatial overlap performance.

2   Materials and Methods

We address a multi-class segmentation task, seeking a meaningful division of the 2D short axis CMR image $\mathbf{X} : \mathbb{R} \times \mathbb{R} \to \mathbb{R}$ into background, right and left ventricular cavities and left ventricular myocardium[1]. We denote the ground truth image segmentation by $\mathbf{Y} : \mathbb{R} \times \mathbb{R} \to \{0,1\}^4$, being made up by four mutually exclusive class label maps: Y_{bg}, Y_{rv}, Y_{my} and Y_{lv}. We consider a deep learning solution, optimising the parameters, θ, of a CNN to infer the probabilistic segmentation, $\hat{\mathbf{Y}} : \mathbb{R} \times \mathbb{R} \to [0,1]^4$, a distribution over the per class segmentation maps: $\hat{Y}_{bg}$, $\hat{Y}_{rv}$, $\hat{Y}_{my}$ and $\hat{Y}_{lv}$. We write segmentation inference as $\hat{\mathbf{Y}} = f(\mathbf{X}; \theta)$. Given the success of CNN-based solutions, we assume that, at least with respect to spatial overlap, $\hat{\mathbf{Y}}$ is a reasonable estimate of $\mathbf{Y}$. In this setting we describe our CNN post-processing framework for the correction of inferred segmentation topology.

---

[1] The semantic classes of this task match those of the ACDC image segmentation challenge [3]. Whilst neither considers the right ventricular myocardium, this could easily be incorporated into the framework set out in Sects. 2.1 and 2.2.

2.1   Multi-class Topological Priors

In 2D, objects with differing topology can be distinguished by the first two Betti numbers: $\mathbf{b} = (b_0, b_1)$. Intuitively, b_0 counts the number of connected components which make up an object, and b_1, the number of holes contained [14]. The Betti numbers are topological invariants permitting the specification of priors for the description of foreground image segments. Consider our short axis example:

$$\mathbf{b}^{rv} = (1,0) \qquad\qquad \mathbf{b}^{rv \cup my} = (1,1)$$
$$\mathbf{b}^{my} = (1,1) \qquad (1) \qquad \mathbf{b}^{rv \cup lv} = (2,0) \qquad (2)$$
$$\mathbf{b}^{lv} = (1,0) \qquad\qquad \mathbf{b}^{my \cup lv} = (1,0)$$

Equation set (1) specifies that each of the right ventricle, myocardium and left ventricle should comprise a single connected component, and that the myocardium should contain a single hole. However, these equations only provide a topological specification in a segment-wise, binary fashion: they fail to capture inter-class topological relationships between cardiovascular anatomy. For instance, they make no specification that the myocardium surround the left ventricular cavity or that the right ventricle and myocardium should be adjacent.

By the inclusion-exclusion principle, the topology of a multi-class image segmentation is characterised by that of all foreground objects and all possible object pairs: see Equation set (2). For convenience, we collect Equation sets (1) and (2) into a 3D Betti array $\mathbf{B} : \{1,2,3\} \times \{1,2,3\} \times \{0,1\} \to \mathbb{R}$. Each element B_d^{ij}[2] denotes the Betti number of dimension d for the ground truth segmentation $Y_{i \cup j}$[3]. Vitally, even in the absence of the ground truth, $\mathbf{B}$ can be determined by prior knowledge of the anatomy to be segmented.

2.2   Topological Loss Function

To expose topological features we apply PH (see [14] for a theoretical background). For a practical understanding of the topological loss described, the results of PH analyses are most easily appreciated by inspection of persistence barcodes. The PH barcode summarises the topological features present within data. However, rather than providing a singular topological description, the barcode returns a dynamic characterisation of the way that the topology evolves as a function of some scale parameter. More concretely, and in our context a barcode reflects the topology of a probabilistic segmentation $\hat{S} : \mathbb{R} \times \mathbb{R} \to [0,1]$,

---

[2] In B_d^{ij} we divide indices between sub and super scripts to make clear the difference between class labels (i, j) and topological dimension (d), without further significance.

[3] We use the union operator ($\cup$) to combine individual classes of a multi-class segmentation. When applied to a binary segmentation, $Y_{i \cup j}$ is the pixel-wise Boolean union of classes i and j. When applied to a probabilistic segmentation, $\hat{Y}_{i \cup j}$ is the pixel-wise probability of class i or j. We consider the union of a class with itself to be the segmentation of the single class: $Y_{i \cup j = i} = Y_i$ and $\hat{Y}_{i \cup j = i} = \hat{Y}_i$.

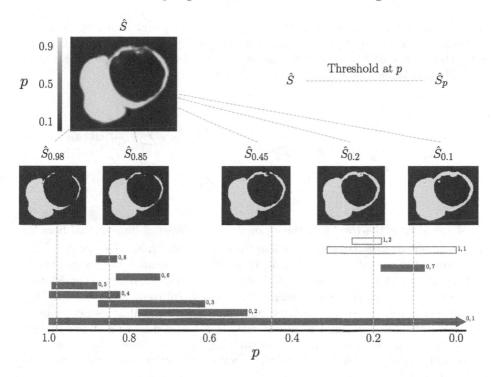

Fig. 1. Construction of the PH barcode. The barcode reflects the topological features of the probabilistic segmentation $\hat{S}$, when binarised at all possible probability thresholds in the interval $[0, 1]$. At a particular p: the number of vertically intersected solid bars counts connected components ($d = 0$); open bars count the number of loops ($d = 1$). Additionally, each bar is labelled with its topological dimension, and its persistence ranking in order of descending lifetime: d, l.

binarised at all possible probability thresholds in the range $[0, 1]$. As the threshold, p, reduces, the barcode diagram tracks the evolving topology of the binarised segmentation $\hat{S}_p$.

Critical values of p admit changes in the topological features of $\hat{S}_p$. In Fig. 1, such values are indicated by the endpoints of each bar. Accordingly, the persistence of a topological feature Δp is the length of its associated bar. Moreover, the presentation of each bar indicates the topological dimension of the feature shown: solid bars are connected components; open bars are loops. Persistent bars are considered robust to small perturbations, suggesting that they are true topological features of the data. Hence, in Figs. 1 and 2, we consider barcodes in order of descending lifetime after grouping by topological dimension. From the persistence barcode of the probabilistic segmentation $\hat{S}$, we write the lifetime of the lth most persistent feature of dimension d as $\Delta p_{d,l}(\hat{S})$.

Importantly, topological persistence can be determined in a fashion that is differentiable and consistent with gradient-based learning [8]. This permits the construction of topological loss functions, exposing the differences between $\hat{\mathbf{Y}}$

and our prior specification $\mathbf{B}$. Key to our formulation is the choice of the probabilistic segmentation S, from which we extract topological features. To align with the theory of Sect. 2.1, we consider the persistence barcode for all foreground class labels and class label pairs (see Fig. 2):

$$L_{topo} = \sum_{d,i,j \geq i} B_d^{ij} - A_d^{ij} + Z_d^{ij} \qquad \begin{aligned} A_d^{ij} &= \sum_{l=1}^{B_d^{ij}} \Delta p_{d,l}(\hat{Y}_{i \cup j}) \\ Z_d^{ij} &= \sum_{l=B_d^{ij}+1}^{\infty} \Delta p_{d,l}(\hat{Y}_{i \cup j}) \end{aligned} \qquad (3)$$

A_d^{ij} evaluates the total persistence of the B_d^{ij} longest, d-dimensional bars for the probabilistic union of segmentations for classes i and j, $\hat{Y}_{i \cup j}$ (see Footnote 2). Assuming that the inferred segmentation closely approximates the ground truth, and recalling that l ranks topological features in descending order of persistence, A_d^{ij} measures the presence of anatomically meaningful topological features. Z_d^{ij} evaluates the persistence of spurious topological features that are superfluous to B_d^{ij}. Summing over all topological dimensions d, and considering all class labels $i, j = i$ and class label pairs $i, j > i$, optimising L_{topo} maximises the persistence of topological features which match the prior specification, and minimises those which do not.

As in the single class formulation presented in [5], L_{topo} is used to guide test time adaptation of the weights of a pre-trained CNN $f(\mathbf{X}; \theta)$, seeking an improvement in inferred segmentation topology. A new set of network parameters θ_n are learned for the individual test case $\mathbf{X}_n$. However, since topology is a global property, there are many segmentations that potentially minimise L_{topo}. Hence, where V_n is the number of pixels in $\mathbf{X}_n$, a similarity constraint limits test time adaptation to the minimal set of modifications necessary to align the segmentation and the topological prior, $\mathbf{B}$:

$$L_{TP} = L_{topo}(f(\mathbf{X}_n; \theta_n), \mathbf{B}) + \frac{\lambda}{V_n}|f(\mathbf{X}_n; \theta) - f(\mathbf{X}_n; \theta_n)|^2 \qquad (4)$$

2.3   Implementation

We apply our loss to a topologically consistent subset of the ACDC [3] training data. Ignoring irregular anatomical appearances at apex and base, we extract three mid ventricular slices from each short axis stack, including end diastolic and systolic frames. We achieve a dataset of 600 examples from 100 subjects. As per the winning submission to the ACDC Challenge, all image - label pairs were resampled to an isotropic pixel spacing of 1.25 mm (less than the mean and median spatial resolution of the training data) and normalised to have zero mean and unit variance [12]. Subjects were randomly divided between training,

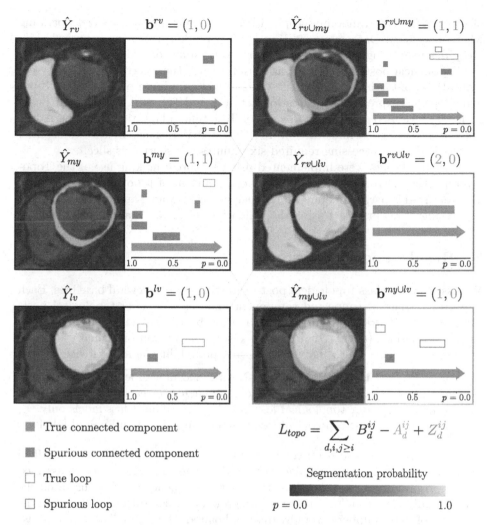

Fig. 2. Construction of the loss L_{topo}. Each probabilistic segmentation (Y_i or $Y_{i \cup j}$), is accompanied by its associated persistence barcode (for clarity, only features with a lifetime $\Delta p_{d,l} \geq 0.05$ are shown). L_{topo} weighs the persistence of topological features which match the topological description (A_d^{ij}; depicted as green bars), against those which do not (Z_d^{ij}; depicted as red bars). To sensitise L_{topo} to multi-class label map topology, the result is repeated for, and summed over all topological dimensions (d), and individual and paired label sets ($i, j \geq i$).

validation and test sets in the ratio 2:1:1, stratified by diagnostic group according to ACDC classification.

A U-Net model [16] was trained using CE loss and the combined training and validation set of 450 examples, for 16,000 iterations. We used the Adam optimiser with a learning rate of 10^{-3}. Each minibatch contained ten patches of size 352 by 352, randomly cropped from ten different patients, zero padding where necessary. Data augmentation applied random rotations between $\pm 15°$. Graphics processing unit (GPU)-accelerated training took nine hours.

Topological post-processing was performed on the inferred multi-class segmentations of the held-out test set. This sought to minimise L_{TP} for the topological priors expressed by Equation sets (1) and (2), and summarised in **B**. In Eq. (4) we used a value of $\lambda = 1000$. Test time adaptation used the Adam optimiser with a learning rate of 10^{-5} for 100 iterations. In the worst case, topological post-processing required six minutes per short axis slice.

All experiments were implemented using PyTorch: making use of the Topolayer package introduced in [8] for the computation of topological persistence. The reported hyperparameters for both supervised training and topological post-processing were optimised using the validation set of 150 examples.

3 Results and Discussion

We assess multi-class topological post-processing against several baselines. Each applies a variant of connected component analysis (CCA) or topological post-processing (TP) to the segmentation inferred by the U-Net trained in a fully supervised fashion ($UNet$). In all cases, a discrete segmentation is finally achieved by the set of labels which maximise inferred probability on a pixel-wise basis:

$UNet$ the output of the U-Net trained in a supervised manner.
$UNet + CCA$ the largest connected components for each foreground label.
$UNet + TP_{i,j=i}$ our topological loss based on individual class labels only.
$UNet + TP_{i,j\geq i}$ our topological loss based on individual and paired labels.

Table 1 presents results on the held-out test set. Spatial overlap is quantified by DSC, averaged across cardiac phases. Topological performance is assessed as the proportion of cases in which the discrete segmentation demonstrated the correct multi-class topology: inferred and ground truth segmentations shared the same set of Betti numbers for individual and paired labels. Finally, we summarise the effect of any post-processing on spatial overlap by the change in DSC.

Table 1 confirms that pixel-wise metrics of spatial overlap do not reliably predict topological performance. Trained by CE, the supervised model infers a segmentation with incorrect topology in almost 15% of cases. Whilst spurious connected components alone account for approximately 50% of U-Net errors, higher dimensional topological errors are not insignificant. As commonly employed, CCA is insensitive to such features, resolving spurious components by their discrete removal. Our method takes a probabilistic approach to the treatment of priors, modifying inferred segmentations by CNN parameter update. This permits expressive topological refinement, as illustrated in Fig. 3.

Optimisation with respect to topological priors for individual labels resolves half of high dimensional errors. This approach, $+TP_{i,j=i}$, reflects the naive extension of the binary segmentation method outlined in [5]. However, by exact

Table 1. Segmentation results on held-out test set. Spatial overlap: Dice Similarity Coefficient (DSC) per class; the average over classes (μ); and the change induced by post-processing ($\Delta\mu$). Topological accuracy: T is the proportion of test images with the correct multi-class topology. σ is the standard deviation.

	$DSC_{(\sigma)}$					$T_{(\sigma)}$
	rv	my	lv	μ	$\Delta\mu$	
$UNet$	$0.891_{(0.115)}$	$0.885_{(0.050)}$	$0.954_{(0.039)}$	$0.910_{(0.045)}$	—	$0.853_{(0.032)}$
$+CCA$	$0.892_{(0.113)}$	$0.886_{(0.049)}$	$0.954_{(0.039)}$	$0.911_{(0.044)}$	$0.001_{(0.004)}$	$0.927_{(0.024)}$
$+TP_{i,j=i}$	$0.889_{(0.130)}$	$0.887_{(0.048)}$	$0.954_{(0.039)}$	$0.910_{(0.049)}$	$0.000_{(0.012)}$	$0.980_{(0.013)}$
$+TP_{i,j\geq i}$	$0.889_{(0.129)}$	$0.888_{(0.048)}$	$0.954_{(0.039)}$	$0.910_{(0.049)}$	$0.000_{(0.011)}$	$1.000_{(0.000)}$

binomial test, and after Bonferroni correction, there was no statistically significant difference (95% confidence) in the proportion of topological errors between $+TP_{i,j=i}$ and $+CCA$ ($p = 0.025 \times 6$). The best performing scheme is our proposed model, $+TP_{i,j\geq i}$, which considers priors for all individual and paired labels. The incremental benefit of $+TP_{i,j\geq i}$ is shown in Fig. 4. Significantly, and without degradation in DSC, our approach resolves all topological errors which remain after CCA ($p = 0.001 \times 6$).

Compared with losses based on a learned latent space of plausible anatomical shapes [13, 15], PH-based loss functions allow optimisation with respect to an explicit topological prior. This is beneficial in terms of interpretability and in low data settings. However, PH-based losses also allow topological prior information to be decoupled from its appearance in training data, on which a learned distribution is necessarily biased. We speculate that additional biases may degrade performance when used to refine the segmentation of out-of-sample test data. Favourably, PH-based priors permit even-handed topological post-processing in the presence of pathology-induced structural variation. At the same time, we recognise the potential complementarity of these approaches if explicit topological specification could be enhanced with learned shape priors.

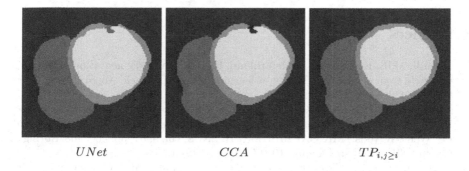

UNet   CCA $TP_{i,j\geq i}$

Fig. 3. Topological post-processing enables expressive correction of U-Net errors.

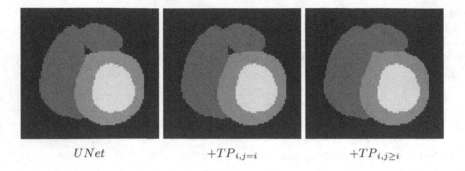

$$UNet \qquad\qquad +TP_{i,j=i} \qquad\qquad +TP_{i,j\geq i}$$

Fig. 4. Multi-class topological priors capture a rich topological description.

More generally, the PH losses described are limited by their reliance on an expert-provided prior. Demonstrating our approach on mid-ventricular slices permitted the consistent application of Equation sets (1) and (2). Allowing for a slice-wise specification, the same approach could equally be applied to apical and basal slices, including associated topological changes. At present this would necessitate operator intervention. However, we also observe that this requirement is an artefact of CMR short axis acquisition. Given 3D cine data, we could specify priors which truly reflect the topology of anatomy rather than its appearance within a 2D acquisition.

4   Conclusion

We have extended PH-based losses to the task of multi-class, CNN-based image segmentation. Leveraging an enriched topological prior, including high dimensional and multi-class features, this approach improved segmentation topology in a post-processing framework. Future work will seek to understand its limits with respect to the fidelity of CNN-based segmentation; its application within weakly supervised learning; and its extension to more challenging targets.

References

1. Avendi, M.R., Kheradvar, A., Jafarkhani, H.: Automatic segmentation of the right ventricle from cardiac MRI using a learning-based approach. Magn. Reson. Med. **78**(6), 2439–2448 (2017)
2. BenTaieb, A., Hamarneh, G.: Topology aware fully convolutional networks for histology gland segmentation. In: Ourselin, S., Joskowicz, L., Sabuncu, M.R., Unal, G., Wells, W. (eds.) MICCAI 2016, Part II. LNCS, vol. 9901, pp. 460–468. Springer, Cham (2016). https://doi.org/10.1007/978-3-319-46723-8_53
3. Bernard, O., et al.: Deep learning techniques for automatic MRI cardiac multi-structures segmentation and diagnosis: is the problem solved? IEEE Trans. Med. Imaging **37**(11), 2514–2525 (2018)

4. Chen, C., et al.: Deep learning for cardiac image segmentation: a review. Front. Cardiovasc. Med. **7**, 25 (2020)
5. Clough, J.R., Byrne, N., Oksuz, I., Zimmer, V.A., Schnabel, J.A., King, A.P.: A topological loss function for deep-learning based image segmentation using persistent homology. IEEE Trans. Pattern Anal. Mach. Intell., 1 (2020). https://doi.org/10.1109/TPAMI.2020.3013679
6. Clough, J.R., Oksuz, I., Byrne, N., Schnabel, J.A., King, A.P.: Explicit topological priors for deep-learning based image segmentation using persistent homology. In: Chung, A.C.S., Gee, J.C., Yushkevich, P.A., Bao, S. (eds.) IPMI 2019. LNCS, vol. 11492, pp. 16–28. Springer, Cham (2019). https://doi.org/10.1007/978-3-030-20351-1_2
7. Duan, J., et al.: Automatic 3D bi-ventricular segmentation of cardiac images by a shape-refined multi- task deep learning approach. IEEE Trans. Med. Imaging **38**(9), 2151–2164 (2019)
8. Gabrielsson, R.B., Nelson, B.J., Dwaraknath, A., Skraba, P.: A topology layer for machine learning. In: International Conference on Artificial Intelligence and Statistics, pp. 1553–1563 (2020)
9. Ganaye, P.-A., Sdika, M., Benoit-Cattin, H.: Semi-supervised learning for segmentation under semantic constraint. In: Frangi, A.F., Schnabel, J.A., Davatzikos, C., Alberola-López, C., Fichtinger, G. (eds.) MICCAI 2018, Part III. LNCS, vol. 11072, pp. 595–602. Springer, Cham (2018). https://doi.org/10.1007/978-3-030-00931-1_68
10. Haft-Javaherian, M., Villiger, M., Schaffer, C.B., Nishimura, N., Golland, P., Bouma, B.E.: A topological encoding convolutional neural network for segmentation of 3D multiphoton images of brain vasculature using persistent homology. In: Proceedings of IEEE/CVF Conference on Computer Vision and Pattern Recognition Workshops, pp. 990–991 (2020)
11. He, Y., et al.: Fully convolutional boundary regression for retina OCT segmentation. In: Shen, D., et al. (eds.) MICCAI 2019, Part I. LNCS, vol. 11764, pp. 120–128. Springer, Cham (2019). https://doi.org/10.1007/978-3-030-32239-7_14
12. Isensee, F., Jaeger, P.F., Full, P.M., Wolf, I., Engelhardt, S., Maier-Hein, K.H.: Automatic cardiac disease assessment on cine-MRI via time-series segmentation and domain specific features. In: Pop, M., et al. (eds.) STACOM 2017. LNCS, vol. 10663, pp. 120–129. Springer, Cham (2018). https://doi.org/10.1007/978-3-319-75541-0_13
13. Oktay, O., et al.: Anatomically constrained neural networks (ACNNs): application to cardiac image enhancement and segmentation. IEEE Trans. Med. Imaging **37**(2), 384–395 (2018)
14. Otter, N., Porter, M.A., Tillmann, U., Grindrod, P., Harrington, H.A.: A roadmap for the computation of persistent homology. EPJ Data Sci. **6**(1), 1–38 (2017). https://doi.org/10.1140/epjds/s13688-017-0109-5
15. Painchaud, N., Skandarani, Y., Judge, T., Bernard, O., Lalande, A., Jodoin, P.M.: Cardiac segmentation with strong anatomical guarantees. IEEE Trans. Med. Imaging **39**, 3703–3713 (2020)
16. Ronneberger, O., Fischer, P., Brox, T.: U-net: convolutional networks for biomedical image segmentation. In: Navab, N., Hornegger, J., Wells, W.M., Frangi, A.F. (eds.) MICCAI 2015, Part III. LNCS, vol. 9351, pp. 234–241. Springer, Cham (2015). https://doi.org/10.1007/978-3-319-24574-4_28
17. Ruijsink, B., et al.: Fully automated, quality-controlled cardiac analysis from CMR. JACC Cardiovasc, Imaging (2019)

Automatic Multiplanar CT Reformatting from Trans-Axial into Left Ventricle Short-Axis View

Marta Nuñez-Garcia[1,2](✉), Nicolas Cedilnik[3], Shuman Jia[3],
Maxime Sermesant[3], and Hubert Cochet[1,4]

[1] Electrophysiology and Heart Modeling Institute (IHU LIRYC), Pessac, France
marta.nunez-garcia@ihu-liryc.fr
[2] Université de Bordeaux, Bordeaux, France
[3] Inria, Université C'ôte d'Azur, Sophia Antipolis, France
[4] CHU Bordeaux, Department of Cardiovascular and Thoracic Imaging
and Department of Cardiac Pacing and Electrophysiology, Pessac, France

Abstract. The short-axis view defined such that a series of slices are perpendicular to the long-axis of the left ventricle (LV) is one of the most important views in cardiovascular imaging. Raw trans-axial Computed Tomography (CT) images must be often reformatted prior to diagnostic interpretation in short-axis view. The clinical importance of this reformatting requires the process to be accurate and reproducible. It is often performed after manual localization of landmarks on the image (e.g. LV apex, centre of the mitral valve, etc.) being slower and not fully reproducible as compared to automatic approaches. We propose a fast, automatic and reproducible method to reformat CT images from original trans-axial orientation to short-axis view. A deep learning based segmentation method is used to automatically segment the LV endocardium and wall, and the right ventricle epicardium. Surface meshes are then obtained from the corresponding masks and used to automatically detect the shape features needed to find the transformation that locates the cardiac chambers on their standard, mathematically defined, short-axis position. 25 datasets with available manual reformatting performed by experienced cardiac radiologists are used to show that our reformatted images are of equivalent quality.

Keywords: Automatic image reformatting · Short-axis view · Deep learning segmentation · Cardiac imaging

1 Introduction

Multiplanar reformatting refers to the process of converting imaging data acquired in a certain plane into another plane. Standard imaging planes are useful to determine normal anatomy and function and to investigate anatomic variants or pathologies. The basic cardiac imaging planes include planes oriented

E. Puyol Anton et al. (Eds.): STACOM 2020, LNCS 12592, pp. 14–22, 2021.
https://doi.org/10.1007/978-3-030-68107-4_2

with respect to the heart (e.g. horizontal and vertical long-axis and short-axis (SAX)), and planes oriented with respect to the major axes of the body (e.g. trans-axial (TA), sagittal, and coronal). Cardiac pathologies are commonly evaluated using cardiac-oriented planes and therefore the position of these planes should be prescribed very accurately. In Computational Tomography (CT) imaging, raw TA images are typically reformatted into SAX view (often the interpretation plane) after manual landmark placement [10]. This manual interaction is operator-dependent and time-consuming as compared to automatic methods. Moreover, the definition of LV SAX plane as going from LV base to apex may differ between observers providing different reformatted images and associated metrics. Marchesseau et al. [12] showed the influence on the evaluation of cardiac function of 4 different SAX acquisition protocols (considering the 2 atrioventricular junctions, considering only the left atrioventricular junction, considering the septum, and considering the LV long-axis).

To the best of our knowledge, no method has been proposed for fully automatic CT image reformatting. However, several methods can be found in the literature for automatic landmark detection and subsequent image reformatting in cardiac MRI. For example, Lu et al. [11] proposed landmark detection using LV segmentations; Le et al. [8] used 3D Convolutional Neural Networks to reformat 4D Flow MR images; and Blansit et al. [3] proposed to use U–Net-based heatmap regression to reformat cine steady-state free precession series. While these methods require manual annotation of anatomical landmarks (tedious and time-consuming) Alansary et al. [2] used acquired standard views for training avoiding the need for any manual labeling. The authors employed a multi-scale reinforcement learning agent framework that enables a natural learning paradigm mimicking operators' navigation steps.

Reformatting to SAX view can be done by resampling the raw TA image along its subject-specific LV SAX plane. In this paper, we propose to first segment the left and right ventricles and use the masks to automatically locate a few anatomical features needed to accurately compute the LV SAX plane. As compared to Deep Learning (DL) methods, our approach does not require annotated data or training a model, usually expensive and time-consuming.

2   Methods

Let $m = \{XYZ\}$ be the coordinate system in original TA view and let $m' = \{X'Y'Z'\}$ be the coordinate system in standard SAX view (see Fig. 1). The standard SAX view can be defined as follows: the LV long-axis is aligned according to the Z'-axis; and the right ventricle (RV) is positioned on the left side of the LV as seen in the X'Y'-plane. The reformatting to SAX view is computed by resampling the input TA view along its subject-specific LV SAX plane. Image resampling involves 4 components [1]:

1. The image that is sampled, given in coordinate system m (i.e. the input image in TA view).

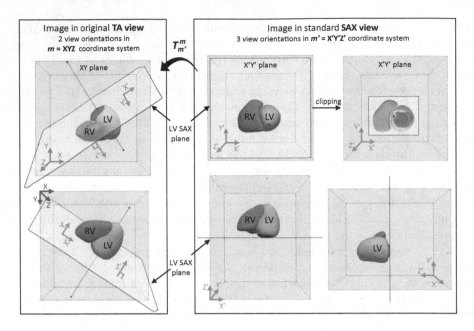

Fig. 1. Schematic representation of multiplanar reformatting to SAX view. On the left, original TA view in m coordinate system. On the right, corresponding SAX view in m' coordinate system. For simplicity, only surface models of the LV and RV are shown. LV = Left Ventricle; RV = Right Ventricle; TA = Trans-axial; SAX = Short-axis.

2. A resampling grid, regular grid of points given in coordinate system m'.
3. The transformation $T_{m'}^{m}$ that maps points from coordinate system m' to coordinate system m.
4. An interpolator method for obtaining the intensity values at arbitrary points in coordinate system m from the values of the points defined by the sampled image.

In our approach, cardiac ventricles are first segmented from the original CT image and surface meshes are computed from the corresponding masks (Sect. 2.1). Mesh coordinates are used to calculate the subject-specific transformation $T_{m'}^{m}$ (Sect. 2.2) that is finally used to resample the input TA image generating its corresponding SAX view (Sect. 2.3).

2.1   Image Segmentation

Automatic segmentation of the cardiac ventricles is performed using a DL-based framework. The method, previously described in [4,7] relies on the use of two successive 3D U-nets [13]. The high spatial resolution of CT images requires high memory resource while at the same time the ventricles take only a fraction of the entire CT volume. In our approach, we input to the first network a low-resolution version of the input data. The output (coarse segmentation) is used to locate the ventricles and keep only the region around them. A high-resolution cropped sub-volume is input to the second 3D U-net. The resulting segmentations are post-processed to obtain clean and non-overlapping masks including up-sampling to the original CT image resolution. The model was trained using 450 CT scans with corresponding expert segmentations of the LV endocardium and epicardium, and the right ventricular epicardium. 50 CT scans were used for validation with a loss function defined as the opposite of the Dice score. Importantly, the trained model was already available to us and we only use it for inference (i.e. prediction or testing) in this work. Therefore, any labeled data or training phase is required for the work presented in this paper.

After image segmentation, triangular surface meshes are computed from the masks using the marching cubes algorithm [9] and they are next uniformly remeshed to improve their quality [14][1].

2.2 Transformation Computation

The transformation $T_{m'}^m$ that maps points from coordinate system m' to coordinate system m can be computed as the inverse of the rotation matrix that rotates points from TA orientation to SAX orientation. Let this rotation matrix be $R_{SAX} \in \mathbb{R}^{3 \times 3}$. It takes into account 3 different rotations that locate 3 different LV features on their target position in SAX view:

1. $R_{MV} \in \mathbb{R}^{3 \times 3}$ aligns the mitral valve (MV) plane.
2. $R_S \in \mathbb{R}^{3 \times 3}$ aligns the LV septum (location of the RV with regard to the LV).
3. $R_{LAX} \in \mathbb{R}^{3 \times 3}$ aligns the LV long-axis.

R_{SAX} is computed as the result of the matrix multiplication:

$$R_{SAX} = R_{LAX} R_S R_{MV} \tag{1}$$

Figure 2 depicts two examples where the alignments are shown consecutively for illustrative purposes only.

All rotation matrices in our approach are computed using the following processing. Given two unit vectors, u_1 and u_2, a rotation matrix R that aligns u_1 to u_2 can be computed as follows:

$$R = I_3 + V_\times + V_\times V_\times \frac{1-c}{s^2} \tag{2}$$

where,
$I_3 = 3 \times 3$ Identity matrix
$v = u_1 \times u_2,$
$s = \|v\|,$
$c = u_1 \cdot u_2,$ and

$V_\times \stackrel{\text{def}}{=} \begin{bmatrix} 0 & -v_3 & v_2 \\ v_3 & 0 & -v_1 \\ -v_2 & v_1 & 0 \end{bmatrix}$ is the skew-symmetric cross-product matrix of $v = (v_1, v_2, v_3)$

---

[1] https://github.com/valette/ACVD.

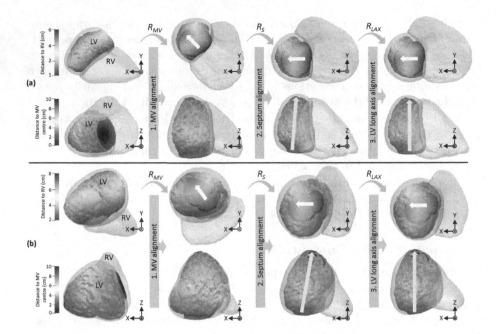

Fig. 2. Two examples of rotation to SAX view using surface meshes. The LV endocardial surface is colored according to its distance to the RV (1st and 3rd row) and according to its distance to the centre of the MV (2nd and 4th row). LV wall and RV epicardial surfaces are shown semitransparent. LV = Left Ventricle; RV = Right Ventricle; MV = Mitral Valve.

MV Plane Alignment. The rotation matrix $\boldsymbol{R}_{MV}$ is computed using Eq. 2 where the unit vectors to be aligned are: $\boldsymbol{u_1}$ = subject-specific MV normal in TA view; and $\boldsymbol{u_2}$ = MV normal in standard SAX view, i.e. opposite of the X'Y'-plane normal ($\boldsymbol{u_2} = (0, 0, -1)$). To compute $\boldsymbol{u_1}$, the region corresponding to the MV orifice is first automatically detected. Distances are computed from each point in the LV endocardial mesh to the LV wall mesh. Similarly, distances are computed from each point in the LV endocardial mesh to the RV epicardial mesh. The MV orifice area is detected as the points in the LV endocardial surface that are far from the LV wall (more than 5 mm), and far from the RV epicardial surface. The latter ensures that the MV orifice is not confused with the aortic orifice and it avoids getting orientations corresponding to this artery. The threshold distance was defined to take into account the LV size and was empirically set to more than half the maximum distance between LV endocardium and RV epicardium. All distances are computed using the distance transform [5,6], that assigns to every point in the mesh a value indicating its distance to the nearest point in the other surface. The MV normal ($\boldsymbol{n_{MV}}$) is computed averaging the normals of all points belonging to the MV orificie area, and $\boldsymbol{u_1}$ is defined as the corresponding unitary vector: $\frac{n_{MV}}{\|n_{MV}\|}$.

LV Septum Alignment. The rotation matrix R_S is computed using Eq. 2 where now the unit vectors to be aligned are: u_1 = unit vector within the MV plane pointing from the RV to the LV; and u_2 = unit vector within the X'Y'-plane pointing horizontally from RV position to LV position in standard SAX view ($u_2 = (1, 0, 0)$). White arrows in Fig. 2 depict these orientations. To compute u_1, the surface corresponding to the MV orifice is first extracted using the same procedure explained above. This surface is uniformly remeshed to increase its spatial resolution and distances to the RV are re-computed. Let p_0 be the point with minimum distance value and let p_1 be the point with maximum distance value. The unitary vector u_1 is computed as $\frac{p_1 - p_0}{\|p_1 - p_0\|}$.

LV Long-Axis Alignment. The rotation matrix R_{LAX} is computed using Eq. 2 where now the unit vectors to be aligned are: u_1 = subject-specific LV long-axis unit vector; and u_2 = standard LV long-axis unit vector in SAX view ($u_2 = (0, 0, 1)$). The LV long-axis is defined as the line connecting the centre of the MV orifice and the LV apex. Yellow arrows in Fig. 2 show schematically these orientations. To compute u_1, the centre of the MV is detected as the centre of mass of the MV orifice surface previously extracted. Let p_{MV} be that point. Distances are computed from p_{MV} to all points in the LV endocardial surface (see Fig. 2, 2nd and 4th rows). The LV apex is detected as the point (p_A) that is farthest from p_{MV}. Then, u_1 is computed as $\frac{p_A - p_{MV}}{\|p_A - p_{MV}\|}$.

2.3 Image Reformatting

Prior to image resampling, a reference image is created with desired origin, size, and resolution. The voxels in this reference image determine the resampling grid that is used to sample the input TA image after mapping the voxel's position from m' to m using $T_{m'}^m$. The final transformation also includes a translation to the area where the image to be sampled is located. In our method we propose to use reference_origin = (0,0,0); reference_image_size = 512 × 512 × 512; and reference_image_resolution = input_image_physical_size / reference_image_size.

3 Results

The method was applied to 25 images with available reformatting to SAX view performed by expert radiologists. Original CT image dimensions were 512 × 512 voxels in the X and Y axes and 402 ± 103 voxels in the Z-axis. The average original voxel size was 0.40 ± 0.05 mm × 0.40 ± 0.05 mm × 0.45 ± 0.15 mm.

The average time to complete the full image reformatting was 28.38 ± 7.73 s (Intel i7 2.60GHz × 12 CPU and 32 GB RAM). Several examples of results are shown in Fig. 3. It can be seen how our method replicates the manual reformatting and it even outperforms it for some cases. For example in (b), (c), and (d) the LV long-axis is more accurately aligned along the Z-axis using our approach than using manual reformatting.

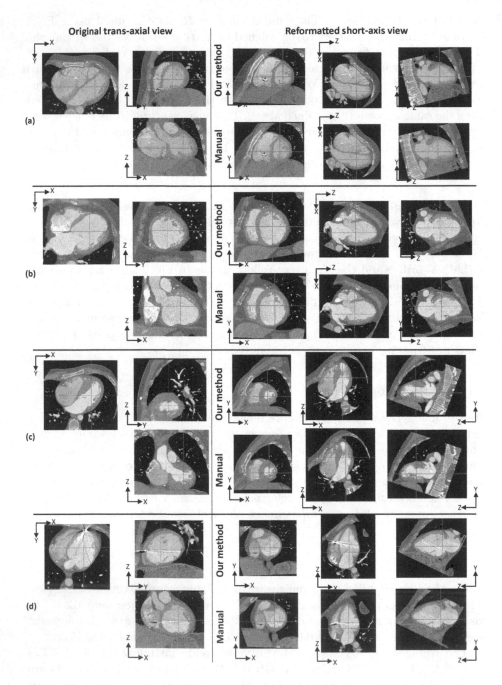

Fig. 3. Example results. On the left, 3 views of original trans-axial images; on the right, 3 views of corresponding short-axis images reformatted with our method (top), and the provided manual reformatting (gold standard) (bottom).

4   Conclusions

Cardiovascular image interpretation often relies on multiplanar reformatted images typically computed manually being time-consuming and lacking reproducibility. Automatic methods such as DL-based techniques often require lots of annotated data and computationally expensive resources. We have presented a method to reformat raw trans-axial CT images to standard SAX view that it is automatic, fast, accurate, reproducible and it does not require labeled data. The main drawback is however the dependency of the method on the quality of the segmentation masks. Previous studies have shown highly precise segmentation results using our proposed method (median Dice score of 0.9 [4]) but the image segmentation step (Sect. 2.1) could be easily replaced by any desired segmentation method (including manual segmentation performed by a radiologist which is yet considered the gold standard). Computing accurate standard SAX images may be useful to compare different image modalities such as CT and MRI since SAX cardiac MRI is currently one of the most common acquisition views in clinical practice. The origin, size, and resolution, of the reformatted SAX image can be specified by the user which is useful in many applications where a standard representation of the data is required (e.g. to use a pre-trained network).

Acknowledgements. Part of this work was funded by the ERC starting grant EC-STATIC (715093), the IHU LIRYC (ANR-10-IAHU-04), the Equipex MUSIC (ANR-11-EQPX-0030) and the ANR ERACoSysMed SysAFib projects. This work was also supported by the French government, through the 3IA Côte d'Azur Investments in the Future project managed by the National Research Agency (ANR) with the reference number ANR-19-P3IA-0002. We would like to thank all patients who agreed to make available their clinical data for research.

References

1. Simple ITK, SPIE2019 COURSE, 02 Images and resampling. https://simpleitk. org/SPIE2019_COURSE/02_images_and_resampling.html. Accessed 15 July 2020
2. Alansary, A., Le Folgoc, L., et al.: Automatic view planning with multi-scale deep reinforcement learning agents. In: International Conference on Medical Image Computing and Computer-Assisted Intervention, pp. 277–285 (2018)
3. Blansit, K., Retson, T., et al.: Deep learning-based prescription of cardiac MRI Planes. Radiol. Artif. Intell. **1**(6), e180069 (2019)
4. Cedilnik, N., Duchateau, J., Sacher, F., Jaïs, P., Cochet, H., Sermesant, M.: Fully automated electrophysiological model personalisation framework from CT imaging. In: Coudière, Y., Ozenne, V., Vigmond, E., Zemzemi, N. (eds.) FIMH 2019. LNCS, vol. 11504, pp. 325–333. Springer, Cham (2019). https://doi.org/10.1007/978-3-030-21949-9_35
5. Danielsson, P.E.: Euclidean distance mapping. Comput. Graph. Image Process. **14**(3), 227–248 (1980)
6. Fabbri, R., Costa, L.D.F., et al.: 2D Euclidean distance transform algorithms: a comparative survey. ACM Comput. Surv. (CSUR) **40**(1), 1–44 (2008)

7. Jia, S., et al.: Automatically segmenting the left atrium from cardiac images using successive 3D U-nets and a contour loss. In: International Workshop on Statistical Atlases and Computational Models of the Heart, pp. 221–229 (2018)

8. Le, M., Lieman-Sifry, J., Lau, F., Sall, S., Hsiao, A., Golden, D.: Computationally efficient cardiac views projection using 3D convolutional neural networks. In: Cardoso, M., et al. (eds.) DLMIA/ML-CDS -2017. LNCS, vol. 10553, pp. 109–116. Springer, Cham (2017). https://doi.org/10.1007/978-3-319-67558-9_13

9. Lorensen, W.E., Cline, H.E.: Marching cubes: a high resolution 3D surface construction algorithm. ACM Siggraph Comput. Graph. **21**(4), 163–169 (1987)

10. Lu, M.T., Ersoy, H., Whitmore, A.G., Lipton, M.J., Rybicki, F.J.: Reformatted four-chamber and short-axis views of the heart using thin section (≤ 2 mm) MDCT images. Acad. Radiol. **14**(9), 1108–1112 (2007)

11. Lu, X., Jolly, M.P., Georgescu, B., et al.: Automatic view planning for cardiac MRI acquisition. In: International Conference on Medical Image Computing and Computer-Assisted Intervention, pp. 479–486 (2011)

12. Marchesseau, S., Ho, J.X., Totman, J.J.: Influence of the short-axis cine acquisition protocol on the cardiac function evaluation: a reproducibility study. Eur. J. Radiol. Open **3**, 60–66 (2016)

13. Ronneberger, O., Fischer, P., Brox, T.: U-net: Convolutional networks for biomedical image segmentation. In: International Conference on Medical Image Computing and Computer-assisted Intervention, pp. 234–241 (2015)

14. Valette, S., Chassery, J.M., Prost, R.: Generic remeshing of 3D triangular meshes with metric-dependent discrete Voronoi diagrams. IEEE Trans. Visual. Comput. Graph. **14**(2), 369–381 (2008)

Graph Convolutional Regression of Cardiac Depolarization from Sparse Endocardial Maps

Felix Meister[1,3]([✉]), Tiziano Passerini[2], Chloé Audigier[3], Èric Lluch[3],
Viorel Mihalef[2], Hiroshi Ashikaga[4], Andreas Maier[1], Henry Halperin[4],
and Tommaso Mansi[2]

[1] Friedrich-Alexander University, Pattern Recognition Lab, Erlangen, Germany
`felix.meister@fau.de`
[2] Siemens Healthineers, Digital Technology and Innovation, Princeton, USA
[3] Siemens Healthineers, Digital Technology and Innovation, Erlangen, Germany
[4] Johns Hopkins University School of Medicine, Cardiac Arrhythmia Service,
Baltimore, USA

Abstract. Electroanatomic mapping as routinely acquired in ablation
therapy of ventricular tachycardia is the gold standard method to identify the arrhythmogenic substrate. To reduce the acquisition time and
still provide maps with high spatial resolution, we propose a novel deep
learning method based on graph convolutional neural networks to estimate the depolarization time in the myocardium, given sparse catheter
data on the left ventricular endocardium, ECG, and magnetic resonance
images. The training set consists of data produced by a computational
model of cardiac electrophysiology on a large cohort of synthetically generated geometries of ischemic hearts. The predicted depolarization pattern has good agreement with activation times computed by the cardiac
electrophysiology model in a validation set of five swine heart geometries
with complex scar and border zone morphologies. The mean absolute
error hereby measures 8 ms on the entire myocardium when providing
50% of the endocardial ground truth in over 500 computed depolarization patterns. Furthermore, when considering a complete animal data
set with high density electroanatomic mapping data as reference, the
neural network can accurately reproduce the endocardial depolarization
pattern, even when a small percentage of measurements are provided as
input features (mean absolute error of 7 ms with 50% of input samples).
The results show that the proposed method, trained on synthetically
generated data, may generalize to real data.

Keywords: Cardiac computational modeling · Deep learning ·
Electroanatomical contact mapping

1 Introduction

Each year, up to 1 per 1,000 North Americans die of sudden cardiac death, among
which up to 50% as a result of ventricular tachycardia (VT) [7]. A well-established

© Springer Nature Switzerland AG 2021
E. Puyol Anton et al. (Eds.): STACOM 2020, LNCS 12592, pp. 23–34, 2021.
https://doi.org/10.1007/978-3-030-68107-4_3

therapy option for these arrythmias is radiofrequency ablation. While success rates between 80–90% have been reported for idiopathic VTs, extinction of recurrent VTs in patients with structural heart disease is only successful in about 50% of the cases and requires additional interventions in roughly 30% of the other half [7]. Crucial to ablation success is the identification of scar regions prone to generate electrical wave re-entry, which is conventionally realized through electroanatomic mapping [7]. This technique poses practical challenges to achieve high spatial resolution and precise localization of the substrate [8]. Moreover, it only provides information about the surface potentials, failing to identify the complex three-dimensional slow conductive pathways within the myocardium [1]. Imaging (MRI or computed tomography) has great potential in helping define the geometry of the substrate [5,17], however electrophysiology assessment of the substrate may not be possible purely based on imaging features.

The use of computational models of cardiac electrophysiology has been explored as a way to combine imaging and catheter mapping information, to accurately estimate the substrate physical properties and potentially use the resulting model to support clinical decisions [2–4,14]. In this work we propose a novel deep-learning based method to enhance sparse left endocardial activation time maps by extrapolating the measurements through the biventricular anatomy. A graph convolutional neural network is trained to regress the local activation times over the left and right ventricle given sparse catheter data acquired on the left ventricular (LV) endocardium, a surface electrocardiogram (ECG), and magnetic resonance (MR) images. The network is trained on data produced by a computational model of cardiac electrophysiology on a large cohort of synthetically generated geometries of ischemic hearts. Biventricular heart anatomies are sampled from a statistical shape model built from porcine imaging data. We evaluate the proposed method by applying it to unseen porcine cases with complex scar morphology. We further illustrate the potential of the method to recover endocardial activation from a reduced set of measurements using one unseen porcine case with high-resolution LV contact maps.

2   Method

We represent the biventricular heart, segmented from medical images, as a tetrahedral mesh [9]. The mesh is an undirected graph $\mathcal{G} = (\mathcal{V}, \mathcal{E}, \mathbf{X})$ comprising a set of N vertices $\mathcal{V}$ and a set of M edges $\mathcal{E}$ connecting pairs of vertices i and j. Trainable graph convolutional filters with shared weights are applied to all vertices $\in \mathcal{V}$, to learn a model of local activation time as a function of vertex-wise features $\mathbf{X} \in \mathbb{R}^{N \times D}$.

Graph Convolutional Layer Definition. In this work GraphSAGE layers with mean aggregation were chosen [6]. Each layer l processes a vector h_j^l, representing any vertex $v_j \in \mathcal{V}$. $h_j^0 \in \mathbf{X}$ denotes the vertex input features. GraphSAGE first aggregates information over a vertex i's immediate neighborhood $\mathcal{N}(i)$ comprising all vertices j that are connected to i via an edge $e_{ij} \in \mathcal{E}$, i.e.

$h_{\mathcal{N}(i)}^{(l+1)} = \text{mean}(h_j^l, \forall j \in \mathcal{N}(i))$. Using learnable weights $\mathbf{W}_{\mathcal{N}(i)}$ and $\mathbf{W}_i$ as well as biases $b_{\mathcal{N}(i)}$ and b_i, the layer output $h_i^{(l+1)}$ is computed from $h_{\mathcal{N}(i)}^{(l+1)}$ and h_j^l according to $h_i^{(l+1)} = \sigma(\mathbf{W}_i \cdot h_i^l + b_i + \mathbf{W}_{\mathcal{N}(i)} \cdot h_{\mathcal{N}(i)}^{(l+1)} + b_{\mathcal{N}(i)})$

Graph Convolutional Regression of Local Activation Times. The feature matrix $\mathbf{X}$ contains geometric features, ECG information, and the measured local activation time (LAT) on the LV endocardial surface. The latter is set to -1 for vertices that are not associated to measurement points. ECG information consists of QRS interval, QRS axis, and 12 features representing the positive area under the curve of the QRS complex in each of the 12 traces. Since each graph convolutional layer is applied to all vertices with shared weights, the same ECG features are appended to each vertex. Geometric features characterize the position of each point in the biventricular heart. For each tetrahedral mesh, we define a local reference system by three orthogonal axes: the left ventricular long axis, the axis connecting the barycenters of the mitral and tricuspid valves, and an axis normal to both. The apex is chosen as the origin of the coordinate system. Each mesh point is then characterized by the cylindrical coordinates radius, angle, and height. In addition, we utilize $[0, 1]$-normalized continuous fields to describe the vertex position between (0 and 1 respectively) apex and base, endocardium and epicardium, and left ventricle and right ventricle. Finally we use mutually exclusive categorical features to denote whether a vertex belongs to the LV/RV endocardial surface or epicardial surface. We also use another set of mutually exclusive categorical features for scar, border zone or healthy tissue region. The numerical features are $[0, 1]$-normalized using the bounds computed from the training data set. The LAT values on the LV endocardium are also normalized as described later.

Network Architecture. Our network architecture (see Fig. 1) is an adaptation of PointNet [15]. The local feature extraction step using spatial transformer networks is replaced by a series of GraphSAGE layers, each leveraging information of the 1-hop neighborhood. The output of each layer is concatenated to form the vertex-wise local features. To obtain a global feature vector, the local features are further processed using multiple fully connected layers with shared weights and max pooling. Given local and global features the local activation times (LATs) are predicted for each vertex using again a series of fully connected layers.

Training Procedure and Implementation. The neural network is trained to optimize a loss function $\mathcal{L} = \mathcal{L}_{LAT} + \mathcal{L}_{qrs}$. $\mathcal{L}_{LAT} = \frac{1}{N} \sum_i \alpha_i \|y_i - \hat{y}_i\|^2$ is a mean-squared error loss, weighted by α_i, on the predicted LATs y_i and the corresponding ground truth $\hat{y}_i$. Since there is no guarantee to perfectly match the information provided at the measurement locations, we set $\alpha_i = 2$ for the vertices with measurements and $\alpha_i = 1$ for all other vertices to guide the network into retaining the measurement data. To preserve QRS duration in the predicted activation pattern, an additional regularization term $\mathcal{L}_{qrs} = \|\text{QRSd} - \hat{\text{QRSd}}\|^2$ is introduced. We approximate the QRS duration by computing the difference

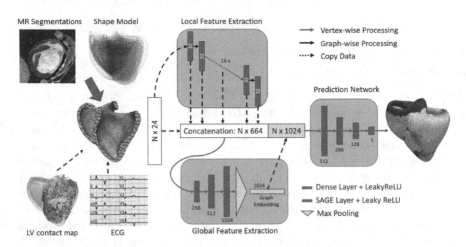

Fig. 1. Illustration of the proposed graph convolutional pipeline. ECG information and endocardial measurements are mapped to the vertices of the mesh. Local features are extracted using a cascade of GraphSAGE layers that gather information about the local neighborhood. A global feature vector is computed using a series of fully connected layers and max pooling taking the concatenated output of each GraphSAGE layer as input. Local activation times (LAT) are regressed using both local features and the global feature vector, and a series of fully connected layers.

between maximum and minimum activation time in the prediction (QRSd) and the ground truth (QR$\hat{\text{S}}$d), respectively.

The proposed network is implemented using PyTorch and the Deep Graph Library [13,16]. Hyperparameters of the network and the optimizer were selected using a grid search on a subset of the training data. In particular, we choose 20 GraphSAGE layers for the local feature extraction, three fully connected layers of size 256, 512, and 1024 for the global feature extraction, and four layers of size 512, 256, 128, and 1 for the final prediction. For all layers leaky rectified linear units are chosen for the activation function. The network is trained for 500 epochs using the Adam optimizer [10] with default parameters and an initial learning rate of 5×10^{-4}. Step-wise learning rate decay is applied every 25 epochs using a decay factor of 0.8. No improvement in the validation loss was seen beyond 500 epochs and the network associated with the epoch of minimal validation loss was used for evaluation.

3   Experiments

Data Generation. For training and testing, 16 swine data sets with MR images (MAGNETOM Avanto, Siemens AG), 12-lead ECG (CardioLab, GE Healthcare), and high-resolution left endocardial contact maps (EnSite Velocity System, St. Jude Medical) were considered. Additional data sets were synthetically generated as follows. A statistical shape model was built from 11 of the 16 swine datasets. A total of 213 swine heart geometries were sampled from it. In short, heart chamber segmentations were extracted from the medical images using a machine learning algorithm [9]. The segmentations were aligned using point correspondences and rigid registration. A mean model was computed over all segmentations and the eigenvectors were computed using principal component analysis. The shape model was obtained from the linear combinations of the five most informative eigenvectors, which captured well the global shape variability such as the variation in ventricle size and wall thickness. Wall thinning as a result of chronic scar has not been considered due to the limited amount of reference data. Biventricular heart anatomical models were generated from the chamber surface meshes using tetrahedralization, mesh tagging, and a rule-based fiber model [9]. A generic swine torso with pre-defined ECG lead positions was first manually aligned to the heart chamber segmentations extracted from one data set, so as to visually fit the visible part of the torso in the cardiac MR images. Using this as a reference, a torso model was automatically aligned to any swine heart anatomical model by means of rigid registration. A fast graph-based computational electrophysiology model [14] was utilized to compute depolarization patterns of the biventricular heart and local activation times on the LV endocardial surface. In essence, the activation time for every vertex was computed by finding the shortest path to a set of activation points. The edge cost w_{ij} between two vertices $\mathbf{v}_i$ and $\mathbf{v}_j$ was computed as $w_{ij} = l_{ij}/c_{ij}$, where c_{ij} refers to the edge conduction velocity, a linear interpolation of the conduction velocity at vertex $\mathbf{v}_i$ and $\mathbf{v}_j$. $l_{ij} = \sqrt{(\mathbf{e}_{ij}^T \mathbf{D} \mathbf{e}_{ij})}$, with $\mathbf{e}_{ij} = \mathbf{v}_i - \mathbf{v}_j$, refers to a virtual edge length that considers the local anisotropy tensor $\mathbf{D}$. Using the fiber direction $\mathbf{f}_{ij}$ from the rule-based fiber model [9], the anisotropy tensor was computed as $\mathbf{D} = (1-r)\mathbf{f}_{ij}\mathbf{f}_{ij}^T + r\mathbf{I}$ with r denoting the anisotropy ratio, i.e. $r = 1/3$, and the identity matrix $\mathbf{I}$. 50 simulations were computed for each anatomical model. For each simulation one region of the biventricular heart was randomly selected from the standard 17 AHA segment model [12] to be either scar or border zone tissue (see Fig. 2a for an example). Conduction velocity was assumed uniform in each of five tissue types: normal myocardium (c_{Myo}), left and right Purkinje system (c_{LV} & c_{RV}), border zone (c_{BZ}) and scar (c_{Scar}). The Purkinje system was assumed to extend sub-endocardially with a thickness of 3 mm, as observed in swine hearts. In each simulation the conduction velocities were randomly sampled within pre-defined physiological ranges: $c_{Myo} \in [250, 750]$ mm/s, c_{LV} and $c_{RV} \in [cMyo, 2,500]$ mm/s, and $c_{BZ} \in [100, cMyo]$ mm/s, with the exception of scar tissue to which a conduction velocity of 0 mm/s was always assigned. The computed endocardial activation time was used to define the endocardial measurement feature, and it was normalized by subtracting the time of earliest ventricular activation of each simulation and dividing by the computed QRS duration. The purely synthetic database comprised 10,650 simulations and was randomly split into 90% for training, 5% for validation, and 5% for evaluation ("Simple Scar Test") ensuring that the same anatomical model was not included in more than one of these partitions. Additional 500 simulations were run using the anatomical models generated for the five left-out swines: 100 simulations per each anatomical model ("Complex Scar Test"). In this case, scar and border zone regions were defined based on segmentation of the MR images, while the set up of the computational model was unchanged (see Fig. 2b for an example).

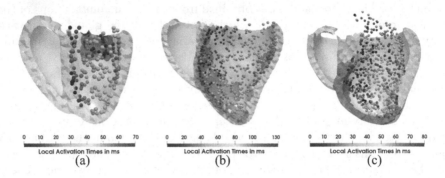

Fig. 2. Three examples for geometry and endocardial measurements for (a) synthetic database with simple uniform region randomly set to be scar or border zone; (b) synthetic database with complex scar derived from images; (c) real dataset with complex scar derived from images and high-resolution contact map. In (a) and (b) 25% of the endocardial ground truth is illustrated.

Comparison to Personalized Computational Model. As a reference method (denoted by DenseEP in this work), we used the same graph-based electrophysiology model with local conduction velocities estimated by the personalization scheme described in [14]. In a first step, local conduction velocities were initialized from the homogeneous tissue conductivities c_{Myo}, c_{LV}, c_{RV}, and c_{BZ}, which were iteratively optimized to match QRS duration and QRS axis. A second step aimed at finding local conductivities to minimize the discrepancy between measurements $\hat{y}_i$ and the computed activation times y_i. In this work, a sum of squared distances $\mathcal{L}_{\mathrm{SSD}} = \sum_i (\hat{y}_i - y_i)^2$ on the N vertices where measurements were available was used. To find a set of edge weights that minimizes $\mathcal{L}_{\mathrm{SSD}}$ an algorithm inspired by neural networks was employed. Hereby, the tetrahedral mesh on which the electrical wave propagation is computed can be seen as a graph that arranges in layers. The activation time at the vertices where measurements are available (output layer) depends on the activation times at the activation points (input layer) and on the path followed by the electrical wave. The wave propagates to a first of set of vertices (first hidden layer), that are connected to the input layer; and recursively propagates to other sets of vertices (hidden layers), each connected to the previous. Only the paths (sets of edges) connecting vertices in the output layer with activation points in the input layer are considered in the optimization step.

For a given path, starting at an activation point j and ending at a measurement point i, the activation time in the end point is computed as $y_i = t_j + \sum_k w_k$ with t_j being the initial activation time in j, and $\{w_k\}_{1,\dots,N}$ being the set of N edges along the path. We seek to find the optimal set of edge weights minimizing $\mathcal{L}_{\text{SSD}}$. The weights are iteratively updated by a gradient descent step $w_k^{t+1} = w_k^t - \gamma g$ with the iteration number t, the step size γ, and the gradient $g = \frac{\partial \mathcal{L}_{\text{SSD}}}{\partial w_k}$. It follows that $\frac{\partial \mathcal{L}_{\text{SSD}}}{\partial w_k} = \frac{\partial \mathcal{L}_{\text{SSD}}}{\partial y_i} \frac{\partial y_i}{\partial w_k}$ with $\frac{\partial \mathcal{L}_{\text{SSD}}}{\partial y_i} = -2(\hat{y}_i - y_i)$ and $\frac{\partial y_i}{\partial w_k} = 1$. Since an edge may be traversed multiple times, gradients are accumulated on the edges. The reader is referred to [14] for more details.

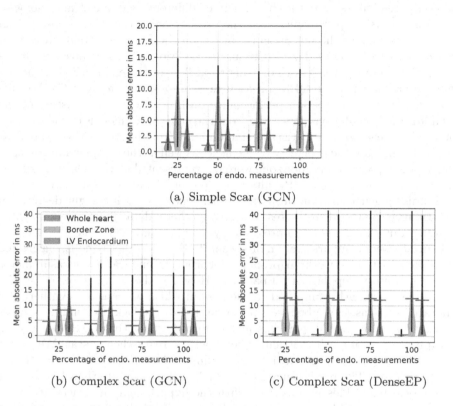

(a) Simple Scar (GCN)

(b) Complex Scar (GCN) (c) Complex Scar (DenseEP)

Fig. 3. Mean absolute error (red: mean, black: 15 to 95 percentile) in local activation time between neural network (GCN) predictions and the ground truth values generated by the computational electrophysiology model for Simple Scar Test (a) comprising simulations with AHA-based scar or borderzone; and Complex Scar Test (b) with chamber geometries and complex scar morphologies not represented by the training database. Good performance is achieved on both test sets. As a reference we report in (c) the same metrics on Complex Scar Test for the computational model with densely estimated conductivities (DenseEP). (Color figure online)

Evaluation on Synthetic Database. The proposed network was trained once on the synthetic training set. We first evaluated the performance of the proposed network on the 5% left out cases from the synthetic database, denoted as "Simple Scar Test" in the figures. We computed the mean absolute error (L1-error) in the predicted vertex-wise LAT for (a) the entire biventricular domain, (b) the LV endocardial surface, (c) all healthy tissue, and (d) border zone tissue. For each case we observed the error variation when the endocardial measurements feature is provided to a decreasing number of vertices (100%, 75%, 50%, and 25% of the total). The results as seen in Fig. 3a indicate good agreement between the prediction and the ground truth even when only using 25% of the measurements. The regressed activation time shows very good agreement with the computed values across a wide range of different anatomical morphologies and tissue physical parameters, and the error remains low when evaluated on the entire volume of the biventricular heart, on the LV endocardial surface, on sub-regions of healthy or diseased tissue. In addition, the mean absolute difference in QRS duration is less than 1 ms regardless of the endocardial sampling. Relative errors w.r.t. the QRS duration measure approximately 2.5%. We further evaluated the prediction performance on the testing set consisting of the 500 additional simulations generated from the anatomical models of the five pigs not used to generate the statistical shape model (referred to as "Complex Scar Test"). In this case the geometry of the heart as well as the complex image-based morphology of scar and border zone were not represented by the relatively simple synthetic training database. Results presented in Fig. 3b and Fig. 4 show overall promising generalization properties of the network, with a slight decrease in performance compared to the test on fully synthetic data. Relative errors w.r.t. the QRS duration measure approximately 7%. Predictions on the LV endocardial surface were comparatively more accurate than on other regions, suggesting the ability of the network to properly incorporate endocardial measurements even from unseen depolarization patterns. However, the accuracy of the prediction decreases slightly compared to the case in which we consider synthetically generated test cases. We believe this to be due to the lack of training samples showing the same type of patterns in the input features that would be seen on a graph representing a biventricular heart with a real ischemic scar. We report in Fig. 3c the errors in activation time computed by DenseEP on the "Complex Scar Test". The results suggest that GCN is less accurate than DenseEP at the endocardium. This is consistent with the hypothesis that a richer training database (in particular for scar and border zone morphology) could help GCN's performance.

Real Data Results. To demonstrate the ability to correctly regress local activation times in clinically relevant scenarios, we applied the network, trained on the synthetic database, to one of the swines from the testing set, which had a high-density left endocardial electro-anatomical map available. Catheter measurements included LAT and peak to peak voltage. The measurement point cloud, which has been curated by an electrophysiologist, was spatially registered to the tetrahedral mesh by manual alignment (see Fig. 2c), so that low voltage signals (<1.5 mV) co-localize with scar and border zone regions. Finally, the LAT data was mapped onto the mesh using nearest neighbor search. The endocardial measurements were randomly subsampled to retain 25%, 50%, 75%, and 100% of the samples and provided as input feature to the network. In all cases,

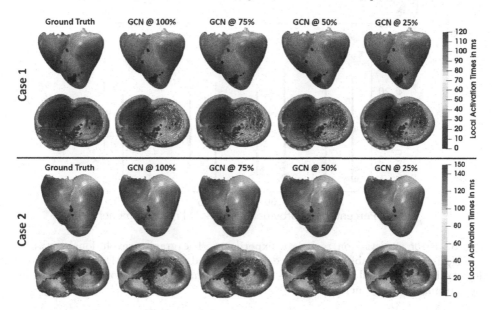

Fig. 4. Comparison of prediction results for two cases when providing 25%, 50%, 75%, and 100% of the LV endocardial measurements (colored dots) to the proposed network (GCN). The main features of the regressed depolarization pattern (i.e. wave front curvature and distribution of isochrones) are stable with respect to the spatial location and density of the measurements and consistent with the ground truth.

the complete set of LV endocardial measurements was used as ground truth and we computed mean absolute error of vertex-wise LAT on the LV endocardium (see Fig. 5). The error increases with decreasing number of provided endocardial measurements, but remains relatively low and similar to the values obtained on unseen geometries with synthetically generated ground truth. The regressed depolarization pattern is qualitatively correct (see Fig. 6) and consistent with the data. The model identifies the main direction of propagation of the electrical wave, with early activation on the septal wall and late activation on the free wall. This qualitative behavior is evident in all solutions provided by the network, regardless of the amount of subsampling in the endocardial measurements, suggesting that the extensive training set including physiologically accurate depolarization patterns produced by a computational model of electrophysiology provides a strong prior for the prediction.

For comparison, DenseEP performs consistently well independently of the amount of data sub-sampling, but with an overall higher average error compared to the network predictions. The depolarization pattern is consistent with the data (see Fig. 6), with a physiologically plausible smooth interpolation in the measurement gaps owing to the solution of an inverse problem using the computational electrophysiology model as a regularizer.

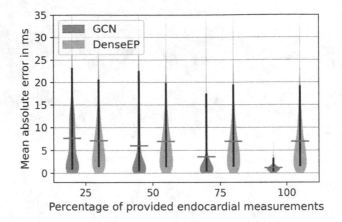

Fig. 5. Error statistics on real data experiment. The graph convolutional network (GCN) has a comparable performance as the method of reference (DenseEP) for low density LV endocardial measurements and outperforms the method of reference as the measurement density increases.

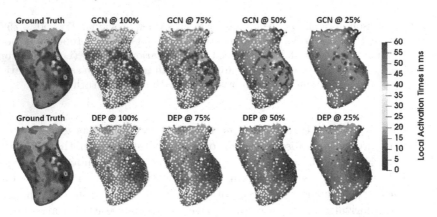

Fig. 6. Comparison of prediction results when providing 25%, 50%, 75%, and 100% of the high-resolution catheter measurements (colored dots) to the proposed network (GCN) and the method of reference (DenseEP). GCN consistently achieves higher accuracy in the neighborhood of the measurement points, while DenseEP provides a smooth and physiologically plausible interpolation of the activation times.

4   Discussion and Conclusion

Graph convolutional networks are based on a definition of neighborhood that adapts naturally to the description of complex physical systems. Topology is one of the main determinants of cardiac electrophysiology, since the function of the organ is intrinsically linked to its structure and the hierarchy of its components. The proposed graph convolutional neural network was trained on a large cohort of synthetic biventricular depolarization patterns that were generated by a graph-based computational model of cardiac electrophysiology. Anatomies were sampled from a statistical shape model. Variability in the depolarization patterns were induced by randomizing conduction velocities and randomly adding uniform regions of scar or border zone on the LV. Our results were obtained on a left out subset from the initial synthetic database as well as on an additional synthetic database that incorporated complex scar and border zone distributions derived from images. The results show that a graph convolutional network can successfully and reliably learn meaningful patterns of activation time as a function of features that are related to local geometrical and functional properties of the heart tissue, when provided with just a sparse set of measurements on the LV endocardial surface. Validation of the method using high density electroanatomic mapping data shows that the main features of the regressed depolarization pattern (i.e. wave front curvature and distribution of isochrones) were stable with respect to the spatial location and density of the measurements, which tend to be incorporated in the depolarization pattern with high accuracy as local spatial discontinuities in the activation time. In general, a decrease in the prediction errors was observed the more endocardial samples were provided. In case of providing 100% of the endocardial samples, however, the network showed to not being able to perfectly match the provided information. We hypothesize that the network is not able to learn to impose the endocardial data points due to the multi-objective loss function and sharing of weights for processing each vertex. Careful construction of a more appropriate loss function and the impact of more complex graph convolutional filters is subject of future work. While the results suggest that the method generalizes to different meshes generated on multiple geometries, the effect of meshing was not studied within the scope of this work and is subject to future research. In addition, the sensitivity of the results on the different input features will be analyzed in future work. Other future directions of this research will focus on improving the model to regress physiologically plausible activation time distribution compatible with the available measurements. A potential path of improvement comprises the extension of the training set, for instance by leveraging a more realistic fiber model, e.g. as presented in [11], or enriching the database with complex scar and border zone morphologies.

Disclaimer. This feature is based on research, and is not commercially available. Due to regulatory reasons its future availability cannot be guaranteed.

References

1. Ashikaga, H., et al.: Magnetic resonance-based anatomical analysis of scar-related ventricular tachycardia: implications for catheter ablation. Circ. Res. **101**(9), 939–947 (2007)
2. Chinchapatnam, P., et al.: Model-based imaging of cardiac apparent conductivity and local conduction velocity for diagnosis and planning of therapy. IEEE Trans. Med. Imaging **27**(11), 1631–1642 (2008)
3. Corrado, C., et al.: A work flow to build and validate patient specific left atrium electrophysiology models from catheter measurements. Med. Image Anal. **47**, 153–163 (2018)

4. Dhamala, J., et al.: Spatially adaptive multi-scale optimization for local parameter estimation in cardiac electrophysiology. IEEE Trans. Med. Imaging **36**(9), 1966–1978 (2017)
5. Dickfeld, T., et al.: MRI-guided ventricular tachycardia ablation: integration of late gadolinium-enhanced 3D scar in patients with implantable cardioverter-defibrillators. Circ. Arrhythm. Electrophysiol. **4**(2), 172–184 (2011)
6. Hamilton, W., et al.: Inductive representation learning on large graphs. In: Advances in Neural Information Processing Systems, pp. 1024–1034 (2017)
7. John, R.M., et al.: Ventricular arrhythmias and sudden cardiac death. Lancet **380**(9852), 1520–1529 (2012)
8. Josephson, M.E., Anter, E.: Substrate mapping for ventricular tachycardia. JACC Clin. Electrophysiol. **1**(5), 341–352 (2015)
9. Kayvanpour, E., et al.: Towards personalized cardiology: multi-scale modeling of the failing heart. PLoS ONE **10**(7), 1–18 (2015)
10. Kingma, D.P., Ba, J.: Adam: a method for stochastic optimization. arXiv preprint arXiv:1412.6980 (2014)
11. Mojica, M., et al.: Novel atlas of fiber directions built from ex-vivo diffusion tensor images of porcine hearts. Comput. Methods Programs Biomed. **187**, 105200 (2020)
12. Cerqueira, M.D., Weissman, N.J., Dilsizian, V., Jacobs, A.K., Kaul, S., Laskey, W.K., Pennell, D.J., Rumberger, J.A., Ryan, T., American Heart Association Writing Group on Myocardial Segmentation and Registration for Cardiac Imaging, et al.: Standardized myocardial segmentation and nomenclature for tomographic imaging of the heart: a statement for healthcare professionals from the cardiac imaging committee of the council on clinical cardiology of the american heart association. Circulation **105**(4), 539–542 (2002)
13. Paszke, A., et al.: PyTorch: an imperative style, high-performance deep learning library. In: Wallach, H., Larochelle, H., Beygelzimer, A., d'Alché-Buc, F., Fox, E., Garnett, R. (eds.) Advances in Neural Information Processing Systems, vol. 32, pp. 8024–8035. Curran Associates, Inc. (2019)
14. Pheiffer, T., et al.: Estimation of local conduction velocity from myocardium activation time: application to cardiac resynchronization therapy. In: Pop, M., Wright, G.A. (eds.) FIMH 2017. LNCS, vol. 10263, pp. 239–248. Springer, Cham (2017). https://doi.org/10.1007/978-3-319-59448-4_23
15. Qi, C.R., et al.: PointNet: deep learning on point sets for 3D classification and segmentation. In: Proceedings CVPR, pp. 652–660 (2017)
16. Wang, M., et al.: Deep graph library: towards efficient and scalable deep learning on graphs. In: ICLR Workshop on Representation Learning on Graphs and Manifolds (2019)
17. Zhang, L., et al.: Multicontrast reconstruction using compressed sensing with low rank and spatially varying edge-preserving constraints for high-resolution MR characterization of myocardial infarction. Magn. Reson. Med. **78**(2), 598–610 (2017)

A Cartesian Grid Representation of Left Atrial Appendages for a Deep Learning Estimation of Thrombogenic Risk Predictors

César Acebes$^{(\boxtimes)}$, Xabier Morales$^{(\boxtimes)}$, and Oscar Camara$^{(\boxtimes)}$

Physense, Department of Information and Communication Technologies,
Universitat Pompeu Fabra, Barcelona, Spain
cesaracebespinilla@gmail.com, {xabier.morales,oscar.camara}@upf.edu

Abstract. Computational fluid dynamics (CFD) simulations have recently been used to assess haemodynamic implications of atrial fibrillation on geometrical meshes built from patient-specific data. Some deep learning architectures, such as Fully Connected Networks (FCN), have demonstrated potential in accelerating CFD simulations, determining the relation between object geometry and model outcomes after finding correspondences with classical surface registration techniques. However, other successful architectures, such as Convolutional Neural Networks (CNN), have not been used yet in this application since geometrical meshes do not present a Euclidean structure, unlike medical images. The primary goal of this study was to estimate a fast surrogate of fluid simulations, based on a CNN architecture, for the prediction of thrombus formation risk in the left atrial appendage (LAA). For this purpose, a new flattened representation of the LAA was achieved by sampling its corresponding mesh in two directions: from the LAA junction to the left atrium (i.e. the ostium) to the tip, using the normalized gradient of the heat flow, producing radial isolines; and along the radial isoline direction, ordering the sampled points by their angle to a reference point. Using the resulting discretization, two FCN and one CNN architectures were tested. The CNN obtained the lowest mean absolute error and better predicted the areas of more elevated thrombogenic risk, while being orders of magnitude more computationally efficient than registration-based methods.

Keywords: Deep learning · Conformal flattening · Thrombus formation · Mesh processing · Left Atrial Appendage · Fluid simulations

1 Introduction

Atrial fibrillation (AF) is the most common sustained arrhythmia in the world, frequently leading to thrombus formation and other heart-related complications. This condition is responsible for about 20% of cardioembolic ischemic

© Springer Nature Switzerland AG 2021
E. Puyol Anton et al. (Eds.): STACOM 2020, LNCS 12592, pp. 35–43, 2021.
https://doi.org/10.1007/978-3-030-68107-4_4

strokes, of which about 99% are caused by thrombi formed in the left atrial appendage (LAA) [2], a tubular hooked structure derived from the left atrium (LA). Such thrombi have been related to (low) blood flow velocities, usually assessed on echocardiography studies. Nevertheless, it can be challenging to interpret echocardiography scans since they provide limited quantitative information about blood flow patterns. Recently, computational fluid dynamics (CFD) have been applied to patient-specific LA geometries to estimate personalized and quantitative thrombogenic risk indices, characterizing the complexity of LA haemodynamics. However, long computational times are required to set up and run patient-specific CFD models, which represents a barrier to processing large numbers of cases and using them in clinical routine; consequently, current CFD studies of the LA are limited to a handful of cases (e.g. 5 in one of the most recent ones [7]).

To reduce the long computational times associated to CFD simulations, deep learning (DL) techniques have been recently proposed as a tool to estimate surrogates of CFD simulation outcomes; DL-based models are able to estimate fluid model results without running simulations, learning the relation between geometry and simulation outcomes. In biomedical applications, Liang et al. [6] proposed a Fully Connected Network (FCN) architecture to directly estimate the wall shear stress of the aorta. Based on that work, Morales et al. [8] also proposed a FCN-based method to estimate the endothelial cell activation potential (ECAP) of the LAA, which is an in-silico CFD-based index that characterizes the thrombogenic susceptibility of a vessel [3]. Correspondences between the processed LAA meshes were obtained by 3D surface registration to a defined template, which was a time-consuming step. Moreover, some of the most efficient DL architectures such as Convolutional Neural Networks (CNN) could not be applied since they require an Euclidean representation of the data. For instance, Li et al. [5] proposed a CNN encoder-decoder architecture to establishing a direct mapping between input features and simulation results for simple reaction-diffusion models run in a small rectangular grid.

In consequence, the objective of this work was to generate a fast surrogate of fluid simulations, based on CNN, to predict the thrombus formation risk. To achieve this goal, we developed a standard representation of the LAA by mapping its surface to a 2D image through a novel flattening algorithm, similarly to techniques previously used in other organs [4,9]. Thereafter, the flattened representation of the LAA was used to train DL models (FCN-based and CNN-based) to learn the relationship between the patient-specific LAA geometry and the ECAP. Additionally, these flat-based methods were benchmarked against the surface registration approach proposed by Morales et al. [8] on a dataset of 206 synthetic LAA, where CFD results were already available.

2   Methods

The general pipeline of the study is shown in Fig. 1. Firstly, the geometry set up and the ECAP computation from CFD simulations was performed as in Morales et al. [8]. Initially, the LAA surfaces were remeshed (1 in Fig. 1) before being introduced to the flattening algorithm, where radial and angular mappings (2 and 3 in Fig. 1, respectively) were computed. The resulting flattened LAA was then represented as a bull's eye plot (4 in Fig. 1). For the CNN architectures, an additional mapping step was required to convert the bull's eye plot circular representation to a rectangular one. Subsequently, the DL networks were trained to estimate the ECAP in new LAA geometries for validation purposes.

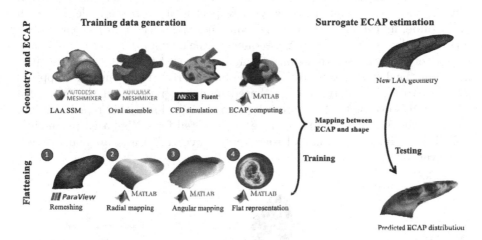

Fig. 1. General pipeline of the study. The four steps related to geometry building and ECAP estimation were performed as in Morales et al. [8]. LAA: Left Atrial Appendage, SSM: Statistical Shape Model, CFD: Computational Fluid Dynamics, ECAP: Endothelial Cell Activation Potential.

A virtual dataset of 206 artificial LAA geometries, with the ECAP values in each node, was used, as computed in [8]. Manual remeshing was applied to the geometries, using Meshmixer[1], increasing the number of nodes in each mesh (from 2466 to ~18,500 nodes).

2.1 Flattening Algorithm

The mapping of the LAA from the 3D geometry to the 2D representation was obtained by calculating polar-based coordinates, employing a method similar to the one of Meuschke et al. [4] for mapping an aneurysm surface to a hemisphere. The flattening algorithm consisted of two successive steps: the radial mapping and the angular mapping, which were used to select the mesh nodes of interest and discretize the 3D geometry. Figure 2 shows a scheme of the flattening algorithm.

To obtain the radial mapping, the geodesic distances between the ostium (i.e. junction of the LAA and the LA) and each mesh node were computed. For this,

---

[1] http://www.meshmixer.com/.

the fast approach proposed by Crane et al. [1] was followed[2], which essentially, consists of three steps: the integration of heat flow for some fixed time; the evaluation of its gradient; and the solving of the Poisson equation. Afterwards, the desired number of radial isolines (i.e. lines with the same geodesic distance to the ostium) was created. Then, a plane was calculated for each isoline from a linear regression with all its nodes as input. After that, each node was projected into the plane corresponding to its isoline, so that each isoline lied in a plane.

To find the angular mapping, the nodes of each radial isoline were ordered depending on the angular distance with respect to a reference node. For each radial isoline, its center was computed, and the furthest node to the circumflex artery was selected as reference. Subsequently, the angular distance of each node to this reference was used to arrange them and to select the nodes of the angular isolines for the discretization. Finally, the obtained radial and angular mappings were used to build a cartesian grid-based mapping, initially in the form of a circle or bull's eye plot, based on polar coordinates. From this representation, a rectangular matrix-like version was easily created for CNN-based processing, in which the image columns and rows represented radial and angular isolines, respectively.

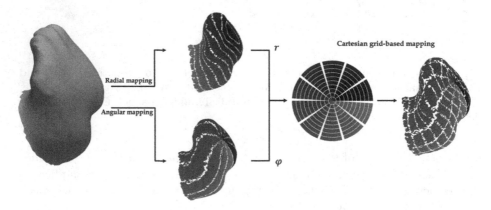

Fig. 2. Scheme of the flattening algorithm, illustrating the calculation of polar-based coordinates. From left to right: 3D mesh; radial (top) and angular (bottom) isolines obtained from radial and angular mapping, respectively; flat representation of the cartesian grid; and corresponding nodes in the 3D geometry.

2.2   Deep Learning Models

Three different DL models were tested to estimate the ECAP map given the flattened representation of the LAA. Firstly, two FCN-based ones based on liter-

---

[2] Available in http://www.numerical-tours.com/matlab/meshproc_7_geodesic_pois son/.

ature (henceforward referred to as Liang-flat-FCN [6] and Morales-flat-FCN [8], respectively) and then, a CNN-based model (Acebes-flat-CNN).

FCN-Based Models. The FCN-based models were adapted from Liang et al. [6] and from Morales et al. [8]. Keras 2.2.4[3], with TensorFlow 1.13.1[4] backend, was chosen as the high-level neural network API. Each model was basically composed of three main steps: the shape encoding, the non-linear mapping and the ECAP decoding. The shape encoding step encoded the spatial coordinates of the LAA shape into a set of scalars. Liang et al. [6] and Morales et al. [8] performed this step using principal component analysis (PCA) and truncated PCA, respectively, allowing reduction of the training set dimensionality.

Next, a fully-connected neural network was utilized to map the shape and ECAP codes. The FCN was comprised of 3 hidden layers in both Liang- and Morales-flat-FCN methods; however, Liang et al. [6] used 64 softplus activation units while Morales et al. [8] chose 256 ReLU activation units. Finally, the ECAP decoding in Liang-flat-FCN was based on a Low Rank Approximation (LRA) algorithm, dividing a quadrilateral version of the 2D flattened LAA into 25 regions, each one composed of 200 nodes. This way, the ECAP of each node was obtained from the ECAP code of the region with the LRA algorithm. On the other hand, the ECAP decoding was carried out by means of truncated PCA in Morales-flat-FCN.

CNN-Based Model. The CNN-based model was inspired by the work of Li et al. [5] to predict the spatial distribution of a reaction-diffusion system. CNN architectures allow multi-scale feature extraction by combining convolution layers with several pooling operations, where the encoder represents each geometry as a condensed feature map. Later, this compacted feature vector can be decoded to produce an output, in this case, a map of ECAP. Since CNN mainly work with Euclidean datasets (e.g.. 2D/3D images), both network input and output must be a matrix. The flattening algorithm presented above allowed the ECAP map of the LAA to be depicted in a 2D plane, as well as the spatial coordinates of each mesh node: the input dataset was divided into three different channels with the spatial coordinates (x, y, z) of the nodes conforming the LAA surface. A graphical representation of the encoder-decoder CNN is displayed in Fig. 3. The input of the network was a three-channel tensor, depicting the spatial (x, y, z) coordinates of the nodes conforming the LAA surface. This tensor was then processed by the encoder part, where some convolutional operations were used to extract the features and max-pooling layers were used to downsample these features. Subsequently, the decoder allowed reversing the process in the encoder and to generate the predicted solution. This model was trained employing the resources provided by Google Colab[5] environment; to be specific, the NVIDIA

---

[3] https://keras.io/.

[4] https://www.tensorflow.org/.

[5] https://colab.research.google.com/.

Tesla P100[6] Graphic Processing Unit or GPU was used, drastically decreasing the training time.

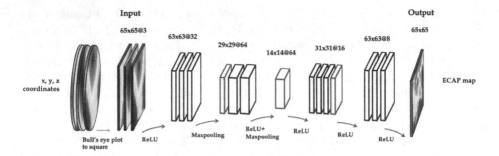

Fig. 3. Overall encoder-decoder CNN architecture. Given the features from the encoder part (red arrows), the ECAP prediction was obtained through multiple decoding operations (blue arrows). (Color figure online)

Network Hyperparameters and Performance Metrics. Adamax and Adam were selected as optimizers for the two FCN-based models and the mean square error (MSE) was selected as the loss function. The DL models were evaluated through a 10-fold Monte-Carlo cross-validation with 100 repetitions and the chosen metric was the mean absolute error (MAE) on ECAP values.

Thrombogenic Areas Detection. Additionally, a binary classification of the ECAP maps was performed to better assess the performance on detecting high thrombogenic risk areas. Following Di Achille et al. [3] we set an ECAP threshold value of $4\,Pa^{-1}$, considering higher values as positive risk and negative otherwise. Afterwards, the rate of true positives and negatives (TPR and TNR, respectively), the accuracy and F_1-score were estimated.

3   Results

The proposed flattening algorithm was successfully implemented in all available 206 LAA geometries. The average processing time of the flattening was 1.24 s per case. Figure 4 shows an example of a 3D LAA mesh and its corresponding flattened 2D representation.

The training of the DL networks was quick, requiring about 1 min and 30 s for each round of cross-validation in the case of the FCN-based models and about 9 min to complete the whole training in the CNN-based model. Considering that, once trained, predicting the ECAP is a near-instantaneous process, there was a reduction in computational time by several orders of magnitude comparing to CFD simulations, which typically take around two hours in the processed data.

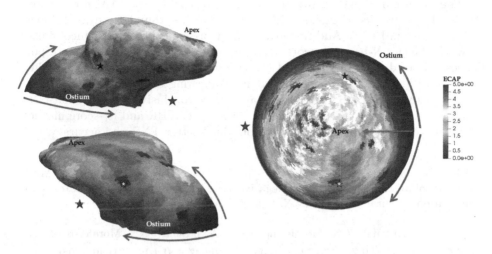

Fig. 4. Example of a LAA flattening result. Direction of radial (in red) and angular (in green and violet) mapping are displayed. The blue star represents the relative position of the circumflex artery; and the black and the orange ones, some correspondences. ECAP: Endothelial Cell Activation Potential. (Color figure online)

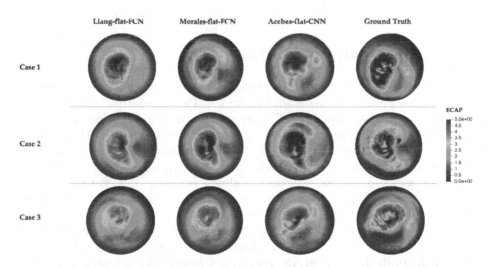

Fig. 5. Set of predicted ECAP values on a random dataset from a single iteration of the Monte Carlo cross-validation, 3 of which are shown in here. Each row corresponds to a given geometry while the columns from left to right display the predicted ECAP bull's eye plot mapping of Liang-flat-FCN, of Morales-flat-FCN and of Acebes-flat-CNN and the respective ground truth ECAP, obtained from CFD simulations. ECAP: Endothelial Cell Activation Potential.

The predicted ECAP 2D mapping of three representative LAAs for the three studied DL networks, alongside with their respective ground truth ECAP map, are displayed in Fig. 5. Additionally, Table 1 presents the performance of each network. The CNN-based network outperformed the FCN ones, using correspondences obtained with the flattened algorithm, in all computed metrics, including around a 10% improvement in MAE. Comparing to the registration-based algorithm (Morales-reg-FCN), results are not as conclusive since the flat-CNN approach achieved lower MAE together with higher TPR and F_1-score metrics, while the registration-based algorithm obtained better TNR and accuracy.

Table 1. Performance of DL models on a 10-fold Monte Carlo cross-validation. MAE: Mean Absolute Error. TNR: True Negative Rate. TPR: True Positive Rate. CNN: Convolutional Neural Network. FCN: Fully Connected Network

	Liang-flat-FCN	Morales-flat-FCN	Acebes-flat-CNN	Morales-reg-FCN
MAE	0.744 ± 0.29	0.727 ± 0.06	$\mathbf{0.627 \pm 0.10}$	0.646 ± 0.05
TNR	0.904	0.889	0.927	**0.976**
TPR	0.455	0.509	**0.719**	0.582
Accuracy	0.810	0.817	0.878	**0.914**
F_1-score	0.501	0.539	**0.725**	0.685

4   Discussion and Conclusions

The developed methodology has successfully been applied to a large dataset of synthetic LAA geometries to obtain a surrogate of a CFD-derived in-silico index for thrombogenic risk estimation. The proposed flattening algorithm allowed the use of state-of-the-art neural networks architectures such as CNN that require Euclidean representations of the data. Node correspondences were achieved in an easier and faster way comparing to surface registration approaches, without issues such as template selection and at almost real-time. The flattening algorithm can also be applied for a swift visual examination of LAA-based maps, without requiring rotations to the 3D structure.

Our results demonstrate that the CNN-based encoder-decoder architecture was a more efficient than FCN-based solutions, using the flattening-based discretization. On the other hand, the registration-based approach provided better accuracy metrics due to a practically perfect rate of true negatives. However, the Acebes-flat-CNN still presented a higher F_1-score and TPR metrics, which are more important for detecting high risk thrombogenic areas.

In conclusion, to the best of our knowledge, this study represents the first successful implementation of a CNN-based surrogate of CFD in a biological

---

[6] https://www.nvidia.com/en-us/data-center/tesla-p100/.

structure. The network obtains accurate ECAP maps based on the LAA geometry that can predict the areas with elevated risk of thrombosis orders of magnitude faster than any conventional CFD approach. Strategies to be studied in the future include the use of graph networks directly on the LAA meshes, therefore omitting the flattening step. Further work will be focused on applying the developed methodology to a large dataset of real left atria, with correspondences achieved with the flattening method proposed by Núñez-Garcia et al. [9].

Acknowledgments. This work was supported by the Spanish Ministry of Economy and Competitiveness under the María de Maeztu Units of Excellence Programme (MDM-2015-0502) and the Retos investigación project (RTI2018-101193-B-I00).

References

1. Crane, K., Weischedel, C., Wardetzky, M.: Geodesics in heat: a new approach to computing distance based on heat flow. ACM Trans. Graph. **32**(5), 1–11 (2013)
2. Cresti, A., et al.: Prevalence of extra-appendage thrombosis in non-valvular atrial fibrillation and atrial flutter in patients undergoing cardioversion: a large transoesophageal echo study. EuroIntervention **15**(3), e225–e230 (2019)
3. Di Achille, P., Tellides, G., Figueroa, C.A., Humphrey, J.D.: A haemodynamic predictor of intraluminal thrombus formation in abdominal aortic aneurysms. Proc. Roy. Soc. A Math. Phys. Eng. Sci. **470**(2172), 20140163–20140163 (2014)
4. Kreiser, J., Meuschke, M., Mistelbauer, G., Preim, B., Ropinski, T.: A survey of flattening-based medical visualization techniques. Comput. Graph. Forum **37**(3), 597–624 (2018)
5. Li, A., Chen, R., Farimani, A.B., Zhang, Y.J.: Reaction diffusion system prediction based on convolutional neural network. Sci. Rep. **10**(1), 1–9 (2020)
6. Liang, L., Liu, M., Martin, C., Sun, W.: A deep learning approach to estimate stress distribution: a fast and accurate surrogate of finite-element analysis. J. Roy. Soc. Interface **15**(138), 20170844 (2018)
7. Masci, A., et al.: The impact of left atrium appendage morphology on stroke risk assessment in atrial fibrillation: a computational fluid dynamics study. Front. Physiol. **9**(January), 1–11 (2019)
8. Morales, X., et al..: Deep learning surrogate of computational fluid dynamics for thrombus formation risk in the left atrial appendage. In: Statistical Atlases and Computational Models of the Heart, pp. 157–166 (2020)
9. Nunez-Garcia, M., et al.: Fast quasi-conformal regional flattening of the left atrium. IEEE Trans. Visual Comput. Graphics **26**, 2591–2602 (2020)

Measure Anatomical Thickness from Cardiac MRI with Deep Neural Networks

Qiaoying Huang[1], Eric Z. Chen[2], Hanchao Yu[3], Yimo Guo[2], Terrence Chen[2], Dimitris Metaxas[1], and Shanhui Sun[2($\boxtimes$)]

[1] Department of Computer Science, Rutgers University, Piscataway, NJ, USA
[2] United Imaging Intelligence, Cambridge, MA, USA
shanhui.sun@united-imaging.com
[3] University of Illinois at Urbana-Champaign, Urbana, IL, USA

Abstract. Accurate estimation of shape thickness from medical images is crucial in clinical applications. For example, the thickness of myocardium is one of the key to cardiac disease diagnosis. While mathematical models are available to obtain accurate dense thickness estimation, they suffer from heavy computational overhead due to iterative solvers. To this end, we propose novel methods for dense thickness estimation, including a fast solver that estimates thickness from binary annular shapes and an end-to-end network that estimates thickness directly from raw cardiac images. We test the proposed models on three cardiac datasets and one synthetic dataset, achieving impressive results and generalizability on all. Thickness estimation is performed without iterative solvers or manual correction, which is 100× faster than the mathematical model. We also analyze thickness patterns on different cardiac pathologies with a standard clinical model and the results demonstrate the potential clinical value of our method for thickness based cardiac disease diagnosis.

1 Introduction

Estimation of shape thickness from images is a fundamental task and has wide applications in practice. Particularly, in the medical domain, anatomical thickness plays a crucial role in disease diagnosis. For instance, abnormal changes of left ventricular myocardial (muscle wall) thickness are signs of many heart diseases [5,8,14,18].

This work specifically focuses on predicting the thickness of left ventricular walls, which can be acquired from cardiac Magnetic Resonance Imaging (MRI) [10,13,17]. There are two common ways to measure such heart anatomical thickness. The first method is to measure regional wall thickness (RWT) [11,12], which divides the heart into several regions and respectively averages the thickness to minimize the effort of manual measurement on the entire heart.

Q. Huang—This work was carried out during the internship of the author at United Imaging Intelligence, Cambridge, MA 02140.

E. Puyol Anton et al. (Eds.): STACOM 2020, LNCS 12592, pp. 44–55, 2021.
https://doi.org/10.1007/978-3-030-68107-4_5

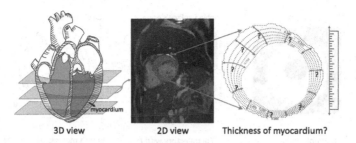

Fig. 1. Overview of 2D multi-slice cardiac MR data acquisition (left) and myocardium thickness measurement (right) from a corresponding 2D MR image (middle).

RWT only provides average values of myocardial regions, and hence has limitations in helping physicians understand precise thickness change for clinical diagnosis. A better solution is dense wall thickness (DWT) [3,5,8,16], which provides detailed thickness measurements at every location of the entire heart. It is also capable of dealing with some convoluted surfaces, which is not solvable by simply calculating the Euclidean distance between corresponding points of the inner and outer borders [3]. However, this approach suffers from a heavy computational burden since numerical methods are used to solve a second-order partial differential equation (PDE). For example, it takes approximately one minute to estimate the thickness of the brain cortex in a 3D volume [16]. Moreover, inner and outer contours need to be defined manually in some disconnected cases, which is time-consuming. Therefore, it is not feasible to apply this method for variously folded and faulted structures. An efficient and effective approach is highly desired (Fig. 1).

To the end, we consider estimating DWT directly from raw images in an automatic manner. However, it is non-trivial to derive an effective approach without utilizing shape information since the model needs to implicitly learn the shape to predict the thickness of the heart. Additionally, segmentation and thickness estimation are highly entangled. Once the segmentation changes, the thickness estimation has to adapt to a new shape from the last segmentation prediction, making the training hard to reach an equilibrium. Therefore, we start with a simpler problem that estimates thickness from a binary image (shape). Inspired by the recently proposed deep PDE solver [2], we introduce a deep learning-based thickness solver. The thickness estimation problem is more challenging than [2], since additional procedures are involved, such as integrating lines and interpolating missing values (refer to Sect. 2.1). Base on this solver, we then propose a novel end-to-end network for the original problem, which effectively decomposes the complex process into relatively easy sub-processes and utilizes shape information learned from the segmentation to benefit the thickness estimation.

Our major contributions are four aspects. (1) We introduce a fast thickness solver that estimates thickness from binary images and train it with a synthetic dataset, making it more generalizable to unseen shapes. (2) We propose a novel end-to-end network that predicts thickness directly from raw images. (3) We

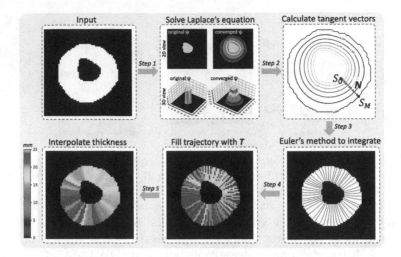

Fig. 2. Overview of dense thickness computation using the mathematical model. Given any shape, e.g., an annular binary mask, we first solve Laplace's equation to obtain the equipotential surfaces (contour lines in step 2). Then we calculate tangent vectors (red arrows in step 2) that are orthogonal to each surface. Euler's method is used to integrate the tangent vectors (red lines in step 3). Finally, an interpolation algorithm is performed to estimate the missing values (purple pixels in step 4) based on the potential surface and adjacent known thickness values.

conduct a comprehensive thickness pathology analysis to show the potential clinical values of the proposed framework. (4) To the best of our knowledge, this is the first DWT deep learning-based approach without hand-engineered steps nor manual annotations.

2 Method

2.1 Background

We first introduce the mathematical model for thickness estimation based on [3], which is used to generate training data for the proposed deep learning models. Suppose the input is a binary shape as in Fig. 2. We consider the electric potential value at the inner contour S to be the maximum (ψ) and that at the outer contour S' to be the minimum (0), where $\psi > 0$. Laplace's equation is a second-order PDE that defines the electric potential function over a region enclosed by boundaries S and S'. Mathematically, for a 2D surface, Laplace's equation is:

$$\nabla^2\psi = \frac{\partial^2\psi}{\partial x^2} + \frac{\partial^2\psi}{\partial y^2} = 0, \tag{1}$$

which is usually solved by numerical methods [9]. Step 1 of Fig. 2 shows the 2D and 3D views of the initial ψ and the converged ψ.

The next step is to compute the vector $N = (N_x, N_y)$ orthogonal to each potential surface, as shown in step 2 of Fig. 2. The path length of these integrated tangent vectors is the thickness between two corresponding points. Suppose $S_0 = (S_0(x), S_0(y))$ is the starting point at the inner border. A curve $S_0 S_1 S_2 \cdots$ is formed by iteratively taking small steps d_s along the tangent line until point S_n touches the outer border. We use Euler's method [4] to integrate:

$$S_{n+1}(x) = S_n(x) + N_x \cdot d_s, \quad S_{n+1}(y) = S_n(y) + N_y \cdot d_s. \tag{2}$$

All integrated lines are plotted in step 3 of Fig. 2. The thickness is computed as $T = \sum_{n=1}^{M} \sqrt{(S_n(x) - S_{n-1}(x))^2 + (S_n(y) - S_{n-1}(y))^2}$ for M points along the trajectory. Since the number of points at the inner boundary must be less than that of the outer boundary, the points with unknown thickness values need to be filled (the dark purple points in step 4 of Fig. 2). We propose to interpolate these points by using the known thickness points and the potential value ψ. Step 5 of Fig. 2 presents the final thickness result after interpolation. The speed of the mathematical model depends on thickness of a structure, the number of points at the inner border and the step size d_s in Eq. (2). It is often slow and needs manual correction. To address the above drawbacks, we propose deep learning-based approaches for thickness estimation.

2.2 Thickness Computation via a Fast Solver

We first focus on the thickness estimation that takes a binary image as input. We propose a fast solver to replace the mathematical model. The solver adopts a U-Net architecture [7], denoted as G. It takes a *binary image* s as input and estimates the thickness $\hat{y}$, represented as $\hat{y} - G(S)$. The goal is to minimize the difference between predicted thickness and ground truth thickness y, which is calculated from the mathematical model. This can be achieved by optimizing over an l_2 loss imposed on $\hat{y}$ and y along with an l_1 regularization on $\hat{y}$:

$$\ell_{\text{thick}}(y, \hat{y}) = \|y - \hat{y}\|_2^2 + \alpha\|\hat{y}\|_1, \tag{3}$$

where the l_1 is used to force sparsity on $\hat{y}$, since the thickness map exhibits many zeros in background. α is a weight to balance two losses. With a large number of diverse training data, the solver generalizes to even unseen shape and achieves $100\times$ speedup compared to the mathematical model.

2.3 Thickness Computation via an End-to-end Network

In this section, we propose an end-to-end network that predicts thickness directly from the *raw image*, as illustrated in Fig. 3. The network perceives the essential shape information during thickness estimation by learning the segmentation mask of the heart. It effectively disentangles shape understanding (segmentation) and thickness estimation tasks. The problem now becomes taking a raw image x as input and predicting both segmentation (shape) s and thickness y. To leverage semantic information from the segmentation s, we enforce the

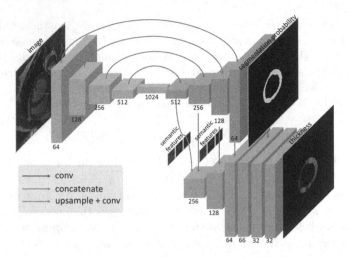

Fig. 3. Illustration of the proposed end-to-end network for thickness estimation directly from the raw image.

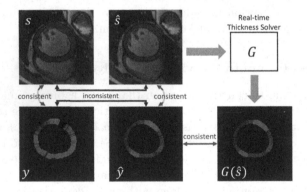

Fig. 4. Illustration of the shape inconsistency problem between predicted thickness $\hat{y}$ and ground truth y. Note that the underline predicted shapes $\hat{s}$ and ground truth shape s are different. During training, we replace the ground truth y using the proposed generic fast thickness solver G to remove this shape discrepancy.

decoder of the thickness estimator to utilize features generated from the segmentation decoder via feature map concatenation. The thickness y is generated based on s that learns features directly from x. The reason why decomposing the thickness estimation task into relatively easy sub-processes is that it is nontrivial to directly predict dense thickness from a raw image since it needs to infer a shape implicitly, which has proven by Sect. 2.1 and 2.2. In Sect. 3, we will show the results of training a U-Net with the loss function in Eq. (3) are blurry with notable artifacts. The U-Net attempts to regress a wide range of float point numbers without using shape information but obtains the mean of them.

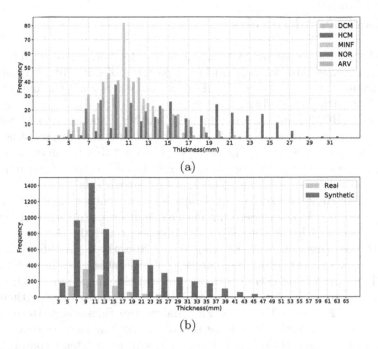

(a)

(b)

Fig. 5. (a) Maximum thickness value distribution of five cardiac diseases from ACDC data.(b) Maximum thickness value distribution from our synthetic data. For a comparison, the corresponding distribution from ACDC data (Real) is also included.

Therefore, to address the above limitations, we propose our novel end-to-end network that utilizes the semantic features from the segmentation task.

Another challenge is to find an optimal loss function. By training the model with a combined segmentation loss and thickness loss, we found that segmentation dominates the training and thickness estimation is not able to converge. This is caused by the shape inconsistency problem (Fig. 4), where the predicted shape $\hat{s}$ is inconsistent with the shape of the thickness ground truth y. Since the thickness decoder utilizes shape features from $\hat{s}$ to predict thickness y, it struggles between two different shapes between $\hat{s}$ and y, making it difficult to reach an equilibrium of training. Therefore we can not use the ground truth thickness y as the guidance for predicting thickness. The solution is to replace y with $G(\hat{s})$ as the ground truth for thickness estimation since $G(\hat{s})$ has the same shape as $\hat{s}$, where $G(\hat{s})$ is the generated thickness from the estimated mask $\hat{s}$ via the proposed fast thickness solver G. We aim to minimize the following loss:

$$\mathcal{L} = \ell_{\text{thick}}(G(\hat{s}), \hat{y}) + \beta \ell_{\text{seg}}(s, \hat{s}), \tag{4}$$

where ℓ_{seg} is cross entropy loss for segmentation and β is to balance two terms.

3   Experimental Results

For model training, we mainly experiment on two following datasets: (1) *ACDC* dataset [1] demonstrates various thickness patterns of different heart diseases, which contains 100 subjects (1,742 2D slices after removing the outliers) and includes five pathologies: dilated cardiomyopathy (DCM), hypertrophic cardiomyopathy (HCM), myocardial infarction (MINF), abnormal right ventricle (ARV) and normal people (NOR). We plot the maximum thickness value distribution of the five categories in Fig. 5(a). We observed that the maximum thickness of each 2D image ranges from 3 to 32 mm. DCM has the largest number of thickness values around 9–11 mm. This is consistent with the fact that DCM is a disease with an enlarged and less functional left ventricular than normal people. HCM patient has thicker myocardium in diastole in several myocardial segments compared to a healthy person. We randomly separated the 100 patients into training (60 patients, 1,016 slices), validation (20 patients, 388 slices) and test (20 patients, 338 slices) set with equal ratio of five pathologies in each dataset. The pixel spacing is normalized to 1.36 mm and the image size is adjusted to 192×192 by center cropping and zero padding. We applied the mathematical model in Sect. 2.1 to ACDC data and analyze the thickness pattern in each disease category. (2) *Synthetic* dataset is built to cover more general and diverse binary annular shapes, as Fig. 5(b) shown. The synthetic dataset contains about 6,000 images and ranges from 0 to 64 mm that has a wider distribution compared to ACDC. We propose the synthetic dataset since the input of the fast thickness solver are binary images, it is easy to synthesize. We also want to show the solver has a good generalizability and can be trained without the burden of annotating data. The synthesis process is to first generate binary annular shape with random radius and position, and then apply elastic transform and piecewise affine transform with random parameter settings. After generating the synthetic dataset, we also applied the mathematical model to generate the ground truth thickness value.

All models are trained with Adam optimizer with learning rate of 10^{-4} and 350 epochs. We set $\alpha = 10^{-3}$ in Eq. (3). The evaluation metrics are Mean Absolute Error (MAE) and Mean Square Error (MSE) inside myocardium regions.

Thickness Estimation on Binary Image. Since there is **NO** other deep learning-based dense thickness estimation method, we consider the Auto-Encoder as a baseline model. U-Net-k as our thickness estimation models (k is

Table 1. Results of the proposed thickness solvers on ACDC and synthetic datasets.

Model	ACDC		Synthetic	
	MAE (mm)	MSE (mm)	MAE (mm)	MSE (mm)
Au-En	1.061 (0.284)	3.472 (2.665)	0.815 (0.156)	2.219 (1.253)
U-Net-2	0.349 (0.067)	0.214 (0.096)	0.335 (0.061)	0.198 (0.090)
U-Net-3	0.355 (0.072)	0.221 (0.109)	0.322 (0.058)	0.185 (0.082)
U-Net-4	0.350 (0.069)	0.212 (0.104)	**0.321 (0.060)**	**0.185 (0.085)**

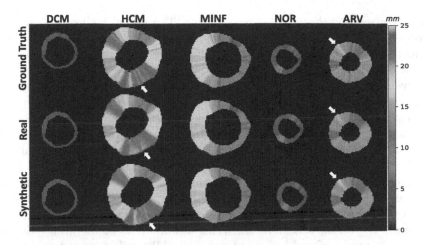

Fig. 6. Examples of thickness prediction by U-Net-4 that are trained on real ACDC data or synthetic data. Tests were performed on ACDC test data including four disease categories (DCM, HCM, MINF, ARV) and one normal category (NOR). The white arrows indicate the area that the U-Net-4 trained with synthetic data performs better than with Real data.

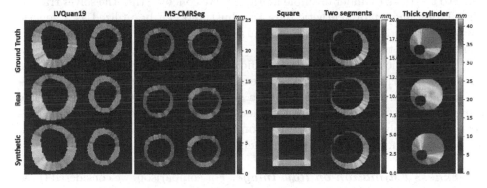

Fig. 7. Examples of thickness prediction on unseen data with the fast thickness solver.

the number of downsampling and upsampling blocks). These models are trained on real data and synthetic data, respectively. For a fair comparison, all models are evaluated on the same ACDC test set.

The results of thickness estimation are reported in Table 1. U-Net-k models outperform Auto-Encoder in MAE and MSE, demonstrating that the skip-connection is essential in thickness estimation. We did not report models with deeper layers since no further improvement compared to the U-Net-4. Furthermore, models trained on synthetic data work better than those with real data, indicating the benefit of including more diverse cases. Figure 6 presents examples of thickness estimation with different cardiac pathologies from models trained on real data and synthetic data, respectively. The model trained on synthetic data produces more similar results to the ground truth, especially in the cases of DCM and MINF. Even in the most challenging case of HCM, the model with synthetic data training recovers better details.

To validate the generalizability of our proposed thickness model, we extended evaluations on two unseen real test datasets and one unseen synthetic test dataset. The real datasets are *LVQuan19* [15] and *MS-CMRSeg* [19], where manual annotations of the myocardium are provided. There are 56 normal subjects with 1,120 2D images in *LVQuan19* and 45 normal subjects with 270 2D images in *MS-CMRSeg*. The synthetic test dataset (*Gen-Data*) includes three additional shapes that are unseen in the training set: square and thick cylinder and two segments. The validation results are reported in Table 2. Both models trained on real data (ACDC) and synthetic data respectively achieve satisfying accuracy, demonstrating that our proposed thickness solver has good generalizability. The model trained on synthetic data outperforms the one on real data in terms of MAE and MSE. Figure 7 illustrates the validation results. The solver trained on the synthetic data performs well on all tested data sets while the solver trained on the ACDC data fails on extremely thick structures (16.0 mm MSE in Table 2, refer to the thick cylinder example in Fig. 7). We conclude that the solver trained on a large amount of synthetic data is robust and can be utilized as a generic thickness solver. This mitigates the data scarcity problem for training a deep learning based thickness solver.

The fast thickness solver is more efficient than the mathematical model. As shown in Fig. 7, for disconnected shapes like the two segments, the mathematical model requires the manual definition of the inner and outer boundaries while our fast thickness solver can directly calculate thickness without the manual processing. Our solver only needs 0.35 seconds on average for thickness prediction, around 100 times faster than the mathematical model.

Thickness Estimation on Raw Image. For comparison, we considered three end-to-end models: (A) The baseline model U-Net-4 trained with loss function Eq. (3). (B) The proposed end-to-end model trained with loss function Eq. (4) but replacing $G(\hat{s})$ with ground truth thickness y (E2ET-gt). (C) The proposed end-to-end model trained with loss function Eq. (4) (E2ET-solver). Note that

Table 2. Results of model generalizability.

Dataset	Model	MAE (mm)	MSE (mm)
LVQuan19	ACDC	0.369 (0.045)	0.240 (0.063)
	Synthetic	**0.332 (0.038)**	**0.194 (0.047)**
MS-CMRSeg	ACDC	0.308 (0.024)	0.164 (0.029)
	Synthetic	**0.285 (0.022)**	**0.143 (0.021)**
Gen-Data	ACDC	1.945 (2.235)	15.969 (22.349)
	Synthetic	**0.438 (0.217)**	**0.445 (0.342)**

Table 3. Comparison of our proposed end-to-end model (E2ET-solver) and two baseline models (U-Net-4 and E2ET-gt) on three cardiac datasets.

Model	*ACDC*	*LVQuan19*	*MS-CMRSeg*
U-Net-4	0.124 (0.060)	0.241 (0.148)	0.151 (0.077)
E2ET-gt	0.071 (0.046)	0.143 (0.103)	0.086 (0.053)
E2ET-solver	**0.067 (0.043)**	**0.121 (0.063)**	**0.085 (0.053)**

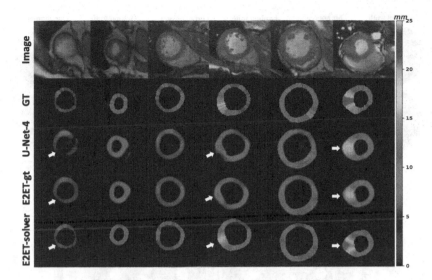

Fig. 8. Examples of direct thickness prediction from images using proposed end-to-end method (E2ET-solver) and other methods (E2ET-gt and U-Net-4). The images are from ACDC data. The white arrows indicate the area that the E2ET-solver performs better than other models.

E2ET-solver utilizes a fast thickness solver G, which is pre-trained on the synthetic dataset (see Sect. 3). In addition, we set $\beta = 15$ in Eq. (4). We take into account the segmentation errors so that MAE is computed on the whole image region. All models are trained on ACDC training set and evaluated on ACDC test set, *LVQuan19* and *MS-CMRSeg* dataset.

Table 3 summarizes the results. Our proposed E2ET-solver achieves the smallest MAE on all three datasets. Figure 8 illustrates examples of thickness prediction from raw images utilizing different algorithms. U-Net-4 shows large errors, which indicates the difficulties of estimating thickness from a raw image without utilizing shape information. E2ET-gt shows purple artifacts at the boundary regions especially for extreme challenging cases (e.g., the thin and thick regions), which may be due to the shape inconsistency as discussed in Sect. 2.3. On the other hand, by disentangling the segmentation task and enforcing shape consistency, E2ET-solver shows no such artifacts and the thickness prediction is very similar to the ground truth even in the difficult cases.

Qualitatively, we use the polar plot of the 17-segment model [6] to visualize different pathologies thickness patterns (can be divided into more segments for disease diagnosis) in Fig. 9. For example, the DCM patient exhibits thin myocardium as overall small thickness values (more purple color than normal people) while the HCM patient exhibits thick myocardium as overall large thickness values (more red color than normal people). Several thin regions can be found in the MINF patient. These observations are clinical symptoms of the corresponding cardiac diseases and often used as diagnostic criteria. Quantita-

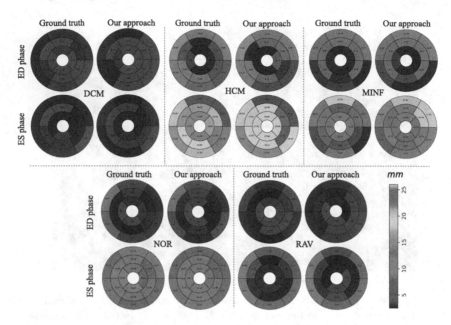

Fig. 9. Examples of 17-segment regional thickness of ground truth and E2ET-solver at ED and ES phases. Zoom in to see the thickness value in the middle of each segment. Five categories of ACDC data are displayed.

tively, the E2ET-solver achieves MAE of 0.64 ± 0.10 mm, 0.73 ± 0.14 mm and 1.08 ± 0.05 mm in the basal, mid-cavity and apical region, respectively. Thickness measurement near the apical region is more challenging due to the small region and the out of plane motion. The results demonstrate the feasibility of these automated measurements for cardiac disease diagnosis.

4   Conclusions

In this work, we propose an efficient and accurate thickness solver to replace the conventional mathematical model as well as a novel end-to-end model that directly estimates thickness from raw images. Our methods demonstrate promising performance on several benchmarks. The fast solver shows its efficiency and good generalizability on unseen shapes. The end-to-end model demonstrates its superior performance compared to other baselines. We also analyze the thickness patterns of different pathologies to show the clinical value of our model in cardiac disease diagnosis.

References

1. Bernard, O., et al.: Deep learning techniques for automatic MRI cardiac multi-structures segmentation and diagnosis: is the problem solved? IEEE Trans. Med. Imaging **37**(11), 2514–2525 (2018)

2. Hsieh, J.T., Zhao, S., Eismann, S., Mirabella, L., Ermon, S.: Learning neural pde solvers with convergence guarantees. arXiv preprint arXiv:1906.01200 (2019)
3. Jones, S.E., Buchbinder, B.R., Aharon, I.: Three-dimensional mapping of cortical thickness using Laplace's equation. Hum. Brain Mapp. **11**(1), 12–32 (2000)
4. Kendall, A.: An Introduction to Numerical Analysis. Wiley, Hoboken (1989)
5. Khalifa, F., Beache, G.M., Gimelrfarb, G., Giridharan, G.A., El-Baz, A.: Accurate automatic analysis of cardiac cine images. IEEE Trans. Biomed. Eng. **59**(2), 445–455 (2012)
6. Cerqueira, M.D., et al.: Standardized myocardial segmentation and nomenclature for tomographic imaging of the heart: a statement for healthcare professionals from the cardiac imaging committee of the council on clinical cardiology of the American heart association. Circulation 105(4), 539–542 (2002)
7. Ronneberger, O., Fischer, P., Brox, T.: U-Net: convolutional networks for biomedical image segmentation. In: Navab, N., Hornegger, J., Wells, W.M., Frangi, A.F. (eds.) MICCAI 2015. LNCS, vol. 9351, pp. 234–241. Springer, Cham (2015). https://doi.org/10.1007/978-3-319-24574-4_28
8. Sliman, H., Elnakib, A., Beache, G.M., Elmaghraby, A., El-Baz, A.: Assessment of myocardial function from cine cardiac MRI using a novel 4D tracking approach. J. Comput. Sci. Syst. Biol. **7**, 169–173 (2014)
9. William, H.P., Saul, A.T., William, T.V., Brian, P.F.: Numerical Recipes in C: The Art of Scientific Computing. Cambridge University Press, Cambridge (1992)
10. Wu, P., Huang, Q., Yi, J., Qu, H., Ye, M., Axel, L., Metaxas, D.: Cardiac MR image sequence segmentation with temporal motion encoding. In: ECCV 2020 Workshop on BioImage Computing (2020)
11. Xue, W., Brahm, G., Pandey, S., Leung, S., Li, S.: Full left ventricle quantification via deep multitask relationships learning. Med. Image Anal. **43**, 54–65 (2018)
12. Xue, W., Nachum, I.B., Pandey, S., Warrington, J., Leung, S., Li, S.: Direct estimation of regional wall thicknesses via residual recurrent neural network. In: Niethammer, M., et al. (eds.) IPMI 2017. LNCS, vol. 10265, pp. 505–516. Springer, Cham (2017). https://doi.org/10.1007/978-3-319-59050-9_40
13. Yang, D., Huang, Q., Axel, L., Metaxas, D.: Multi-component deformable models coupled with 2D-3D U-Net for automated probabilistic segmentation of cardiac walls and blood. In: 2018 IEEE 15th International Symposium on Biomedical Imaging, pp. 479–483 (2018)
14. Yang, D., Huang, Q., Mikael, K., Al'Aref, S., Axel, L., Metaxas, D.: MRI-based characterization of left ventricle dyssynchrony with correlation to CRT outcomes. In: 2020 IEEE 17th International Symposium on Biomedical Imaging, pp. 1–4 (2020)
15. Yang, G., Hua, T., Xue, W., Li, S.: Lvquan19, September 2019. https://lvquan19.github.io/
16. Yezzi, A.J., Prince, J.L.: An Eulerian PDE approach for computing tissue thickness. IEEE Trans. Med. Imaging **22**(10), 1332–1339 (2003)
17. Yu, H., Chen, X., Shi, H., Chen, T., Huang, T.S., Sun, S.: Motion pyramid networks for accurate and efficient cardiac motion estimation. arXiv preprint arXiv:2006.15710 (2020)
18. Yu, H., et al.: FOAL: fast online adaptive learning for cardiac motion estimation. In: Proceedings of the IEEE/CVF Conference on Computer Vision and Pattern Recognition, pp. 4313–4323 (2020)
19. Zhuang, X.: Multivariate mixture model for myocardial segmentation combining multi-source images. IEEE Trans. Pattern Anal. Mach. Intell. **41**, 2933–2946 (2018)

Modelling Cardiac Motion via Spatio-Temporal Graph Convolutional Networks to Boost the Diagnosis of Heart Conditions

Ping Lu[1]([✉]), Wenjia Bai[2,3], Daniel Rueckert[2], and J. Alison Noble[1]

[1] Institute of Biomedical Engineering, University of Oxford, Oxford, UK
ping.lu@eng.ox.ac.uk
[2] Department of Computing, Imperial College London, London, UK
[3] Department of Brain Sciences, Imperial College London, London, UK

Abstract. We present a novel spatio-temporal graph convolutional networks (ST-GCN) approach to learn spatio-temporal patterns of left ventricular (LV) motion in cardiac MR cine images for improving the characterization of heart conditions. Specifically, a novel GCN architecture is used, where the sample nodes of endocardial and epicardial contours are connected as a graph to represent the myocardial geometry. We show that the ST-GCN can automatically quantify the spatio-temporal patterns in cine MR that characterise cardiac motion. Experiments are performed on healthy volunteers from the UK Biobank dataset. We compare different strategies for constructing cardiac structure graphs. Experiments show that the proposed methods perform well in estimating endocardial radii and characterising cardiac motion features for regional LV analysis.

Keywords: Spatio-temporal graph convolutional networks · Cardiac MR · Endocardium · Epicardium · Motion-characteristic features

1 Introduction

Cardiac motion estimation plays an important role in the diagnosis of heart condition [6–8,10]. Motion-characteristic features, such as time series of the endocardial radius, thickness, and radial strain (Err), can be evaluated by the sampling nodes of the endocardium and the epicardium from magnetic resonance imaging (MRI). These features are related to cardiac disease and they are easy to explain as characteristics of pathological cardiac motion [6]. The motion trajectory estimation of these sampling nodes can be regarded as motion analysis in computer vision.

In recent years, geometric deep learning-based methods have achieved promising results in motion recognition. Yan et al. [9] proposed a spatial-temporal graph neural network (ST-GCN) for human action recognition via

© Springer Nature Switzerland AG 2021
E. Puyol Anton et al. (Eds.): STACOM 2020, LNCS 12592, pp. 56–65, 2021.
https://doi.org/10.1007/978-3-030-68107-4_6

motion classification. Li et al. [5] suggested actional-structural GCN for human action prediction using skeleton and joint trajectories of human bodies. Huang et al. [3] predicted human trajectories with a spatio-temporal graph attention network (STGAT). These applications of GCNs [4] extend convolutional neural networks (CNNs) to graphs of arbitrary structures and model dynamic graphs over time sequences, which can be applied to cardiac motion analysis.

For the diagnosis of heart conditions, especially for early stage characterisation, a major challenge is estimating the effect of cardiac functional changes via automated cardiac motion analysis. Before more significant cardiac structural changes occur, functional changes such as motion and strain may occur, which are subtle but often indicates the early onset of cardiac disease symptoms. But the early onset of symptoms already causes an increased strain on the heart [6].

In this paper, we propose a geometric deep learning-based architecture with a self-supervised strategy. This method characterises the spatio-temporal patterns of left ventricular (LV) cardiac motion in cine MR image sequences to boost the diagnosis of heart conditions. We investigate two strategies of graph construction and cardiac motion estimation. The proposed method is to predict node locations on the endocardium and the epicardium separately. This requires cardiac structure graphs constructed from the endocardium and the epicardium respectively. The other method is node locations prediction on the endocardium and the epicardium simultaneously, which needs a graph constructed from both endocardium and epicardium. Moreover, we extract motion-characteristic features and time series of the endocardial radius, thickness, and radial strain (Err), based on the output cardiac motion trajectory, and compare these features between the prediction and the ground truth in a group of healthy volunteers.

Contributions. The contributions of this work are as follows. (1) To our knowledge, this is the first time that cardiac motion has been modelled with a geometric deep learning-based architecture. (2) The predicted cardiac structure trajectory of this method can be used to evaluate motion characteristics, namely a time series of the endocardial radius, thickness, and Err. (3) We demonstrate that spatio-temporal patterns achieve good performance for cardiac motion estimation and regional analysis of LV function.

2   Method

2.1   Cardiac Structure Graph Construction

The high-level steps in cardiac structure graph construction are summarised in Fig. 1. The cardiac structure sequence is represented by 2D coordinates of nodes on both the endocardium and epicardium in each cardiac MR frame. These nodes are chosen by the left and right ventricle geometry, according to the mid-slice 6-segments model of the 17-Segment AHA model [2]. Firstly, we select the barycenter of the left ventricle (LV) and the right ventricle (RV) in the middle slice of the short axis view image. Secondly, we define the straight line between these two nodes as the initial line [6]. Thirdly, we rotate this initial line around the barycenter node of the LV by every 15 degrees. The intersection nodes of the rotated line and the endocardial and epicardial borders are chosen. These intersection node locations, 2D pixel coordinates (x, y), are obtained in every single frame. Morphological transformations and finding the barycenter location of the LV and the RV are implemented using OpenCV. Later, these nodes will be classified into the mid-slice 6-segments model for the regional analysis of left ventricular function. Moreover, these selected node locations are the ground truth in our work.

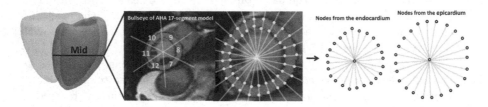

Fig. 1. Overview of the proposed framework for the cardiac structure graph construction in the mid-ventricular of short-axis view cardiac MR image sequences. The barycenter of the left ventricle (LV) and the 24 node locations from both the endocardium and epicardium are considered as cardiac structure graphs.

We construct two undirected spatio-temporal graphs $G = (V, E)$ on the cardiac structure with N nodes and T frames. As shown in Fig. 2, one graph features both intra-endocardium and inter-frame connection, the other graph includes both intra-epicardium and inter-frame connection. These two graphs have the same structure. In the graph, the node set $V = \{v_{it} | i = 1, ..., N, t = 1, ..., T\}$ includes all the 25 nodes in the cardiac structure sequence. The feature vector on a node $F(v_{it})$ is the 2D pixel coordinate (x, y) of the i-th node on frame t.

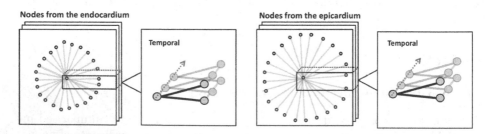

Fig. 2. Overview of the cardiac structure as a spatio-temporal graph. One graph (left side) represents nodes from the endocardium, the other graph (right side) represents nodes from epicardium. In both graphs, the inter-frame edges connect the same nodes between consecutive sampling frames. Node locations, 2D pixel coordinates (x,y), are defined as inputs to the spatio-temporal graph convolutional networks.

The spatio-temporal graphs on the cardiac structure sequences are constructed from the endocardium and the epicardium respectively. In the graph constructed from the endocardium, nodes from the endocardium in the same frame are connected to the barycenter of the LV with edges, which is presented in Fig. 1. After that, each node will be connected to the same node in the consecutive sampling frame, as described in Fig. 2. The graph constructed from the epicardium follows the same steps as the graph constructed from the endocardium. Based on the validation performance, every third frame is selected.

The edge set E consists of two subsets. The first subset describes the intra-structural connection at each cardiac MR frame, denoted as $E_s = \{(v_{it}, v_{jt})|(i,j) \in H\}$, where H is the set of nodes from both the endocardium and epicardium separately. The second subset includes the inter-frame edges, $E_f = \{(v_{it}, v_{i(t+1)})\}$, which connect the same nodes in consecutive sampling frames. Hence, for every node n, the edge in E_f records its trajectory over the time period.

2.2 Spatio-Temporal Graph Convolutional Neural Network

Let $A \in \{0,1\}^{N \times N}$ represent the adjacency matrix of the cardiac structure graph. If the i-th and the j-th nodes are connected, $A_{i,j} = 1$. Otherwise, $A_{i,j} = 0$. Let $D \in \mathbb{R}^{N \times N}$ indicate the diagonal degree matrix where $D_{i,i} = \sum_j A_{i,j}$.

Spatio-Temporal GCN. Spatio-temporal GCN (ST-GCN) contains a range of the ST-GCN blocks [9]. Each block includes a spatial graph convolution and then a temporal convolution, which extracts spatial and temporal features respectively. The crucial element of ST-GCN is the spatial graph convolution operation, which provides a weighted average of neighboring features for each node.

Let $F_{in} \in \mathbb{R}^{C \times T \times N}$ be the input features, where C is the number of channels, T is the temporal length and N is the number of nodes in one frame. Let $F_{out} \in \mathbb{R}^{C \times T \times N}$ be the output features obtained from the spatial graph convolution. The spatial graph convolution is described as

$$F_{out} = \sum M \circ \widetilde{A} F_{in} W.$$

Here M is the edge weight matrix, W is the feature importance matrix, $\widetilde{A} = D^{-\frac{1}{2}} A D^{-\frac{1}{2}}$ is the normalized adjacent matrix, $\circ$ represents the Hadamard product.

In the temporal dimension, there are 2 corresponding nodes in the adjacent frames. Therefore, the number of neighbors for each node is 2. A $1D$ temporal convolution ($K_t \times 1$ convolution) is applied to extract features, K_t is the kernel size of the temporal dimension.

Encoder. First of all, we normalize input cardiac structures with a batch normalization layer. The ST-GCN model contains 9 layers of spatial-temporal graph convolution blocks (ST-GCN blocks). The output channel for these layers are

64, 64, 64, 128, 128, 128, 256, 256, 256 (see Fig. 3). Each layer has a temporal kernel size of 9. The ResNet mechanism is applied to each ST-GCN block. The strides of the 4-*th* and the 7-*th* temporal convolution layers are set to 2 as the pooling layer.

Decoder. There are 5 layers of ST-GCN blocks and the output channel of each layer is 128 (see Fig. 3). The output of these blocks and the previous motion status are concatenated and put into an ST-GCN block. The output channel of this block is 2. The sum of the output of the decoder and the previous motion status predicts the future 2D node locations which represent the future cardiac motion trajectory. The loss function for future prediction in one frame is the mean squared error (MSE) of the node locations.

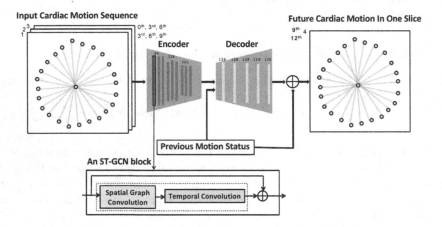

Fig. 3. Network overview. The cardiac motion sequence is given as input to the ST-GCN encoder-decoder framework. The output of the encoder and the previous motion status are fed into the decoder. The sum of the output of the decoder and the previous motion status represents the output future cardiac motion trajectory (predicted node locations), which is used in the left ventricular function evaluation. ⊕ denotes the element-wise addition.

3   Experiments

3.1   Data Acquisition

In this study, we use short-axis view cardiac MR image sequences from the UK BioBank[1]. A stack of short-axis images, around 12 slices, cover the entire left and right ventricles. The 3 slices from the middle slice of short-axis images are selected for experiments. In-plane resolution is 1.8×1.8 mm^2, while the

---

[1] UK BioBank https://www.ukbiobank.ac.uk/.

slice gap is 2.0 mm and the slice thickness is 8.0 mm. Each sequence contains 50 consecutive time frames per cardiac cycle. We randomly selected image sequences of 1071 subjects for training, 270 subjects for validation and 270 subjects for testing.

3.2 Implementation Details

Pre-processing. The segmentation of the LV endocardial and epicardial borders and the RV was generated from using the FCN method proposed by Bai et al. [1] and used for nodes extraction. For training and testing, we obtained 1 barycenter node location of the LV and 24 node locations from both the endocardial and epicardial borders of the myocardium from the sampling frames shown in Fig. 4. The detail of nodes extraction are described in Sect. 2.1 and Fig. 1. The input features are described with tensors (C, T, N). Here C denotes 2 channels for the 2D pixel coordinate (x, y). T denotes 3 consecutive sampling time frames. N denotes 25 nodes in each time frame.

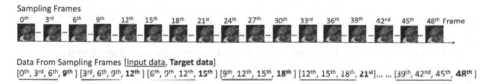

Fig. 4. Data preparation. Every third frame is selected from a cardiac MR image sequence. After sampling, 17 selected time frames denote a cardiac cycle for a sequence. The input cardiac structure data are extracted from 3 consecutive sampling frames and are used for cardiac motion estimation in the next sampling frame.

Training. The model is trained over 500 epochs using stochastic gradient descent with a learning rate of 0.01. The learning rate is decayed by 0.1 after every 10 epochs until the $50th$ epoch, after which the learning rate is constant. We set the input cardiac structure sequence length to 3 frames. As shown in Fig. 4, 3 consecutive sampling time frames are selected for each sample. The proposed network was implemented using Python 3.7 with Pytorch. Experiments are run with computational hardware GeForce GTX 1080 Ti GPU 10 GB.

3.3 Results

Quantitative Results. We compared three cardiac structure graphs and measured the predicted node locations with the ground truth using the mean square error (MSE). In the first cardiac structure graph, 24 nodes from the endocardium connect the centre node of the LV with edges. In the second graph, 24 nodes from the epicardium connect the centre node of the LV with edges. In the third graph,

48 nodes from both the endocardium and epicardium connect the centre node of the LV with edges. As shown in Fig. 5, the second graph, graph nodes from the epicardium, have the least MSE and achieved better-predicted performance.

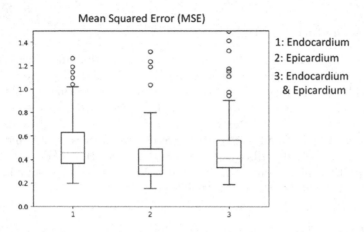

Fig. 5. Box plot of the mean square error (MSE) for three different cardiac structure graphs using the proposed method. Graph nodes are from the endocardium, the epicardium, and both endocardium and epicardium. Each cardiac structure graph includes a node from the centre of the left ventricle.

Representative Examples. There are 14 predicted sampling frames (frames 9–48, in increments of 3 frames (i.e 9, 12, 15 etc.)) of node locations in a cardiac cycle for each sequence. Figure 6 shows an example of cardiac motion estimation on frames 9, 18, 27, 36, 45 of the MRI sequence between the proposed method with cardiac structure graph constructed from the endocardium and the epicardium, and a graph constructed from both the endocardium and epicardium. Both methods better predict the location of the epicardium nodes than the endocardium nodes. The node locations are predicted very accurately on the epicardium by both methods, especially on the 36th and 45th frame.

Left Ventricular Function Evaluation. Based on the 17-Segment AHA model, the prediction of 24 node locations from both the endocardium and epicardium in each frame is classified into 6 segments [2]. Figure 7 shows an example of a time series of the endocardial radius, thickness, radial strain (Err) [6] in the six segments of myocardium from the prediction and the ground truth of nodes' locations of a healthy volunteer. Compared to the ground truth, the prediction of the endocardial radius achieves better performance. The thickness and the Err has a similar plot shape.

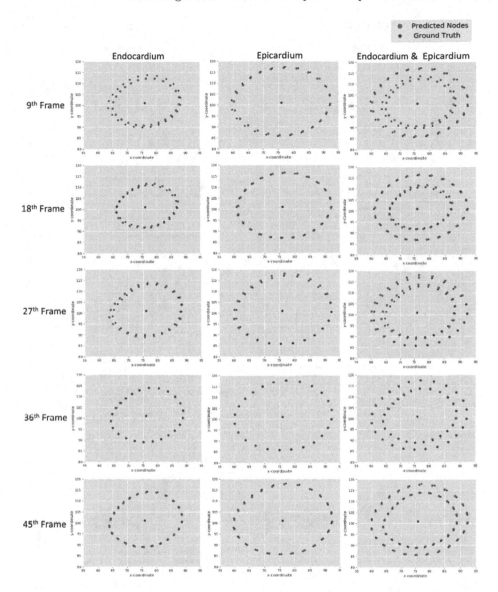

Fig. 6. Cardiac motion estimation comparison between the proposed method with different cardiac structure graphs and the ground truth. Both of these cases are modelled by three cardiac structure graphs constructed from the endocardium, the epicardium and both the endocardium and the epicardium. The predictions and the ground truth on the 9th, 18th (end-systolic), 27th, 36th and 45th frame of the MRI sequence are shown there.

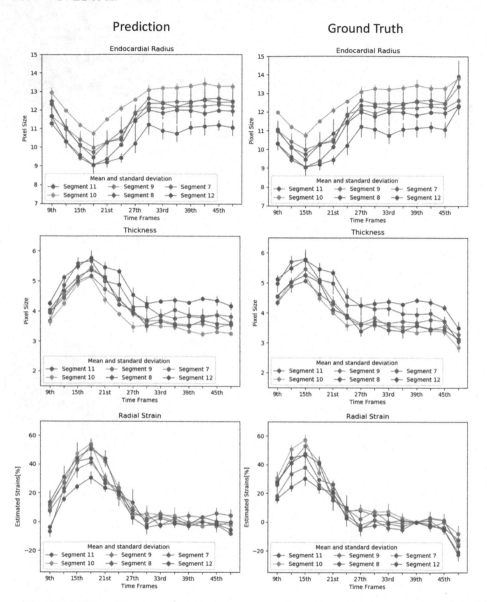

Fig. 7. Example results of the estimated endocardial radius (mean and standard deviation shown), thickness (mean and standard deviation shown) and radial strain (mean and stan- dard deviation shown) for cardiac segments (7–12) plotted on 9th, 12th, 15th, 18th (end-systolic), 21th, 24th, 27th, 30th, 33rd, 36th, 39th, 42th, 45th and 48th frame over a cardiac cycle. Results for prediction (top), and ground truth (bottom) in a healthy volunteer.

4   Discussion and Conclusion

In this work, we propose a spatio-temporal graph convolutional network (ST-GCN) to characterise cardiac motion. We evaluated and compared the effect of different cardiac structure graphs on cardiac motion estimation. In terms of accuracy, all the cardiac structure graphs yield good performance, especially the prediction of the node locations on the epicardium. In the next step, we will compare the proposed method with other deep learning-based architectures, such as [6]. Future work will extend this method to analyse the basal and apical slices of the LV for both healthy heart and ischaemic heart disease.

Acknowledgements. This work is supported by SmartHeart. EPSRC grant EP/P001009/1. This research has been conducted using the UK Biobank Resource under Application Number 40119.

References

1. Bai, W., et al.: Automated cardiovascular magnetic resonance image analysis with fully convolutional networks. J. Cardiovasc. Magn. Reson. **20**(1), 65 (2018)
2. Cerqueira, M.D., et al.: Standardized myocardial segmentation and nomenclature for tomographic imaging of the heart: a statement for healthcare professionals from the cardiac imaging committee of the council on clinical cardiology of the American heart association. Circulation **105**(4), 539–542 (2002)
3. Huang, Y., Bi, H., Li, Z., Mao, T., Wang, Z.: STGAT: modeling spatial-temporal interactions for human trajectory prediction. In: Proceedings of the IEEE International Conference on Computer Vision, pp. 6272–6281 (2019)
4. Kipf, T.N., Welling, M.: Semi-supervised classification with graph convolutional networks. arXiv preprint arXiv:1609.02907 (2016)
5. Li, M., Chen, S., Chen, X., Zhang, Y., Wang, Y., Tian, Q.: Actional-structural graph convolutional networks for skeleton-based action recognition. In: Proceedings of the IEEE Conference on Computer Vision and Pattern Recognition, pp. 3595–3603 (2019)
6. Lu, P., Qiu, H., Qin, C., Bai, W., Rueckert, D., Noble, J.A.: Going deeper into cardiac motion analysis to model fine spatio-temporal features. In: Papież, B.W., Namburete, A.I.L., Yaqub, M., Noble, J.A. (eds.) MIUA 2020. CCIS, vol. 1248, pp. 294–306. Springer, Cham (2020). https://doi.org/10.1007/978-3-030-52791-4_23
7. Qin, C., et al.: Joint learning of motion estimation and segmentation for cardiac MR image sequences. In: Frangi, A.F., Schnabel, J.A., Davatzikos, C., Alberola-López, C., Fichtinger, G. (eds.) MICCAI 2018. LNCS, vol. 11071, pp. 472–480. Springer, Cham (2018). https://doi.org/10.1007/978-3-030-00934-2_53
8. Qiu, H., Qin, C., Le Folgoc, L., Hou, B., Schlemper, J., Rueckert, D.: Deep learning for cardiac motion estimation: supervised vs. unsupervised training. In: Pop, M., et al. (eds.) STACOM 2019. LNCS, vol. 12009, pp. 186–194. Springer, Cham (2020). https://doi.org/10.1007/978-3-030-39074-7_20
9. Yan, S., Xiong, Y., Lin, D.: Spatial temporal graph convolutional networks for skeleton-based action recognition. In: Thirty-Second AAAI Conference on Artificial Intelligence (2018)
10. Zheng, Q., Delingette, H., Ayache, N.: Explainable cardiac pathology classification on cine MRI with motion characterization by semi-supervised learning of apparent flow. Med. Image Anal. **56**, 80–95 (2019)

Towards Mesh-Free Patient-Specific Mitral Valve Modeling

Judit Ros[1]([✉]), Oscar Camara[1], Uxio Hermida[2], Bart Bijnens[1,3],
and Hernán G. Morales[2]

[1] Physense, BCN MedTech, Department of Information and Communication
Technologies, Universitat Pompeu Fabra, Barcelona, Spain
juditros01@estudiant.upf.edu
[2] Philips Research Paris, Suresnes, France
[3] ICREA, Barcelona, Spain

Abstract. Computational modeling is a tool that has gained importance recently to better understand valve physiopathology, to assess
safety and efficacy of cardiovascular devices and as a supporting tool
for therapy planning. Mesh-based methods, such as the Finite Element
Method (FEM), have shown high accuracy and application modeling
the mitral valve (MV). However, when it comes to irregular and complex geometries, FEM techniques suffer from well-documented limitations, such as the labor and time-consuming intrinsic need of a mesh. In
this work, novel structural models of the MV and mitral valve regurgitation (MVR) dynamics are presented using a mesh-free method. Obtained
results show that the developed models are capable of reproducing MV
and MVR behaviour with good agreement with respect to both in-vivo
and in-silico studies, in terms of valve closure and opening, valve deformation, as well as stress magnitudes. This paper shows that mesh-free
methods have the potential to become a powerful alternative to the currently most used modeling approaches.

Keywords: SPH · Smoothed particles hydrodynamics ·
Computational modeling · Mitral valve · Mitral valve regurgitation ·
Mesh-free

1 Introduction

The mitral valve (MV) is a complex cardiac valve located between the left atrium
and ventricle. Its structure involves the mitral annulus ring, the anterior and
posterior leaflets, both papillary muscles (PM) and some tendinous chords named
chordae tendineae that link the leaflets with the PM. In light of this complexity,
the development of computational models is a powerful instrument to quantify
and further understand valve function both in healthy and diseased conditions.
The predictive behaviour of modeling has also the potential to become a strong
tool to assist medical decisions, as well as playing an important role in regulatory

E. Puyol Anton et al. (Eds.): STACOM 2020, LNCS 12592, pp. 66–75, 2021.
https://doi.org/10.1007/978-3-030-68107-4_7

pathways, speeding up the incorporation of new cardiovascular devices to the market. To that end, models not only need to be accurate, reliable and robust, but also efficient enough to be used by industry and in a clinical environment.

In the field of cardiac biomechanics, numerical mesh-based approaches, such as the Finite Element Method or Finite Volume Method, have been largely used to solve solid and fluid dynamics problems related to the MV, showing great performance and reliability [1]. Nevertheless, these techniques suffer from well-documented limitations, including the need of high computational cost and time algorithms to handle contact between structures or to represent large deformations [2]. Besides, when coping with irregular geometries such as the MV, mesh generation demands for laborious and manual editing workload, not only being one of the most time-consuming steps of the modeling procedure, but also obstructing the processing of large amount of data.

Contrarily to mesh-based techniques, mesh-free methods, such as Smoothed Particles Hydrodynamics (SPH), may offer some advantages due to the no need of a mesh, such as the calculation of the interaction between implicit surfaces, including solid-solid (contact) and solid-fluid. Also more irregular segmentations can be taken as geometrical input since the neighbour search implicitly smooths those local irregularities. These features are especially interesting for cardiac valve modeling.

Although mesh-free approaches are not as mature as mesh-based ones and have been mostly used to solve fluid dynamics equations, recent developments in SPH have allowed the modeling of a wider range of problems, including cardiac electromechanics [3] and biomechanics [4], being a promising alternative to mesh-based methods. Nevertheless to our knowledge, there is no SPH structural 3D model of the MV nor any of its pathological behaviours. Therefore, the main objectives of this work were: 1) developing a 3D image-based mesh-free structural model of the MV using SPH; and 2) proposing different strategies to model diseased mitral valve regurgitation (MVR) cases.

2   Materials and Methods

2.1   SPH Continuum Mechanics

Total lagrangian SPH discretization was used in order to solve the continuous mechanics equations describing the MV dynamics. The interested reader is referred to [4] for further details. In SPH, the domain is divided in a set of moving particles with no connectivity among them. A function f_i is defined for each particle, which can be approximated by Eq. (1). In this work, the kernel used was the C2-Wendland Kernel, with the kernel correction detailed in [5].

$$f_i = \sum_{j=N_i} f_j V_j W_h(r_{ij}), \tag{1}$$

where the sub-index i is the particle of interest and the j corresponds to a particle inside the neighbourhood of i, denoted as N_i. Let V_j be the volume of the particle j and $W_h(r_{ij})$ the kernel function, which depends on h, corresponding to the size of the support domain around particle i, and is evaluated in the vector $r_{ij} = x_i - x_j$, with x being the particle position.

To model MV solid mechanics four equations were used. Firstly, the momentum conservation in the reference configuration is expressed as follows:

$$\frac{\partial^2 x_i}{\partial t^2} = \frac{1}{\rho_0} \sum_{j \in N_i} V_j (\mathbf{P}_{ji} - \mathbf{\Pi}_{0ij}) \nabla_0 \tilde{W}_h(r_{ij}) + b_i, \tag{2}$$

where the sub-index 0 denotes the quantity in the reference configuration (at time $t = 0$), ρ is the density, $\mathbf{P}_{ji}$ indicates the difference between the first Piola-Kirchoff stress tensor of particles i and j, and b_i is the sum of external body forces being applied on the particle i, such as gravity, contact (if the particle lays on the surface) or the force exerted by the chordae apparatus, explained in detail in Sect. 2.3. The tensor $\mathbf{\Pi}_{0ij}$ is introduced as an artificial viscosity to compensate local particle oscillations [6]. Secondly, the continuity equation is determined as $\rho_i = \dfrac{\rho_0}{J_i}$, being J the determinant of the tensor $\mathbf{F}$. Finally, the deformation gradient tensor $\mathbf{F}$ and the neo-Hookean constituve law are given as:

$$\mathbf{F}_i = \frac{\partial x_i}{\partial X_i} = \sum_{j \in N_i} V_j r_{ij} \nabla_0 \tilde{W}_h(r_{ij}), \tag{3}$$

$$\mathbf{P_i} = \mu J_i^{-\frac{2}{3}} (\mathbf{F}_i - \mathbf{F}_i^{-T}(\frac{1}{3} tr(\mathbf{F}_i \cdot \mathbf{F}_i^T))) + \mathbf{F}_i^{-T}(p_i J_i), \tag{4}$$

where the second term on the right is included to reinforce incompresibility with the Lagrange multiplied $p_i = K(J_i - 1)$, being K the bulk modulus.

2.2   Contact with SPH

The frictional contact algorithm based on ideal plastic collision developed by Wang et al. [7] was used. In contrast to most mesh-free contact approaches [8,9], this algorithm avoids particle penetration without the need of a penalization coefficient.

Briefly, contact occurs when a particle i is near a non-neighbour particle k within a given distance δ_c. Once contact is detected, the relative velocity v_{ik} is computed and weighted by their mass ratio $\zeta_m = \dfrac{m_i}{m_i + m_k}$, being m the particle mass. A contact force $f_{ik,c}$ is then applied from particle k to particle i calculated as follows:

$$f_{ik,c} = \frac{\zeta_m(v_{ik,n} + v_{ik,t})}{\Delta t}, \tag{5}$$

where $v_{ik,n}$ and $v_{ik,t}$ are the normal and tangential relative velocities between particle i and the colliding particle k. The normal component $v_{ik,n}$ is calculated by projecting v_{ik} in the average normal direction $\bar{n}$, determined following the formulation from Seo and Min [10]. Finally, the tangential component $v_{ik,t}$ is computed as follows:

$$v_{ik,t} = \min\{v_{ik} - v_{ik,n}, \quad \mu|v_{ik,n}|\frac{v_{ik} - v_{ik,n}}{|v_{ik} - v_{ik,n}|}\} \qquad (6)$$

where μ is the friction coefficient.

2.3 Mitral Valve Simulations

A MV geometry extracted from the segmentation of a Computed Tomography (CT) scan of a patient diagnosed with MVR was used as the starting point for this model. Specifically, the valve used was *case 03 user 01* retrieved from *zenodo.org* [11]. A manual remodeling was performed to the original geometry in order to reduce leaflets stenosis. Two extra particle layers were added to generate a constant total valve thickness of 1 mm. The resulting geometry included 2699 regularly distributed particles and a valve surface area of 4.65 cm^2.

The MV leaflets were modeled as an hyperelastic material following a neo-Hookean law, as presented in Equation (4). Previously reported Young's Modulus for modeling the MV leaflets ranged from 0.8 to 9 MPa [12]. In this work, both leaflets were assumed to have a Young's Modulus of 1.5 MPa. The Poisson's Ratio was set to 0.49 to model leaflets nearly incompressible behaviour [13].

Chordae tendineae were modeled as a set of 156 straight springs linking the PM tips with the leaflets free edges, shown in Fig. 2a. Each spring offers a resistance f_s against the motion towards the atrium, applied only when the spring is stretched and defined with the following equation:

$$f_s = \begin{cases} (L - pL_0)K & if\, L \geq pL_0 \\ 0 & if\, L < pL_0 \end{cases} \qquad (7)$$

K is the stiffness coefficient and L is the current distance from the corresponding PM to the particle anchored. L_0 is the end-diastolic distance between the PM and the anchored leaflet particle, and p is a percentage pre-multiplying this distance. Parameters K and p were empirically tuned until a complete valve closure was ensured. Since the assumption of linear elasticity of the chords, its distribution and number are unknown; it is not possible to compare the values of these parameters with previous studies.

PM tips were represented as two moving 3D Cartesian coordinates. The motion was described following [14]. Besides, a time-varying relative transvalvular physiological pressure was uniformly applied to particles facing the ventricle, defined accordingly to [15]. Both are shown in Fig. 1 and were assumed to be the same for all models. Finally, mitral annulus ring particles were considered to be fixed.

2.4 Mitral Valve Regurgitation Simulations

Three strategies to model different cases of MVR were proposed. Firstly, primary MVR (PMVR) was modeled. Chordae degeneration is the most prevalent cause

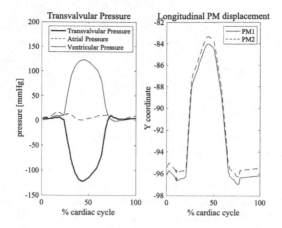

Fig. 1. Physiological transvalvular pressure and PM motion over one cardiac cycle in the longitudinal (Y) direction, corresponding to the ventricular apex to mitral annulus direction.

of PMVR and it is characterized by a decreased stiffness and increased elongation of a specific group of chordae. To mimic this pathological mechanism, the chordae stiffness parameter K from 50 chordae attached to the P1-P2 scallops was decreased 20%. Following Carpentier's classification, P1-P2 scallops corresponded to the anterolateral and middle segments of the posterior MV leaflet. The remaining chordae parameters were the same as in the healthy case.

The second modeled pathological case, named Traumatic MVR (TMVR), corresponded to a more severe case of PMVR posterior to a traumatic injury. A total rupture of the same 50 chords considered in PMVR was simulated, corresponding with a valve with no chordae linked to the P1-P2 scallops.

Finally, the last pathological condition modeled was secondary MVR (SMVR), which is caused by mitral annulus dilatation and PM displacement secondary to a global remodeling of the left ventricle. Therefore, the strategy followed in this case was a dilatation of the valve annulus, together with an apically and laterally translation of the PM. The SMVR annular and PM dimensions can be appreciated in Fig. 2b.

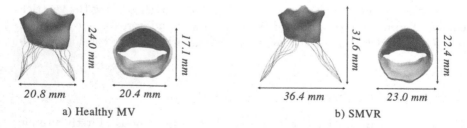

Fig. 2. PM and annulus dimensions in a) healthy and b) SMVR case.

3   Results

The structural healthy MV dynamics in relation to the transvalvular pressure applied during one cardiac cycle are shown in Fig. 3a. The valve remained opened between the 0–20% of the cardiac cycle, corresponding to diastole. Valve closing started at the 25% of the cycle, coinciding with an increase in the ventricular pressure. Valve reopened at the 70% of the cycle, when a decrease in the ventricular pressure is applied.

Under SMVR condition, the valve showed a loss of leaflets coaptation during systole, preventing the valve from closing. The leaflets non-coaptation can be appreciated in Fig. 3b between the 30–60% of the cardiac cycle. It appeared as a consequence of the mitral annulus and PM remodeling.

In the case of PMVR and TMVR, a prolapse into the left atrium was observed during the 45–60% of the cardiac cycle. This prolapse is shown in Fig. 4. The prolapse occurred in the area where the chords were damaged or removed. As expected, in TMVR a more severe prolapse compared to the PMVR was appreciated, due to the chordae rupture.

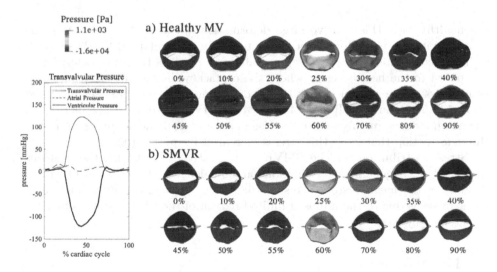

Fig. 3. Valve dynamics over time [%] for healthy and SMVR cases.

In Fig. 5, peak systolic maximum principal stresses are shown. The healthy valve condition presented a mean stress of 127.7 kPa. Higher stresses up to 1 MPa were appreciated in the valve commissures and P1 leaflet scallop, resulting from the non-symmetry of the geometry.

An increase of stresses in the anterior leaflet was clearly visible in the PMVR and TMVR cases, highlighting the peak stresses reached in the annular part of the same leaflet up to 1 MPa. The stress increase was consequence of the larger local deformation that occurred in that area when the prolapse was created. In

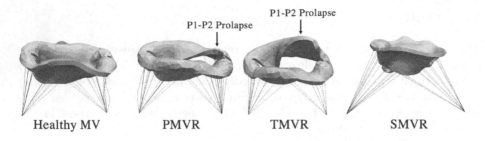

Fig. 4. Anterior view of all cases during systole.

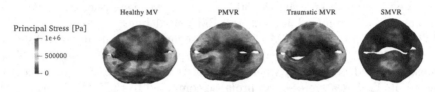

Fig. 5. Maximum peak systolic principal stress.

the healthy case, this relative large deformation did not take place, since the contact with the posterior leaflet prevented it. In the prolapsed P1-P2 scallops, a slight decrease on stresses was appreciated as a result of the lower deformation compared to the healthy case, where the contact with the anterior leaflet prevented tissue free motion. Globally, the overall mean systolic stresses increased up to 145.5 kPa in PMVR and 143.6 kPa in TMVR conditions. The increase in the mean stress is mainly due to the rise in stresses near the annumus. In Fig. 4, the deformation that leads this stresss increment is appreciated.

Stress distribution for the SMVR case greatly differed with the previous states. Here, a generalized reduction on both leaflets stresses was observed, decreasing the mean systolic stress to 68.3 kPa. This significant reduction in the stresses is due to the imposed limited motion of the SMVR state compared to the healthy case.

4   Discussion

The first aim of our work was to develop a 3D image-based mesh-free MV model using SPH. This model was verified with respect to physiological valve dynamics, in term of the overall shape, opening and closure, for given boundary conditions, naming pressure, PM motion and chordae modeling. The contact algorithm properly worked, preventing particle penetration. Moreover, it is coherent with previous mesh-based simulations in terms of stress range. Difference in the stress distribution and values with respect to previous mesh-based simulations are due to the differences in the anatomy, as well as boundary condition.

The second objective was developing some strategies to simulate MVR. The mechanisms responsible for diseased valves were also well modeled as shown with the PMVR, TMVR and SMVR simuations. In the first two, a damage/rupture of the chords did lead to a prolapse, which is indicative of MVR pathology [16,17]. This prolapse was located in the region where the chords properties were modified. Similarly to [18], the prolapse in the PMVR condition was less severe than in the case of TMVR, since chords were only damaged in the first case and removed in the second one. In the case of SMVR, a dilated heart did influence the performance of the whole valve apparatus even if the valve was healthy. A general leaflets loss of coaptation was observed, commonly presented in SMVR disease [19].

Regarding peak systolic principal stresses, they were found to be between 10–370 KPa in the P2-P3 posterior leaflet scallops and 100–520 KPa in the anterior leaflet of the healthy case, respectively. These ranges agree with previously reported mesh-based analysis [18,20]. For instance, in [20], reported peak systolic stresses ranged from 60 to 270 KPa and 130 to 540 KPa in the posterior and anterior leaflets, respectively. Nevertheless, high stress regions were observed in the valve commissures and P1 scallop of the posterior leaflet, which seems to be consequence of the geometry used.

In PMVR and TMVR cases, an increase on the the anterior leaflet stresses was found. High stress values were observed near the annular region, as in [17]. Besides, a slight decrease on the stresses of the prolapsing scallop was found, also observed, for example, in [18]. SMVR condition showed a 53% decrease of the mean leaflets stresses compared to the healthy case, as a result of the limited deformation of both leaflets due to the annulus dilatation and restricted PM motion.

This work can be complemented with previous efforts where SPH is used to simulate fluid flow in complex cardiac cavities and around valves [21,22], aiming at a pure mesh-free solver for valve dynamics. Although, the use of SPH has been already proposed for FSI problems, including mitral valve opening [23,24], those studies serve as a base since they used 2D idealized geometries, elastic materials and do not consider neither contact mechanics nor PM influence on valve dynamics.

4.1 Limitations

This study is not excluded from limitations. Firstly, this work is a proof of concept for mitral valve modeling using a 3D mesh-free approach and therefore only one image-based geometry was study. A larger set of geometries will be used in the future to test the robustness of the method when changing the geometries. Secondly, only the open valve was available from the patient and therefore, clinical claims cannot be derived from this study. Additional patient data, such as papillary tips, annulus motion and pressure condition are required. Finally, a sensitivity analysis of particle resolution and SPH parameters, including frictional coefficient and kernel size needs to be conducted for complex anatomies.

5   Conclusion

In this work, the first 3D image-based mesh-free structural model of the MV was proposed and developed. Additionally, strategies to reproduce MVR mechanisms were proposed and deeply analyzed. Valve dynamics for healthy and diseased scenarios were qualitative and quantitative evaluated. Results were verified against previous studies with good agreement, in terms of valve closure and opening, as well as deformation and stress ranges. We have demonstrated that SPH-based models have the potential to become an alternative to mesh-based approaches in the modeling of both healthy and diseased MV cases.

Acknowledgments. We gratefully acknowledge clinicians from the Cardiology Service of Hospital Clínic (Barcelona) for clinical consultation.

References

1. Gao, H., et al.: Modelling Mitral Valvular Dynamics - current trend and future directions: review of MV modelling. Int. J. Numer. Meth. Biomed. Eng. **33**(12) (2016)
2. Zhang, L., Ademiloye, A., Liew, K.: Meshfree and particle methods in biomechanics: prospects and challenges. Arch. Comput. Meth. Eng. **26**, 1547–1576 (2019)
3. Lluch, E., et al.: Calibration of a fully coupled electromechanical meshless computational model of the heart with experimental data. Comput. Meth. Appl. Mech. Eng. **364**, 112869 (2020)
4. Lluch, È., De Craene, M., Bijnens, B., Sermesant, M., Noailly, J., Camara, O., Morales, H.G.: Breaking the state of the heart: meshless model for cardiac mechanics. Biomech. Model. Mechanobiol. **18**(6), 1549–1561 (2019). https://doi.org/10.1007/s10237-019-01175-9
5. Chen, J., Beraun, J., Carney, T.: A corrective smoothed particle method for boundary value problems in heat conduction. Int. J. Num. Meth. Eng. **46**(2), 231–252 (1999)
6. Gray, J., Monaghan, J., Swift, R.: SPH elastic dynamics. Comput. Meth. Appl. Mech. Eng. **49**(50), 6641–6662 (2001)
7. Wang, J., Chan, D.: Frictional contact algorithms in SPH for the simulation of soil-structure interaction. Int. J. Num. Anal. Meth. Geomech. **38**(7), 747–770 (2014)
8. Xiao, Y.H., et al.: Simulation of normal perforation of aluminum plates using axisymmetric smoothed particle hydrodynamics with contact algorithm. Int. J. Comput. Meth. **10**(3), 1350039 (2013)
9. Liu, M.B., Liu, G.R.: Smoothed particle hydrodynamics (SPH): an overview and recent developments. Arch. Comput. Meth. Eng. **17**, 25–76 (2010)
10. Seo, S., Min, O.: Axisymmetric SPH simulation of elasto-plastic contact in the low velocity impact. Comput. Phys. Commun. **175**, 583–603 (2006)
11. Tautz, L., et al.: CT segmented mitral valves in open state. Zenodo (2019)
12. Lim, K., Yeo, J., Duran, C.: Three-dimensional asymmetrical modeling of the mitral valve: a finite element study with dynamic boundaries. J. Heart Valve Dis. **14**(3), 386–392 (2005)
13. Maisano, F.: An annular prosthesis for the treatment of functional mitral regurgitation: finite element model analysis of a dog bone-shaped ring prosthesis. Ann. Thorac. Surg. **79**(4), 1268–1275 (2005)

14. Sanfilippo, A., et al.: Papillary muscle traction in mitral valve prolapse: quantification by two-dimensional echocardiography. J. Am. Coll. Cardiol. **19**(3), 1268–1275 (1992)
15. Kaiser, A., McQueen, D., Peskin, C.: Modeling the Mitral Valve. Wiley, New York (2019)
16. Gabriel, V., Kamp, O., Visser, C.: Three-dimensional echocardiography in mitral valve disease. Eur. J. Echocardiogr. **6**(6), 443–454 (2005)
17. Rim, Y., et al.: Personalized computational modeling of mitral valve prolapse: virtual leaflet resection. PLoS ONE **10**(6), e0130906 (2015)
18. Caballero, A., et al.: New insights into mitral heart valve prolapse after chordae rupture through fluid-structure interaction computational modeling. Sci. Rep. **8**, 17306 (2018)
19. Izumi, S., et al.: Mechanism of mitral regurgitation in patients with myocardial infarction: a study using real-time two-dimensional doppler flow imaging and echocardiography. Circulation **76**(4), 777–785 (1987)
20. Votta, E., et al.: Mitral valve finite-element modelling from ultrasound data: a pilot study for a new approach to understand mitral function and clinical scenarios. Philos. Trans. R. Soc. A Math. Phys. Eng. Sci. **366**(1879), 3411–3434 (2008)
21. Yuan, Q., et al.: Fluid structure interaction study of bioprosthetic heart valve with FESPH method. Int. J. Adv. Comput. Technol. **8**, 695–702 (2013)
22. Mao, W., et al.: Fully-coupled fluid-structure interaction simulation of the aortic and mitral valves in a realistic 3D left ventricle model. PLOS ONE **12**(9), e0184729 (2017)
23. Antonci, C., et al.: Numerical simulation of fluid-structure interaction by SPH. N. Engl. J. Med. **339**(24), 1725–1733 (1998)
24. Amanifard, N., Rahbar, B., Hesan, M.: Numerical simulation of the mitral valve opening using smoothed particle hydrodynamics. In: Proceedings of the World Congress on Engineering, vol. 3 (2011)

PIEMAP: Personalized Inverse Eikonal Model from Cardiac Electro-Anatomical Maps

Thomas Grandits[1,4]($\boxtimes$), Simone Pezzuto[2], Jolijn M. Lubrecht[2], Thomas Pock[1,4], Gernot Plank[3,4], and Rolf Krause[2]

[1] Institute of Computer Graphics and Vision, Graz University of Technology, Graz, Austria
{thomas.grandits,pock}@icg.tugraz.at
[2] Center for Computational Medicine in Cardiology, Institute of Computational Science, Università della Svizzera italiana, Lugano, Switzerland
{simone.pezzuto,jolijn.marieke.lubrecht,rolf.krause}@usi.ch
[3] Institute of Biophysics, Medical University of Graz, Graz, Austria
gernot.plank@medunigraz.at
[4] BioTechMed-Graz, Graz, Austria

Abstract. Electroanatomical mapping, a keystone diagnostic tool in cardiac electrophysiology studies, can provide high-density maps of the local electric properties of the tissue. It is therefore tempting to use such data to better individualize current patient-specific models of the heart through a data assimilation procedure and to extract potentially insightful information such as conduction properties. Parameter identification for state-of-the-art cardiac models is however a challenging task.

In this work, we introduce a novel inverse problem for inferring the anisotropic structure of the conductivity tensor, that is fiber orientation and conduction velocity along and across fibers, of an eikonal model for cardiac activation. The proposed method, named PIEMAP, performed robustly with synthetic data and showed promising results with clinical data. These results suggest that PIEMAP could be a useful supplement in future clinical workflowss of personalized therapies.

1 Introduction

Patient-specific modeling in cardiac electrophysiology has nowadays reached the status of a clinically feasible tool for assisting the cardiologist during the therapeutic intervention. As models became more mature, and thanks to the increasingly

This research was supported by the grants F3210-N18 and I2760-B30 from the Austrian Science Fund (FWF) and BioTechMed Graz flagship award "ILearnHeart", as well as ERC Starting grant HOMOVIS, No. 640156. This work was also financially supported by the Theo Rossi di Montelera Foundation, the Metis Foundation Sergio Mantegazza, the Fidinam Foundation, the Horten Foundation and the CSCS–Swiss National Supercomputing Centre production grant s778.

© Springer Nature Switzerland AG 2021
E. Puyol Anton et al. (Eds.): STACOM 2020, LNCS 12592, pp. 76–86, 2021.
https://doi.org/10.1007/978-3-030-68107-4_8

availability of high-resolution data such as high-density electroanatomic maps (EAMs), parameter identification has emerged as a key topic in the field.

A high-density EAM is composed by a large number of contact recordings (1000 points or more), each with local electrogram and spatial information. Activation and conduction velocity maps, for instance, can be derived by combining electric and geometric data. Conductivity parameters in a propagation model may therefore be adapted to reproduce such maps for model personalization.

The reconstruction of conduction velocity maps is generally based on local approaches [4, 7]. In these methods, the local front velocity is estimated from an appropriate interpolation of the local activation time (LAT). Anisotropic conductivity can be deduced from front velocity and prior knowledge on fiber structure (rule-based or atlas-based), or by combining multiple activation maps [11].

Although being computationally cheap, these local methods may miss effective mechanisms for global consistency in electric wave propagation models, and may introduce artefacts in conduction velocity due to front collisions or breakthroughs. A different approach, also adopted in this work, relies instead on the (possibly strong) assumption that a calibrated model for the cardiac activation can reproduce the measured activation with sufficient accuracy. The electric conductivity in the model is eventually identified through an optimization procedure aiming at minimizing the mismatch between the model output and the collected data. The model can either be enforced pointwise, yielding for instance PDE-constrained optimization [2], or act as a penalization term [12].

To the best of our knowledge, however, the problem of estimating *simultaneously* distributed fiber architecture and conduction velocities from sparse contact recordings has never been attempted before with either approaches. In this work we aim to bridge this gap by proposing a novel method to extract from a single EAM the full electric conductivity tensor, with the only assumption of symmetric positive-definiteness (s.p.d.) of the tensor field. Local fiber orientation and conduction velocities are then deduced from the eigendecomposition of the conductivity tensor. As forward model, we consider the anisotropic eikonal model, which is a good compromise between physiological accuracy and computational cost [6]. The corresponding inverse model, employing Huber regularization, a smooth total variation approximation, to stabilize the reconstruction and log-Euclidean metric in the parameter space to ensure s.p.d. of the tensor field, is solved by an iterative quadratic approximation strategy combined using a Primal-Dual optimization algorithm. Finally, we extensively test the algorithm with synthetic and clinical data for the activation of the atria, represented as a 2-D manifold, showing promising results for its clinical applicability.

2 Methods

2.1 Forward Problem

The anisotropic eikonal equation describes the activation times u of a wave propagating with direction-dependent velocity. Given a smooth 2-D manifold

$\Omega \subset \mathbb{R}^3$, the equation reads as follows

$$\sqrt{\langle D(\mathbf{x}) \nabla_\mathcal{M} u(\mathbf{x}), \nabla_\mathcal{M} u(\mathbf{x}) \rangle} = 1 \ \text{s.t.:} \ \forall \mathbf{x} \in \Omega : D_3(\mathbf{x}) \in \mathcal{P}(3)$$
$$\forall \mathbf{x} \in \Gamma_0 \subset \partial\Omega : u(\mathbf{x}) = 0, \tag{1}$$

with Γ_0 representing the domain of fixed activation times, $\mathcal{P}(n)$ being the space of $n \times n$ symmetric positive definite matrices and $\nabla_\mathcal{M}$ being the surface gradient. The conductivity tensor $D(\mathbf{x})$ specifies the conduction velocity in the propagation direction, that is $\sqrt{\langle D(\mathbf{x})\mathbf{k}, \mathbf{k} \rangle}$ is the velocity of the wave at $\mathbf{x} \in \Omega$ in direction $\mathbf{k}$, $\mathbf{k}$ unit vector.

To solve Eq. (1), we based our algorithm—purely implemented in Tensor-Flow to allow for automatic gradient computation through back-propagation—on the Fast Iterative Method (FIM) for triangulated surfaces [8]. The only fixed assumed point in PIEMAP $\{\mathbf{x}_0\} = \Gamma_0$ is the chosen earliest activation site, which is assumed to be the earliest of all measured activation points. We use a slightly altered fixed-point iteration $\mathbf{u}^{k+1} = F(\mathbf{u}^k)$ which iteratively updates the activation times. For sake of simplicity, we keep denoting by $u(\mathbf{x})$ the piecewise linear interpolant of the nodal values u_i and by D a piecewise constant tensor field on the triangulated surface. The approximated solution of the Eq. (1), henceforth denoted by $\text{FIM}_D(\mathbf{x})$ is then the unique fixed-point of the map F. Specifically, the map F updates each of the nodal values as follows:

$$u_i^{k+1} = \begin{cases} 0, & \text{if } \mathbf{x}_i \in \Gamma_0, \\ \underset{T_j \in \omega_i}{\text{smin}} \underset{\mathbf{y} \in e_{i,j}}{\text{smin}} \left\{ u(\mathbf{y}) + \sqrt{\langle D_j^{-1}(\mathbf{y} - \mathbf{x}_i), \mathbf{y} - \mathbf{x}_i \rangle} \right\}, & \text{otherwise.} \end{cases} \tag{2}$$

where ω_i is the patch of triangles T_j connected to the vertex $\mathbf{x}_i$, $e_{i,j}$ is the edge of the triangle T_j opposite to the vertex $\mathbf{x}_i$, $D_j = D|_{T_j}$ and smin^k being the soft-minimum function, defined as $\text{smin}^\kappa(\mathbf{x}) = -\frac{1}{\kappa} \log(\sum_i \exp(-\kappa x_i))$.

Differently from the classic FIM method [8], we concurrently update *all* the nodes, that is the map F is applied in parallel to each node and not just on a small portion of "active" nodes. We also replaced the min-function of the original FIM-algorithm by the soft-minimum smin^κ to ensure a limited degree of smoothness of the function and avoid discontinuities in the gradient computation.

2.2 PIEMAP and Inverse Problem

PIEMAP implements an inverse problem in which the optimal conductivity tensor field D for Eq. (1) is selected such that the mismatch between recorded activation times $\hat{u}(\mathbf{x})$ and the simulated activation times $\text{FIM}_D(\mathbf{x})$ on the measurement domain $\Gamma \subset \Omega$ is minimized in the least-squares sense.

In principle, after accounting for symmetry, $D(x)$ has 6 component to be identified for every $x \in \Omega$. Since the dynamic of the wave propagation is bound to the 2-D manifold, however, the component normal to the surface does not influence the solution. We therefore define $D(\mathbf{x})$ as follows:

$$D(\mathbf{x}) = P(\mathbf{x}) \begin{pmatrix} \tilde{D}(\mathbf{x}) & \mathbf{0} \\ \mathbf{0}^\mathsf{T} & 1 \end{pmatrix} P(\mathbf{x})^\mathsf{T}, \tag{3}$$

where $\tilde{D}(\mathbf{x}) \in \mathcal{P}(2)$ and $P(x)$ is a rotation from the canonical base in $\mathbb{R}^3$ to a local base $\{\mathbf{v}_1(\mathbf{x}), \mathbf{v}_2(\mathbf{x}), \mathbf{n}(\mathbf{x})\}$. The local base at $\mathbf{x} \in \Omega$ is such that $\mathbf{n}(\mathbf{x})$ is the normal vector to the surface and $\mathbf{v}_1(\mathbf{x})$, $\mathbf{v}_2(\mathbf{x})$ span the tangent space. In such way, the dimension of the parameter space is reduced from 6 to only 3. Any basis $\mathbf{v}_1(\mathbf{x})$ in the tangent space is valid, but we compute a smooth basis by minimizing the variation across the manifold:

$$\min_{\mathbf{v}_1} \int_\Omega \|\nabla \mathbf{v}_1(\mathbf{x})\|_2^2 \, d\mathbf{x} \ \text{s.t.:} \ \forall \mathbf{x} \in \Omega : \|\mathbf{v}_1(\mathbf{x})\| = 1,$$

to ensure a meaningful result through the later introduced regularization term. The computed local bases, used in all experiments throughout this paper, are shown in Fig. 1.

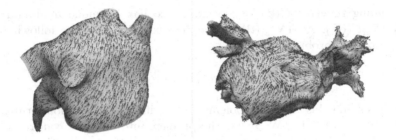

Fig. 1. Generated local bases on the atria manifold models.

Finally, we consider the Log-Euclidean metric [1] for ensuring a s.p.d. tensor field: given $\mathbf{d}(\mathbf{x}) \in \mathbb{R}^3$, $\mathbf{x} \in \Omega$, we set $\tilde{D}$ as follows:

$$\tilde{D}(\mathbf{x}) := \exp \begin{pmatrix} d_1(\mathbf{x}) & d_2(\mathbf{x}) \\ d_2(\mathbf{x}) & d_3(\mathbf{x}) \end{pmatrix}, \tag{4}$$

where the matrix exponential is computed from the eigendecomposition. In particular, the admissible set $\mathbb{R}^3$ is mapped through (3) and (4) to $\mathcal{P}(3)$.

The inverse problem, therefore, consists in finding the vector field $\mathbf{d} \in \mathbb{R}^3$, which minimizes the following objective function:

$$\min_{\mathbf{d}} \underbrace{\frac{1}{2} \int_\Gamma \left(\mathrm{FIM}_{D(\mathbf{d})}(\mathbf{x}) - \hat{u}(\mathbf{x}) \right)^2 d\mathbf{x}}_{U(\mathbf{d})} + \underbrace{\lambda \int_\Omega H_\varepsilon(\nabla_{\mathcal{M}} \mathbf{d}) \, d\mathbf{x}}_{\mathrm{TV}_{\varepsilon,\lambda}(\mathbf{d})}, \tag{5}$$

where $\mathrm{TV}_{\varepsilon,\lambda}(\mathbf{d})$ is a smooth total variation (TV) regularization term which alleviates the ill-posedness of the problem. Specifically, H_ε is the Huber function:

$$H_\varepsilon(\mathbf{x}) = \begin{cases} \frac{1}{2\varepsilon} |\mathbf{x}|^2, & \text{if } |\mathbf{x}| \le \varepsilon, \\ |\mathbf{x}| - \frac{1}{2}\varepsilon, & \text{else} \end{cases} \tag{6}$$

We set $\epsilon = 5 \times 10^{-2}$ for our experiments, while the optimal choice of regularization parameter λ is obtained by using a cross-validation approach.

2.3   Forward-Backward Splitting and Numerical Solution

The computational complexity of solving Eq. (5) is dominated by the time for computing FIM_D and $\nabla_\mathbf{d}\mathrm{FIM}_D$. The implementation of FIM_D in TensorFlow allows for an efficient computation of $\nabla_\mathbf{d}\mathrm{FIM}_D$ via backpropagation on a graphical processing unit (GPU). While the minimization of the residual $U(\mathbf{d})$ is usually achieved very quickly, at least when $\hat{u}$ is a (possibly corrupted) solution of Eq. (1), the TV term tends to increase the number of needed iterations for convergence. In order to increase the convergence rate, we apply the principle of the Fast Iterative Shrinking and Thresholding Algorithm (FISTA) [3], quadratically bounding the non-linear, non-convex function U around the current point $\mathbf{d}_k$:

$$U(\mathbf{d}) \leq U(\mathbf{d}_k) + \langle \nabla_\mathbf{d} U(\mathbf{d}_k), (\mathbf{d} - \mathbf{d}_k)\rangle + \frac{L}{2}\|\mathbf{d} - \mathbf{d}_k\|_2^2 =: G(\mathbf{d}). \qquad (7)$$

The bounding function $G(\mathbf{d})$ is convex, hence has a unique minimum $\bar{\mathbf{d}} = \mathbf{d}_k - L^{-1}\nabla_\mathbf{d} U(\mathbf{d}_k)$. As $\mathrm{TV}_{\varepsilon,\lambda}(\mathbf{d})$ is also convex, we obtain the following convex minimization problem:

$$\min_\mathbf{d} \frac{L}{2}\|\mathbf{d} - \bar{\mathbf{d}}\|_2^2 + \mathrm{TV}_{\varepsilon,\lambda}(\mathbf{d}).$$

Iteratively solving this class of problems along with an acceleration term is usually referred to as FISTA. We recast the problem into a convex-concave saddle-point problem:

$$\min_\mathbf{d} \max_\mathbf{p} \frac{L}{2}\|\mathbf{d} - \bar{\mathbf{d}}\|_2^2 + \langle \nabla_\mathcal{M}\mathbf{d}, \mathbf{p}\rangle - \mathrm{TV}_{\varepsilon,\lambda}^*(\mathbf{p}) \qquad (8)$$

which can be solved using the Primal-Dual algorithm [5] given by:

$$\begin{cases} \mathbf{d}^{i+1} = \mathrm{prox}_{\tau G}(\mathbf{d}^i - \tau\nabla_\mathcal{M}^*\mathbf{d}^i) \\ \mathbf{d}_\Theta = \mathbf{d}^{i+1} + \theta\left(\mathbf{d}^{i+1} - \mathbf{d}^i\right) \\ \mathbf{p}^{i+1} = \mathrm{prox}_{\sigma\mathrm{TV}_{H_\epsilon,\mathcal{M}}^*}\left(\mathbf{p}^k + \sigma\nabla_\mathcal{M}\mathbf{d}_\Theta\right) \end{cases} \qquad (9)$$

with

$$\hat{\mathbf{d}} = \mathrm{prox}_{\tau U}(\tilde{\mathbf{d}}) = \left(\tilde{\mathbf{d}} + \tau L\bar{\mathbf{d}}\right)/(\tau L + 1)$$

$$\hat{\mathbf{p}} = \mathrm{prox}_{\sigma\mathrm{TV}_{\varepsilon,\mathcal{M}}^*}(\mathbf{p}) \Leftrightarrow \hat{\mathbf{p}}_j = \begin{cases} \frac{\bar{\mathbf{p}}_j}{|\bar{\mathbf{p}}_j|/\lambda} & \text{if } |\bar{\mathbf{p}}_j| > 1 \\ \bar{\mathbf{p}}_j & \text{else} \end{cases}$$

for $\bar{\mathbf{p}}_j = \frac{\mathbf{p}_j}{\sigma\epsilon/\lambda+1}$, $\theta = 1$ and $\tau\sigma\|\nabla_\mathcal{M}\|_2^2 \leq 1$. The parameter L in Eq. (7), usually challenging to evaluate, is computed through a Lipschitz backtracking algorithm [3].

3   Experiments

For the evaluation of PIEMAP, we first assessed its effectiveness on reconstructing known conduction velocity and fibers on a realistic human left atrium (LA) model, also in the presence of white noise and heterogeneity. The LA model was generated from MRI data of a patient, with the fibers semi-automatically assigned as described previously [9]. Fiber and transverse velocity were set to 0.6 $\frac{m}{s}$ and 0.4 $\frac{m}{s}$ respectively for the entire LA, except for the low conducting region, where we used 0.2 $\frac{m}{s}$ for both fiber and transverse velocity. We tested PIEMAP both in the case of fully anisotropic and in the case of isotropic conduction. In the latter case, in particular, we compared PIEMAP to existing methods for the evaluation of conduction velocity, namely a local method [4] and Eikonal-Net [12], a Physics Informed Neural Network (PINN) method. In a second set of experiments, we eventually applied PIEMAP to clinically acquired data, in the form of high-density EAM.

All experiments were run on a desktop machine with an Intel Core i7-5820K CPU with 6 cores of each 3.30GHz, 32GB of working memory and a NVidia RTX 2080 GPU. All examples were optimized for 2000 iterations, with each iteration taking about 1.8 seconds, totalling into a run-time of approximately 1 h for one optimization.

3.1 Numerical Assessment

All the experiments were performed on a human, cardiac magnetic resonance (CMR)-derived left atrium model, with semi-automatically placed fiber directions based on histological studies. The ground-truth (GT) solution was computed with a single earliest activation site using Eq. (1), and with a low-conducting area being close to the left atrial appendage. Different levels of independent and identically distributed (i.i.d.) Gaussian noise with standard deviation σ_N were tested. The measurement domain was a set of 884 points uniformly distributed across the atrium. The reconstruction root-mean-square error (RMSE) with respect to GT was evaluated in terms of conduction velocity (m/s), propagation direction and, only for PIEMAP, fiber-angle error. To evaluate the results, we compute the front direction and fiber direction unit vectors, denoted as $\mathbf{e}$ and $\mathbf{f}$ respectively. The front and fiber angle-errors are then defined as $\alpha_{\mathbf{e}} = \arccos \langle \mathbf{e}, \mathbf{e}_{GT} \rangle \in [0, 180°)$ and $\alpha_{\mathbf{f}} \arccos |\langle \mathbf{f}, \mathbf{f}_{GT} \rangle| \in [0, 90°)$. The velocity errors in propagation direction are then $v \cos(\alpha_{\mathbf{x}}) - v_{GT}$ for computed velocity v and exact velocity v_{GT}, both in the front and fiber direction.

Results are reported in Table 1. All methods correctly captured the low conduction region. PIEMAP compared favourably to the local method at all noise levels in terms of absolute conduction velocity. EikonalNet shows a slightly more accurate front angle error, which is counteracted by the considerably high front velocity error, both compared to our and the local method.

Overall, PIEMAP had the benefit over EikonalNet that the GT was generated with the anisotropic eikonal model, and thus it is in theory possible to reproduce the data exactly with a zero noise level. In the local method no model assumption is made. Interestingly, the error in front direction for the local method could be linked to the fact that, in the presence of anisotropic conduction, propagation direction and $\nabla_{\mathcal{M}} u$ differ. For instance, a circular propagation from the source x_0 satisfies Eq. (1) with $u(x) = \sqrt{\langle D^{-1}(x - x_0), (x - x_0) \rangle}$, thus

Table 1. Comparison of the front-velocity/front-angle error of PIEMAP with the local approach [4] and EikonalNet [12], assuming different noise levels for the in-silico models. Errors in $\frac{m}{s}$/degree. In the last column, we compare the fiber velocity error and fiber angle error.

		Error in Propagation Direction			Fiber Error
		PIEMAP	Local Method	EikonalNet	PIEMAP
σ_N/PSNR	0ms/∞ dB	**0.20**/10.58	**0.20**/22.95	0.53/**9.20**	0.25/38.34
	0.1ms/64.1 dB	**0.19**/10.61	0.20/23.17	0.40/**9.10**	0.25/38.46
	1ms/43.9 dB	**0.20/11.03**	0.21/23.57	0.49/14.60	0.25/38.59
	5ms/29.9 dB	**0.25/19.94**	0.29/30.20	1.24/49.40	0.26/40.14

$\nabla_{\mathcal{M}} u$ differs from $x - x_0$, which is the propagation direction. In the local method, $\nabla_{\mathcal{M}} u$ is used to establish such direction. In EikonalNet, results were less robust to noise. A plausible explanation is that training Neural Networks does not always yield the same results, as multiple local minima might be present. Therefore, error can be slightly lower or higher depending on the initial conditions. In terms of computational time, PIEMAP was comparable to EikonalNet, but significantly slower than the local method.

Regarding the reconstruction of fiber directions (see Fig. 2), we observed a very good performance for the fiber and cross-fiber velocity, and a reasonable reconstruction for the direction. In particular, reconstruction in fiber direction was poor around the boundaries (mitral ring and pulmonary veins, where fibers run parallel to the opening) and in the scarred region, which attribute the most to the fiber angle error in Table 1. The distribution of fiber angle errors is a slightly left-skewed uniform distribution (not shown), indicating that the chosen smooth basis along with a simple TV prior can provide resonable results with respect to the activation timings. Still, it may not be sufficient to account for the partly complicated fiber orientation, especially in areas of high-curvature of the mesh or sudden changes of fiber orientation on the endocardium as an effect of the volumetric structure of the atria, such as is the case for the mitral valve. Physiological priors will need to be considered in the future for this purpose.

3.2   Application to Real Clinical Data

In a patient candidate to ablation therapy, a high-density activation map along with a 3D patient-specific atrial model was acquired with an EAM system (Catheter: Pentaray® System: CARTO® 3 System, Biosense Webster). The recordings encompassed roughly 850 "beats" of 2.5 sec including both the electrode position in 3D space and the unipolar electrogram (1 kHz). Recordings that were deemed to be untrustworthy due to 1) insufficient contact, 2) sliding of the electrode in 3D space >1 cm, 3) correspondence to a inconsistent surface P-wave, 4) minimal unipolar amplitude, were excluded automatically from the study. To avoid degenerated triangles with acute angles, sometimes created by

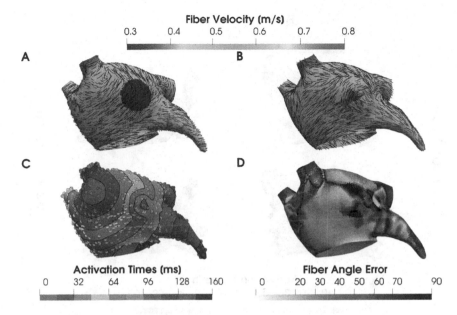

Fig. 2. Results of our method on the noise-less LA-model with known ground-truth fiber orientation and velocity (A). The scarred region is correctly activated at a later time by a combination of reducing fiber-velocity, as well as aligning fibers along the contour lines (B/C). The activation map can faithfully capture the observed measurement points $u(\mathbf{x} \subset \Gamma)$, marked as colored dots (C) and matches the GT-model's activation closely (not shown). Fiber alignment of our model (B) shows mostly errors around the mitrial valves, the scarred region and pulmonary veins (D), as well as regions of high curvature. Best viewed online.

the EAM recordings, we used PyMesh[1] to postprocess the mesh. A further manual pre-processing of the signals was eventually performed for a correct detection of the local activation time (steepest negative deflection in the unipolar signal) in the last beat and compared to local bipolar signals for confirmation. Distribution of points was uneven across the LA, as many points were located around the pulmonary veins (PVs).

Of the remaining valid 565 beats, randomly chosen 80% (452 points) were used to optimize Eq. (5), while the remaining 20% were used as a cross-validation set to find the optimal regularization parameter λ. Figure 3 shows the fiber velocity, orientation and activation map after the optimization. The cross-validation error over several values of λ is shown in Fig. 4, which lead us to the used value of λ. The best cross-validation error lead to a relatively smooth fiber velocity field, with velocities ranging up to $1.5\frac{m}{s}$ in the initiation region, probably a consequence of choosing only one mesh node as an initiation site when in reality the initiation site is larger or composed of multiple sites. A speed-up of propagation

---

[1] https://github.com/PyMesh/PyMesh.

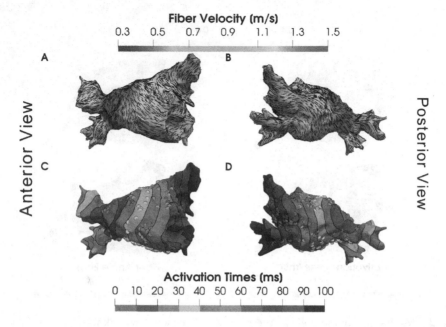

Fig. 3. Anterior (A/C) and posterior view (B/D) of PIEMAP's results on a patient's left atria. The panels A and B show the found fiber direction and fiber velocity, while the panels C and D show the activation map along with the actual measured points on top, similar to Fig. 2.

near the atrial wall can often be witnessed and is compensated in our model by an overall higher fiber-velocity.

4   Discussion and Conclusion

In this paper, we proposed PIEMAP, a global method to reconstruct the conductivity tensor (fiber direction, fiber- and cross-fiber velocity) of an anisotropic eikonal model from sparse measurement of the activation times. We compared our method to existing approaches for determining the conduction velocity map from the same data (a local method and a PINN method) and we demonstrated its effectiveness in a real application.

Our method showed promising results on atrial electrical data, acquired using an EAM system, but may be used with any electrical measurements, mapped to a manifold. In Sect. 3.1, we demonstrated the possibility to infer low conducting regions, as sometimes witnessed for scarred regions, but future studies could apply the algorithm to analyze different pathologies, such as fibrosis.

With special care for registration, PIEMAP could also be combined with high-resolution 3D imaging, such as CT or MRI, to improve anatomical accuracy. An interesting question is whether PIEMAP could also be applied to ventricular

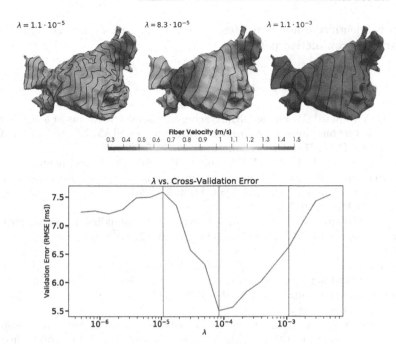

Fig. 4. Influence on the final cross-validation error when varying over λ for the optimization on the EAM recordings. The black contours are the isochrones of the modelled activation. Shown are the results of three different λ values: The left result is the least regularized with $\lambda = 1.08 \times 10^{-5}$, while the result on the right side is heavily regularized with $\lambda = 1.08 \times 10^{-3}$. A compromise is the figure in the middle with $\lambda = 8.34 \times 10^{-5}$, but finding the physiologically most plausible value for λ is a non-trivial task since we do not know the true distribution of velocities and activations in the atria.

activation. A major difference between ventricular and atrial activation is transmural propagation in the former, which is not accessible by contact mapping. Moreover, endocardial activation in the ventricles of healthy subjects, due to the Purkinje network, is extremely complex and may overshadow myocardial propagation. Under specific pathological conditions, such as ventricular tachycardia or bundle branch block, myocardial activation becomes relevant and heterogeneity in conduction of potential interest, justifying the applicability of PIEMAP. While it is true that no transmural data would be available, it is also known that fibers in ventricles follow a peculiar pattern in the transmural direction with low inter-patient variability [13]. Such *prior knowledge* may be used in the inverse procedure by appropriately changing the regularization term. In a recent work [10], we actually applied an inverse method similar to PIEMAP in the ventricles by using epicardial data, obtaining convincing results also in the transmural direction.

In light of the presented results, we believe that PIEMAP can assist future medical interventions by estimating cardiac conduction properties more robustly and help in identifying ablation sites, as well as in better understanding atrial and ventricular conduction pathways.

References

1. Arsigny, V., Fillard, P., Pennec, X., Ayache, N.: Geometric means in a novel vector space structure on symmetric positive-definite matrices. SIAM J. Matrix Anal. Appl. **29**(1), 328–347 (2007). https://doi.org/10.1137/050637996
2. Barone, A., Gizzi, A., Fenton, F., Filippi, S., Veneziani, A.: Experimental validation of a variational data assimilation procedure for estimating space-dependent cardiac conductivities. Comput. Method Appl. M. **358**, 112615 (2020). https://doi.org/10.1016/j.cma.2019.112615
3. Beck, A., Teboulle, M.: A fast iterative shrinkage-thresholding algorithm for linear inverse problems. SIAM J. Imag. Sci. **2**(1), 183–202 (2009). https://doi.org/10.1137/080716542
4. Cantwell, C., Roney, C., Ng, F., Siggers, J., Sherwin, S., Peters, N.: Techniques for automated local activation time annotation and conduction velocity estimation in cardiac mapping. Comput. Biol. Med. **65**, 229–242 (2015). https://doi.org/10.1016/j.compbiomed.2015.04.027
5. Chambolle, A., Pock, T.: A first-order primal-dual algorithm for convex problems with applications to imaging. J. Math. Imag. Vis. **40**(1), 120–145 (2011). https://doi.org/10.1007/s10851-010-0251-1
6. Colli Franzone, P., Guerri, L.: Spreading of excitation in 3-d models of the anisotropic cardiac tissue. I. validation of the eikonal model. Math. Biosci. **113**(2), 145–209 (1993). https://doi.org/10.1016/0025-5564(93)90001-Q
7. Coveney, S., et al.: Gaussian process manifold interpolation for probabilistic atrial activation maps and uncertain conduction velocity. Philos. T. R. Soc. A. **378**(2173), 20190345 (2020). https://doi.org/10.1098/rsta.2019.0345
8. Fu, Z., Jeong, W., Pan, Y., Kirby, R., Whitaker, R.: A fast iterative method for solving the eikonal equation on triangulated surfaces. SIAM J. Sci. Comp. **33**(5), 2468–2488 (2011). https://doi.org/10.1137/100788951
9. Gharaviri, A., et al.: Epicardial fibrosis explains increased endo-epicardial dissociation and epicardial breakthroughs in human atrial fibrillation. Front. Phys. **11**, 68 (2020). https://doi.org/10.3389/fphys.2020.00068
10. Grandits, T., et al.: An inverse Eikonal method for identifying ventricular activation sequences from epicardial activation maps. J. Comput. Phys. **419**, 109700 (2020). https://doi.org/10.1016/j.jcp.2020.109700, http://www.sciencedirect.com/science/article/pii/S0021999120304745
11. Roney, C.H., et al.: A technique for measuring anisotropy in atrial conduction to estimate conduction velocity and atrial fibre direction. Comput. Biol. Med. **104**, 278–290 (2019). https://doi.org/10.1016/j.compbiomed.2018.10.019
12. Sahli Costabal, F., Yang, Y., Perdikaris, P., Hurtado, D.E., Kuhl, E.: Physics-informed neural networks for cardiac activation mapping. Front. Phys. **8**, 42 (2020). https://doi.org/10.3389/fphy.2020.00042
13. Streeter, D.D., Spotnitz, H.M., Patel, D.P., Ross, J., Sonnenblick, E.H.: Fiber orientation in the canine left ventricle during diastole and systole. Circ. Res. **24**(3), 339–347 (1969). https://doi.org/10.1161/01.RES.24.3.339

Automatic Detection of Landmarks for Fast Cardiac MR Image Registration

Mia Mojica[1]([envelope]), Mihaela Pop[2], and Mehran Ebrahimi[1]

[1] Faculty of Science, University of Ontario Institute of Technology,
Oshawa, ON, Canada
{mia.mojica,mehran.ebrahimi}@uoit.ca
[2] Department of Medical Biophysics, Sunnybrook Research Institute,
University of Toronto, Toronto, ON, Canada
mihaela.pop@utoronto.ca

Abstract. Inter-subject registration of cardiac images is a vital yet challenging task due to the large deformations influenced by the cardiac cycle and respiration. Various intensity-based cardiac registration methods have already been proposed, but such methods utilize intensity information over the entire image domain and are thus computationally expensive. In this work, we propose a novel pipeline for fast registration of cardiac MR images that relies on shape priors and the strategic location of surface-approximating landmarks. Our holistic approach to cardiac registration requires minimal user input. It also reduces the computational runtime by 60% on average, which amounts to an 11-min speedup in runtime. Most importantly, the resulting Dice similarity coefficients are comparable to those from a widely used elastic registration method.

Keywords: Classification · Landmark detection · Cardiac MR image registration

1 Introduction

Cardiac image registration can generally be classified under two main categories: intensity- and point-based registration. The latter deforms images by minimizing a distance measure that compares voxel similarity patterns over the entire image domain [3]. As such, these methods are typically computationally expensive. On the other hand, point-based registration can be viewed simply as an interpolation problem, with the optimal transformation being the function passing through each control point while satisfying other constraints, such as minimizing the oscillation of the interpolant.

Point-based cardiac registration can be challenging and prone to errors since the selection of landmarks highly depends on the ability of the physician to mentally integrate information from different images [5]. In addition, the heart only has few spatially accurate and repeatable anatomical landmarks [10]. For instance, in [12], only the two papillary muscles and the inferior junction of

© Springer Nature Switzerland AG 2021
E. Puyol Anton et al. (Eds.): STACOM 2020, LNCS 12592, pp. 87–96, 2021.
https://doi.org/10.1007/978-3-030-68107-4_9

the right ventricle were used to rigidly transform cardiac PET and US images. To circumvent the aforementioned issues, point-based registration is typically used either only as a preliminary step to correct cardiac scaling and orientation, or in conjunction with an intensity-based approach to improve image overlap, as in [10].

Here, we propose a mathematical approach to automatically detect landmarks for fast registration of cardiac images. Our approach incorporates shape priors in the segmentation and classification of epicardial and endocardial contours, as well as in the determination of natural edge-bisectors that make it possible to generate an ordered set of contour-approximating landmarks. We first discuss the landmark detection pipeline on short-axis slices of the heart and then present its 3D analogue. The method is validated on 2D and 3D *ex-vivo* MR images of explanted porcine hearts.

2   MRI Data

In this work, we used eight explanted healthy porcine hearts which were fixed in formalin. For MR imaging, the hearts were placed in a Plexiglass phantom filled with Fluorinert (to avoid susceptibility artefacts at air-tissue interface) and scanned on a 1.5T GE Signa Excite using a high-resolution head coil, as described in a previous study [11]. Specifically, here, we obtained the heart anatomy from T2-weighted images (3D FSE MR pulse sequence, TR = 700 ms, TE = 35 ms, NEX = 1, ETL = 2, FOV = 10–12 cm) which were reconstructed and interpolated at $0.5 \times 0.5 \times 1.6$ mm spatial resolution.

3   Methods

We now present each step involved in the detection of contour-approximating landmarks in short-axis slices. The pipeline (summarized in Fig. 1) starts with the segmentation and classification of the epicardial and endocardial regions. These steps utilize prior shape knowledge about the structures of the myocardial segments. In the final step, an ordered sampling of each contour is obtained through the location of points of interest (POIs) in the ventricles. The process of locating the POIs will be discussed in detail in Sect. 3.2.

We will also extend our proposed landmark detection scheme to generate surface-approximating landmarks for 3D cardiac MR volumes.

3.1   Segmentation and Classification

Short-axis slices were binarized to facilitate the segmentation and classification of epicardial, LV endocardial, and RV contours. The Moore-Neighbor tracing algorithm [4] was implemented to detect the holes in the image, the exterior boundaries, and the maximal region of connected pixels tracing each segment.

The following assumptions on myocardial segments were employed to label the segmented regions:

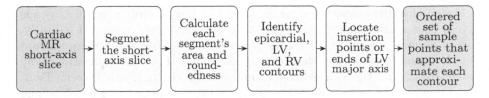

Fig. 1. Landmark detection in short-axis cardiac images

a. Healthy epicardial and LV endocardial regions possess elliptical contours.
b. RV contours are lune-shaped.

In conjunction with these shape priors, we also calculated the area and roundedness of each segment. The largest region is automatically classified as the epicardium. In basal and mid-cavity slices (see Fig. 2), the LV endocardium is differentiated from the RV through the roundedness measure R. It is given by

$$R = \frac{4\pi A}{P^2},$$

where A and P denote the area and perimeter of a region, respectively. Note that R-values range from 0 to 1 (circle). The segment with the smallest R-value is labeled as the RV because RVs are typically crescent-shaped, while LV cross sections resemble ellipses. On the other hand, in binarized apical slices where only the LV endocardial and the epicardial contours are visible, the smaller region is classified as the LV endocardium.

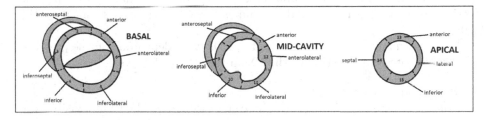

Fig. 2. Classifications of short-axis slices. Image adapted from [2].

3.2 Landmark Detection in 2D Short-Axis Images

To generate an ordered set of contour-approximating landmarks, we begin by identifying two points of interest in each slice. These 2 points will partition each myocardial segment into halves, which in turn, will allow us to define a starting and end point for each partition, divide each partition into shorter arcs of equal length, and identify an ordered set of pixels that ultimately approximate each contour. Here, we define the POIs to be the points along the ventricular contours where the maximum curvature occurs.

Recall that the curvature κ in $\mathbb{R}^2$ measures how sharply a curve C bends [1] and that maximizing κ is equivalent to minimizing its reciprocal $\rho = 1/\kappa$, more commonly known as the radius of curvature. Finding κ is straightforward provided that the equation describing C is known. For our application, only the pixels tracing the epicardial and endocardial contours are known. Thus, computing the curvature entails the additional step of performing a polynomial fit at every point on each contour. On the other hand, calculating ρ involves finding the center of the osculating circle at every $x \in C$ and then calculating the distance of the center from x.

To simplify the problem of finding the POIs, note that the sharpest bends along the myocardial contours occur where the contours are narrowest (Figs. 3a–c). Therefore, we can instead focus on measuring the narrowing at every contour point.

Given an ordered list of connected pixels $C = \{x_k\}_{k=0}^{N-1}$, we define the interior angle of C at pixel x_k as a function of the dot product of two unit vectors

$$\phi(x_k) = \arccos \left[\frac{\overrightarrow{x_k x_{(k+\Delta k) \bmod N}}}{\left\| \overrightarrow{x_k x_{(k+\Delta k) \bmod N}} \right\|} \cdot \frac{\overrightarrow{x_k x_{(k-\Delta k) \bmod N}}}{\left\| \overrightarrow{x_k x_{(k-\Delta k) \bmod N}} \right\|} \right],$$

where the window size Δk ($0 < \Delta k < N - 1$ and $\Delta k \in \mathbb{N}$) dictates the points along the curve that make up the interior angle $\phi(x_k)$. More specifically, the interior angle $\phi(x_k)$ at pixel x_k is made up of two rays: one extending from x_k to $x_{k-\Delta k}$, and the other from x_k to $x_{k+\Delta k}$, the pixels that, respectively, precede and succeed x_k in the ordered list of connected pixels by Δk places. See Figs. 3b–c.

A smaller interior angle corresponds to a smaller radius of curvature or, equivalently, to a larger curvature. Thus, we hypothesize that minimizing the interior angle along a myocardial contour is a good alternative to locating the POIs.

As a consequence of the definition of κ, POIs vary depending on the presence (or absence) of a cross section of the RV in the slice as follows.

A. Basal and mid-cavity slices

On the upper slices of the heart where cross sections of the RV are visible, the POIs are given by the RV insertion points. Visually, these points correspond to the cusps of the crescent-shaped RV contour and consequently divide it into two arcs. After finding the insertion points using the method described above, we then bisect the LV and the epicardial contours by locating the centers of the LV and RV, and then passing a straight line through their centers. The points along the LV (epicardial) contour that intersect this line are labeled as the first pixels in the ordered sampling for their respective halves of the LV (epicardial) contour. These steps are shown in Fig. 3e.

B. Apical slices

On apical slices where only the LV and epicardial contours are present, the bisection of the myocardial contours begins with the identification of the ends of the major axis of the elliptic LV contour. We remark that the curvature-based POI detection method described above is still applicable in this case because the maximum curvature in every ellipse occurs at the ends of its major axis. In this case, a line passing through the two LV POIs is used to bisect the epicardial and endocardial contours, again enabling the identification of a starting and an end point for each contour. See Fig. 3f.

Finally the contour-approximating landmarks in each bisected arc are obtained by sampling the ordered set of connected pixels. The number of sample points is user-specified.

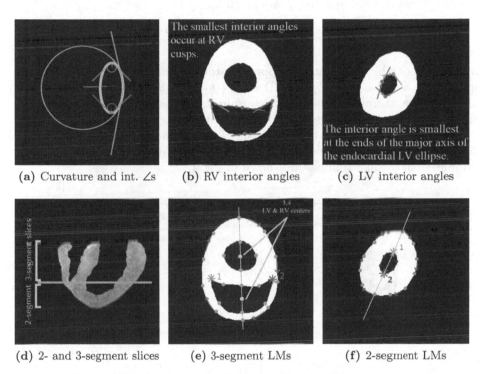

(a) Curvature and int. ∠s   (b) RV interior angles   (c) LV interior angles

(d) 2- and 3-segment slices   (e) 3-segment LMs   (f) 2-segment LMs

Fig. 3. (a) Comparison of curvature, radius of curvature, and interior angles. Osculating circles drawn at different points on an ellipse demonstrate that a high curvature κ corresponds to a small radius of curvature ρ and a small interior angle ϕ, (b)–(c) interior angles in 3- and 2-segment slices, (d) 3- and 2-segment slices in lateral view, (e)–(f) POIs and a sampling of each segment/contour.

4   Landmark Detection in 3D Cardiac Volumes

Provided that the hearts were scanned as specified in Sect. 2, the 3D analogue of the landmark detection method presented in the previous section merely involves the extra step of identifying the short-axis slices from which contour-approximating landmarks are to be generated.

Suppose that

a. T denotes the slice thickness,
b. the two-segment (apical, see Fig. 3d) slices occupy levels $z \in [a, b]$,
c. the three-segment (basal and mid-cavity) slices are on levels $z \in [b + T, c]$,
d. m and n refer to the user-specified number of sub-intervals in the levels spanned by the two-segment and the three-segment slices.

Then the sampling increments in the vertical direction for the two-segment and three-segment slices, respectively, are given by

$$\Delta m = \frac{b - a}{m} \quad \text{AND} \quad \Delta n = \frac{c - (b + T)}{n},$$

and the slices where the contour-approximating landmarks are to be identified are $z \in \{a, a + \Delta m, \ldots, b\}$ AND $z \in \{(b + T), (b + T) + \Delta n, \ldots, c\}$.

Once the sampling slices are determined, the same landmark detection pipeline discussed in the previous section is implemented on each sampling slice. Collectively, the landmarks from these slices approximate the surface of the cardiac volume. Examples demonstrating the placement of automatically detected landmarks in cardiac volumes are shown in Fig. 4.

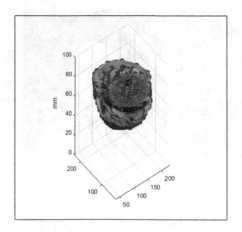 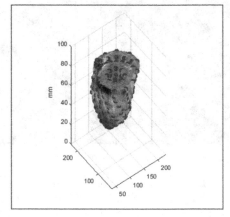

Fig. 4. Examples of automatically detected surface-approximating landmarks in cardiac volumes.

5   Experiments

Prior to registration, the hearts were manually segmented in ITK Snap (www.itksnap.org, [14]) and the contours were checked by an expert. Both 2D and 3D cardiac MR images were used to validate the proposed pipeline. The use of images of explanted hearts in our experiments was advantageous because it helped avoid motion artefacts. It also simplified the heart segmentation step since no other structures were present.

For both 2D and 3D experiments, landmarks were located using the method discussed in Sect. 3. Note that in order for the proposed fast registration pipeline to work, a one-to-one correspondence between the sets of reference and template landmarks has to be defined. To do this, the user-specified number of sampling slices (m and n) and the number of contour-approximating points along the short-axis myocardial contours (epicardium, LV, RV) have to be the same for the reference and template hearts.

Thin Plate Spline (TPS) [13] registration was then used to align the hearts, with the automatically detected landmarks serving as the control points. To compute the accuracy of the registration results, Dice coefficients were compared before and after registration. Elastic registration was also implemented using the FAIR toolkit [6] in Matlab to provide a benchmark for our registration results.

6 Results and Discussion

Shown in Figs. 5a c are exemplary results obtained after TPS registration was applied on pairs of short-axis slices. The top row shows the pre- and post-registration image overlap of the template with the reference image. Note that the use of automatically detected landmarks resulted in better post-registration image similarity, demonstrated by the difference image $|\mathcal{R} - \mathcal{T}[\theta]|$ being close to null.

The proposed fast registration approach also proved to be effective in registering 3D volumes of porcine hearts. All fifty-six possible reference-template pairings in our dataset of eight hearts were considered. Some of the results are displayed in Figs. 5d–e, and the complete tabulation of DSCs is in Table 2. Our method yielded consistently accurate results despite the high variability in cardiac size and surface curvatures present in our dataset. Statistically, 46 of the 56 (82.14%) image pairings that were registered using our proposed pipeline had resulted in DSCs that were comparable (i.e., within a 5% margin of error) to the baseline elastic registration results [7–9]. Notably, an additional 8/56 (14.28%) of our results improved on their corresponding baseline DSC by at least 10%.

We also calculated the average change in runtime (measured in seconds) over all reference-template pairings between the two registration methods. We found that the TPS, when paired with the automatically detected landmarks on the original cardiac volumes, was approximately 60% faster compared to implementing multi-level elastic registration [6] on downsampled versions of the reference and template volumes. This amounts to an 11-min runtime speedup (Table 1).

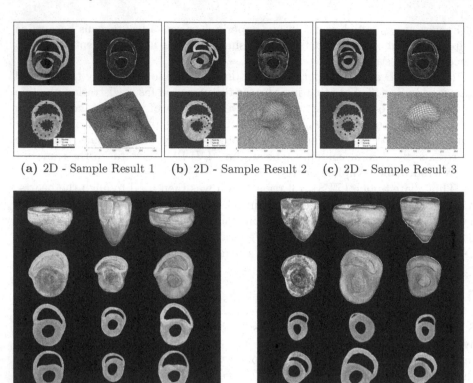

(a) 2D - Sample Result 1   **(b)** 2D - Sample Result 2   **(c)** 2D - Sample Result 3

(d) 3D - Sample Result 1   **(e)** 3D - Sample Result 2

Fig. 5. 2D and 3D TPS registration using automatically detected landmarks. (a)–(c) Top row: Difference images before and after registration, Bottom row: Registered template image and optimal transformation; (d)–(e) First col: Different views of the reference volume, Second col: Template, Third col: Registered template.

Table 1. Reference-template image similarity before registration, measured in terms of the Dice similarity coefficient.

		Pre-Registration Dice Similarity Coefficient							
		Template Heart							
		Heart1	Heart2	Heart3	Heart4	Heart5	Heart6	Heart7	Heart8
Reference Heart	Heart1	1.00	0.57	0.58	0.27	0.44	0.46	0.42	0.55
	Heart2	0.57	1.00	0.69	0.21	0.34	0.30	0.28	0.41
	Heart3	0.58	0.69	1.00	0.26	0.43	0.31	0.28	0.59
	Heart4	0.27	0.21	0.26	1.00	0.26	0.24	0.34	0.20
	Heart5	0.44	0.34	0.43	0.26	1.00	0.43	0.36	0.49
	Heart6	0.46	0.30	0.31	0.24	0.43	1.00	0.73	0.34
	Heart7	0.42	0.28	0.28	0.34	0.36	0.73	1.00	0.28
	Heart8	0.55	0.41	0.59	0.20	0.49	0.34	0.28	1.00

Table 2. Dice similarity coefficient after elastic (ER) and TPS registration.

		Post-Registration Dice Similarity Coefficient															
		Template Heart															
		Heart1		Heart2		Heart3		Heart4		Heart5		Heart6		Heart7		Heart8	
		ER	TPS	ER	TPS	ER	TPS	ER	TPS	ER	TPS	ER	TPS	ER	TPS	ER	TPS
Reference Heart	Heart1	1.00	1.00	0.95	0.93	0.95	0.93	0.96	0.93	0.96	0.94	0.96	0.94	0.94	0.93	0.92	0.94
	Heart2	0.97	0.93	1.00	1.00	0.97	0.94	0.44	0.93	0.95	0.94	0.97	0.93	0.88	0.92	0.95	0.93
	Heart3	0.97	0.93	0.97	0.93	1.00	1.00	0.59	0.94	0.96	0.94	0.97	0.93	0.96	0.93	0.95	0.93
	Heart4	0.97	0.93	0.97	0.93	0.97	0.94	1.00	1.00	0.97	0.94	0.97	0.93	0.97	0.94	0.94	0.93
	Heart5	0.97	0.92	0.97	0.91	0.97	0.92	0.92	0.94	1.00	1.00	0.98	0.92	0.96	0.92	0.94	0.92
	Heart6	0.96	0.92	0.95	0.92	0.95	0.94	0.96	0.95	0.96	0.95	1.00	1.00	0.96	0.94	0.93	0.94
	Heart7	0.95	0.93	0.95	0.93	0.95	0.94	0.95	0.95	0.96	0.95	0.98	0.95	1.00	1.00	0.92	0.93
	Heart8	0.82	0.89	0.89	0.89	0.83	0.89	0.16	0.90	0.81	0.92	0.78	0.90	0.53	0.89	1.00	1.00

Future work includes the improvement of the segmentation and classification steps to allow the proposed pipeline to accommodate the co-registration of *ex-vivo* and *in-vivo* cardiovascular MR images. We also plan to generalize the landmark detection scheme to other clinical images that exhibit one or more structures with prominent curvature.

7 Conclusions

In this work, we presented the mathematical framework behind each step in a fast cardiac image registration pipeline. Our proposed method uses shape priors to determine the strategic location of landmarks on the reference and template images, which ultimately guides the registration process. Our holistic approach was able to successfully carry out the segmentation and classification of myocardial regions, the detection of surface-approximating landmarks, and the implementation of a point-based registration method to align pairs of cardiac images with minimal user input.

Experiments using the automatically detected landmarks together with Thin Plate Spline registration demonstrated that our proposed method consistently yields accurate registration results that are comparable to those from an intensity-based elastic registration scheme. In addition, the computational runtime was reduced. We conclude that our pipeline provides a fast but still an effective alternative to intensity-based registration methods.

Acknowledgments. This work was supported in part by a Natural Sciences and Engineering Research Council of Canada (NSERC) grant for Dr. Mehran Ebrahimi and a Canadian Institutes of Health Research (CIHR) project grant for Dr. Mihaela Pop. Mia Mojica is supported by an Ontario Trillium Scholarship (OTS).

References

1. Anton, H., Bivens, I., Davis, S., Polaski, T.: Calculus: Early Transcendentals. Wiley, Hoboken (2010)
2. Cerqueira, M.D., et al.: Standardized myocardial segmentation and nomenclature for tomographic imaging of the heart. Circulation **105**(4), 539–542 (2002)
3. Fischer, B., Modersitzki, J.: Combining landmark and intensity driven registrations. In: PAMM: Proceedings in Applied Mathematics and Mechanics, vol. 3, pp. 32–35. Wiley Online Library (2003)
4. Gonzalez, R.C., Woods, R.E., Eddins, S.L.: Digital Image Processing Using MATLAB. Pearson Education India (2004)
5. Makela, T., et al.: A review of cardiac image registration methods. IEEE Trans. Med. Imaging **21**(9), 1011–1021 (2002)
6. Modersitzki, J.: FAIR: Flexible Algorithms for Image Registration, vol. 6. SIAM, Philadelphia (2009)
7. Mojica, M., Pop, M., Sermesant, M., Ebrahimi, M.: Multilevel non-parametric groupwise registration in cardiac mri: application to explanted porcine hearts. In: Pop, M., et al. (eds.) STACOM 2017. LNCS, vol. 10663, pp. 60–69. Springer, Cham (2018). https://doi.org/10.1007/978-3-319-75541-0_7
8. Mojica, M., Pop, M., Sermesant, M., Ebrahimi, M.: Constructing an average geometry and diffusion tensor magnetic resonance field from freshly explanted porcine hearts. In: Medical Imaging 2019: Image Processing, vol. 10949, p. 109493C. International Society for Optics and Photonics (2019)
9. Mojica, M., Pop, M., Sermesant, M., Ebrahimi, M.: Novel atlas of fiber directions built from ex-vivo diffusion tensor images of porcine hearts. Comput. Methods Programs Biomed. **187**, 105200 (2020)
10. Peyrat, J.M., et al.: A computational framework for the statistical analysis of cardiac diffusion tensors: application to a small database of canine hearts. IEEE Trans. Med. Imaging **26**(11), 1500–1514 (2007)
11. Pop, M., et al.: Quantification of fibrosis in infarcted swine hearts by ex vivo late gadolinium-enhancement and diffusion-weighted MRI methods. Phys. Med. Biol. **58**(15), 5009 (2013)
12. Savi, A., et al.: Spatial registration of echocardiographic and positron emission tomographic heart studies. Eur. J. Nucl. Med. **22**(3), 243–247 (1995). https://doi.org/10.1007/BF01081520
13. Wahba, G.: Spline Models for Observational Data, vol. 59. SIAM, Philadelphia (1990)
14. Yushkevich, P.A., et al.: User-guided 3D active contour segmentation of anatomical structures: significantly improved efficiency and reliability. Neuroimage **31**(3), 1116–1128 (2006)

Quality-Aware Semi-supervised Learning for CMR Segmentation

Bram Ruijsink[1,2,3]($\boxtimes$), Esther Puyol-Antón[1], Ye Li[1], Wenjia Bai[4],
Eric Kerfoot[1], Reza Razavi[1,2], and Andrew P. King[1]

[1] School of Biomedical Engineering and Imaging Sciences, King's College London,
London, UK
[2] St Thomas' Hospital NHS Foundation Trust, London, UK
[3] Department of Cardiology, University Medical Centre Utrecht,
Utrecht, The Netherlands
jacobus.ruijsink@kcl.ac.uk
[4] Biomedical Image Analysis Group, Imperial College London, London, UK

Abstract. One of the challenges in developing deep learning algorithms
for medical image segmentation is the scarcity of annotated training data.
To overcome this limitation, data augmentation and semi-supervised
learning (SSL) methods have been developed. However, these methods
have limited effectiveness as they either exploit the existing data set only
(data augmentation) or risk negative impact by adding poor training
examples (SSL). Segmentations are rarely the final product of medical
image analysis - they are typically used in downstream tasks to infer
higher-order patterns to evaluate diseases. Clinicians take into account a
wealth of prior knowledge on biophysics and physiology when evaluating
image analysis results. We have used these clinical assessments in previous works to create robust quality-control (QC) classifiers for automated
cardiac magnetic resonance (CMR) analysis. In this paper, we propose
a novel scheme that uses QC of the downstream task to identify high
quality outputs of CMR segmentation networks, that are subsequently
utilised for further network training. In essence, this provides quality-
aware augmentation of training data in a variant of SSL for segmentation networks (semiQCSeg). We evaluate our approach in two CMR
segmentation tasks (aortic and short axis cardiac volume segmentation)
using UK Biobank data and two commonly used network architectures
(U-net and a Fully Convolutional Network) and compare against supervised and SSL strategies. We show that semiQCSeg improves training of
the segmentation networks. It decreases the need for labelled data, while
outperforming the other methods in terms of Dice and clinical metrics.
SemiQCSeg can be an efficient approach for training segmentation networks for medical image data when labelled datasets are scarce.

Keywords: CMR · Segmentation network · Quality control · Data
augmentation

E. Puyol-Antón—Joint first authors.

© Springer Nature Switzerland AG 2021
E. Puyol Anton et al. (Eds.): STACOM 2020, LNCS 12592, pp. 97–107, 2021.
https://doi.org/10.1007/978-3-030-68107-4_10

1   Introduction

Automated segmentation and analysis of medical imaging data using deep learning (DL) has been shown to greatly improve analysis of a wide range of diseases. Using the output of segmentation networks, functional processes in the body can be quantified. For example, in cardiac magnetic resonance (CMR) imaging, the segmentations obtained from individual frames of a cine acquisition of the beating heart can be used to calculate the volumes of the ventricles throughout the cardiac cycle, providing a detailed description of the pumping function of the heart.

One of the main challenges facing the use of DL models in medical imaging is the scarcity of training data. To createa complete training dataset for cardiac volume segmentation over the full cardiac cycle from short-axis CMR, multiple slices (typically 10–12) and multiple time-frames (typically 30–50) need to be manually segmented for each case by an experienced CMR cardiologist. This labour intensive task cannot be feasibly done at scale. As a result, segmentations are often available in only one or two time-frames.

Data augmentation has become a vital part of network training. Most augmentation techniques focus on the existing training data only, typically applying spatial and intensity transformations. Semi-supervised learning (SSL) techniques have also been proposed [4], in which the training data are augmented by new cases labelled by a DL model trained in a supervised way [1]. However, if the new labels are unreliable this has the potential to degrade model performance.

So far, information about downstream tasks has mostly been ignored when producing extra training data for segmentation networks. However, such knowledge is potentially very useful: the behaviour of biological segmentation targets is constrained by their biophysical properties. For example, the shape and volume of the aorta and/or cardiac left and right ventricles (LV and RV), and their changes during the cardiac cycle follow certain patterns, even in disease. Clinicians intuitively use this biophysical knowledge when evaluating results of CMR analysis. In previous work, we showed that a clinician's assessment of shape and volume changes during CMR analysis can be used to train quality control (QC) classifiers to identify erroneous segmentations [14]. In this paper, we aim to utilise these QC classifiers to increase the training data of segmentation models in a semi-supervised setting. Specifically, we propose to identify highly accurate segmentation results on unlabelled cases (i.e. 'best-in-class' cases) using the QC classifiers and use these cases to increase the available training data.

We evaluate our proposed approach in two different segmentation tasks (aortic and cardiac segmentation) using two commonly used architectures (fully convolutional network (FCN) [2] and U-Net [13]). Our results show that the proposed method outperforms state-of-the-art SSL and fully supervised methods, while significantly reducing the initial (manually labelled) training set needed.

Related Work: Data augmentation techniques can be used to create new and realistic-looking training data. Most augmentation methods involve alteration of existing training examples, for instance by using spatial and intensity transformations or applying kernel filters [16]. More recently, DL approaches have been employed to generate new training data. Several works have proposed synthesis of image segmentation training data using Generative Adversarial Networks (GANs), for example for computed tomography [15] and chest X-rays [3]. Another approach, demonstrated by Oksuz et al. [9], is to use knowledge of CMR acquisition to synthesise a specific image type of interest. The authors created images that contained motion artefacts by reverse modelling non-corrupted images to k-space and inserting irregularities that would have occurred if motion were present during acquisition.

SSL methods represent another way of improving training in the face of limited training data. Examples include self-training [8], bootstrapping [11] and generative modeling [6]. Most common approaches focus on weakly supervised learning [11,12], which consists of generating pseudo pixel-level labels to increase the training database and then train a segmentation network in a supervised manner. For generative modeling, extensions of the generic GAN framework have been used to obtain pixel-level predictions that can be used as training samples [6]. In this case the discriminator is jointly trained with an adversarial loss and a supervised loss over the labelled examples.

In the cardiac imaging domain, Bai et al. [1], proposed a SSL approach for CMR image segmentation that combines labelled and unlabelled data during training. This technique used a conditional random field (CRF) based post-processing step to improve the quality of new training samples produced by a FCN. Another SSL approach was proposed by Kervadec et al. [8], who used a teacher-student network approach to segment 3D atrial volumes. A more general approach to improve segmentation outcomes is by applying multiple networks simultaneously and select only the best result [5].

The benefits of standard data augmentation are limited by the variation present in the existing data, while DL-generated augmented examples and SSL potentially suffer from generation of unrealistic or inaccurate data. These are significant drawbacks, particularly in medical image analysis. Our proposed method utilises aspects of data augmentation and SSL alongside robust QC. It creates an approach for semi-supervised DL model training that is informed by an independent QC network, trained using prior knowledge of biophysical behaviour in a downstream task. To the best of our knowledge, no previous works have exploited downstream analysis of model output for a QC-informed SSL approach to image segmentation.

2   Materials

We evaluate our proposed approach in two different segmentation tasks using two separate segmentation models. The first task entails segmentation of the aortic area (1 label) from a cine CMR acquisition of the ascending aorta using a FCN [2]. The second task entails segmentation of the LV and RV blood pools, and LV myocardium (3 labels) from a short-axis (SAX) cine CMR acquisition using a U-net [13]. The image datasets were obtained from the UK Biobank cohort. The aortic cine CMR acquisitions consists of a single slice and 100 temporal frames. The SAX acquisition contains multiple slices covering the full heart. This acquisition contains 50 temporal frames. Details of the image acquisition protocol can be found in [10].

For training of the models, we used manual segmentations obtained by experienced CMR operators. The training set consisted of 138 subjects with 3 segmented timeframes (randomly chosen throughout the cardiac cycle) for the aortic images, and 500 subjects with 2 segmented timeframes each for the SAX data (at the end-diastolic and end-systolic phases). This approach is in keeping with common practice for training DL segmentation models for cine CMR. Both datasets consisted of a mix of healthy subjects and patients with a range of cardiovascular diseases.

For evaluation, we used a set of test cases with full cardiac cycle ground truth segmentations in a total of 102 subjects for both the aortic images and SAX images. These datasets consist of 50% healthy subjects and 50% patients with cardiomyopathies.

3   Methods

In the following sections we describe the automated segmentation algorithms, the QC step and the process of identifying the 'best-in-class' segmentations. Figure 1 summarises these steps and gives an overview of our proposed method.

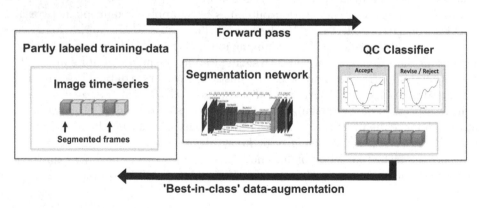

Fig. 1. Overview of the proposed framework for QC semi-supervised learning.

3.1   Proposed Framework

Automatic Segmentation Network: We used a FCN with a 17 convolutional layer VGG-like architecture for automatic segmentation of the aorta in all frames through the cardiac cycle [2,14]. For segmentation of the LV blood pool, RV blood pool and LV myocardium from the cine SAX images, we used a U-net with the layers of the encoding and decoding stages defined using residual units [7].

Output Measures: From each segmentation output, the aortic area and LV and RV blood pool volumes were calculated. By combining the measures from all time-frames, curves of aortic area and ventricular blood volume change were formed.

Quality Control: For QC, we used a 3-layer Long-Short-Term-Memory (LSTM) classifier, preceded by a single dense-layer convolution. The aim of this network is to classify cases into 'good' and 'erroneous' based on the area/volume curves. Separate classifiers were trained for each of the aortic area and ventricular volume data. To train the networks, a clinician reviewed 1,000 cases using plots of the LV and RV blood volume curves and aortic area change in combination with an animation of the obtained segmentations, see also [14]. The 1,000 training cases for QC were randomly selected from the UK Biobank dataset and were not used for training or evaluating our segmentation models. This way, overfitting of the QC network to the segmentation training data was avoided.

The loss functions of these networks were optimised for sensitivity of detection of erroneous results. This was done by obtaining a receiver operating characteristic (ROC) curve after training and selecting the classifier that maximised sensitivity (i.e. recall) using a weighted Youden index. The QC classifiers used during the experiments described in this paper had a sensitivity of 96% (aorta) and 95% (SAX) for detection of erroneous results.

Identifying 'Best-in-Class' Cases: After QC, we used the probability outputs of the QC classifiers to rank cases from highest probability of being accurate to lowest probability of being accurate. We identified the 30 'best-in-class' cases and from these cases, the full-cardiac cycle segmentations and their corresponding images were used to augment the training dataset of the segmentation model.

3.2 Experiments

Network Training: We trained both segmentation models in four scenarios: (1) using all cases originally available for training of the network (i.e. the 'full dataset') - this acts as a 'best achievable' scenario, (2) a baseline approach using only half the available training data in a fully supervised manner ('half dataset'), (3) using a SSL approach, initially using half of the training data, but feeding the outputs of 30 random cases back into the training set without QC (semi-supervised no QC), (4) a state-of-the-art SSL approach [1] that utilises CRF postprocessing to improve the quality of the new training segmentations, and finally (5) using our proposed approach (semiQCSeg), initially using half of the training data, and then after feeding the 30 'best-in-class' outputs back into the training data. Table 1 shows a summary of the data used for each experiment. To achieve a fair comparison, we trained all networks in a similar way: from scratch and for a fixed number of 2,000 epochs.

Table 1. Summary of the training data (number of subjects and number of labelled images) used for each experiment. Note that, for the SSL approaches (3)-(5), the number of images increases dramatically because segmentations are available for all timeframes.

Methods	Aorta FCN		SAX U-net	
	Ssubjects (n)	Images (n)	Subjects (n)	Images (n)
(1) Full dataset	138	414	500	1000
(2) Half dataset	69	207	250	500
(3) Semi-supervised no QC	99	2160	280	2000
(4) Semi-supervised no QC + CRF	99	2160	280	2000
(5) SemiQCSeg	99	2160	280	2000

Evaluation: We evaluated the models' segmentation performances using the independent test set (described above), which was not used during training. We compared segmentation performance using mean Dice scores for aorta and SAX segmentations and additionally evaluated the performance of the different approaches in the downstream task for the SAX dataset, i.e. the calculation of ventricular volume and ejection fraction.

4   Results

Table 2 shows the mean Dice scores obtained using the different approaches. The Dice score shown for the SAX data is a pooled mean of the individual LV and RV blood pool and LV myocardium values.

The semiQCSeg approach achieved a significantly higher Dice score than the 'half dataset' baseline approaches for both SAX and aorta experiments. More importantly, semiQCSeg outperformed the best achievable model ('full dataset') for SAX, while achieving Dice scores similar to full dataset training for Aorta. SSL without QC performed significantly worse than our semiQCSeg approach. Adding the CRF postprocessing to this model improved performance, but not to the level of our proposed approach. See Fig. 3 for an example of the performance of the different approaches for segmentation of a SAX stack of a selected subject.

Selecting output cases for training augmentation might bias a network towards learning to segment 'easy' cases only, while decreasing accuracy in 'hard' cases. Figure 2 shows that there was no bias in improvement of Dice scores in easy cases alone using semiQCSeg.

Table 2. Mean Dice and standard deviation (between brackets) of the tested models. For SAX, the pooled mean dice of LV and RV blood pool and LV myocardium is shown. * denotes a p-value < .001 with respect to semi-supervised QC approach (semiQCSeg), using a paired t-test.

Training strategy	Aorta FCN	SAX U-net
	Dice	Dice
(a) Full database	95.45 (1.45)	88.17 (4.35)*
(b) Half database	93.71 (1.51)*	82.57 (10.79)*
(c) Semi-supervised no QC	92.86 (2.95)*	85.30 (6.97)*
(d) Semi-supervised no QC CRF	94.89 (1.71)*	85.97 (6.00)*
(e) SemiQCSeg	95.56 (1.53)	89.44 (4.01)

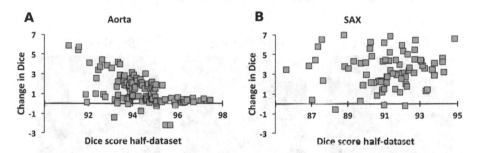

Fig. 2. Change in Dice score using semiQCSeg with respect to half dataset training for each individual case. A) for aorta, B) for SAX (mean pooled dice of LV and RV blood pool, and myocardium).

4.1 Downstream Task

A comparison of the different approaches in terms of downstream clinical metrics (ventricular volumes and ejection fraction) for the SAX dataset is shown in Table 3. Of all the methods, semiQCSeg yields results closest to those based on manual segmentations. We do not show data for the aortic cases, as the calculation of the clinical metric (aortic distensibility) necessitates additional blood pressure data.

5 Discussion

In this paper, we have proposed a novel method for quality-aware SSL for medical image segmentation. We show that this approach is able to boost network performance, while markedly reducing the amount of annotated datasets needed.

Importantly, our approach does not infer quality from the individual segmentations directly (for example using voxel probability). Instead, it utilises clinical characteristics obtained from a downstream analysis of the segmentations to

Table 3. Mean and standard deviation (in brackets) of clinical metrics produced by the tested models. LVEDV: Left Ventricular End Diastolic Volume, LVEF: Left Ventricular Ejection Fraction, RVEDV: Right Ventricular End Diastolic Volume, RVEF: Right Ventricular Ejection Fraction.

Training strategy	SAX U-net			
	LVEDV	LVEF	RVEDV	RVEF
(a) Full database	151 (36)	56 (8)	163 (42)	52 (6)
(b) Half database	153 (38)	54 (6)	171 (46)	47 (8)
(c) Semi-supervised no QC	152 (39)	55 (6)	165 (44)	47 (8)
(d) Semi-supervised no QC CRF	151 (38)	56 (6)	164 (42)	48 (7)
(e) SemiQCSeg	150 (36)	57 (6)	160 (43)	54 (6)
(f) Manual segmentation	147 (36)	58 (5)	156 (42)	55 (6)

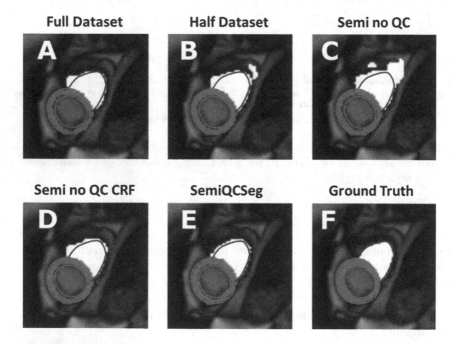

Fig. 3. Examples of segmentations obtained using the different approaches for a CMR short axis image, with our proposed method (SemiQCSeg) in panel E. The contour overlay in each image represents the ground truth segmentation. QC: quality control, CRF: conditional random field.

make use of biophysical knowledge used by clinicians when judging CMR segmentation output. We show that our approach allows segmentation models to learn more effectively; it uses less initial data, while yielding similar or even better results compared to full dataset training. The latter illustrates the strength of our approach; it delivers extra full cardiac cycle datasets while tightly controlling their quality using metrics not directly obtained from the segmentation network.

Augmenting data based on direct inference from the segmentations may bias further learning towards a certain group of cases. As can be appreciated in Fig. 2, this bias seems not to be present using our proposed approach. However, more robust evaluation is needed to ensure the impact of our approach on bias after multiple iterations of the semiQCSeg process. A large number of iterations is likely to result in some bias in learning. Therefore this method is most suitable for boosting performance using a single or limited number of iterations. To detect increasing bias over iterations, we propose to monitor changes in loss of the validation set for a potential increase to trigger early stopping of the semiQCSeg process.

While training of our QC step necessitates some extra effort from clinicians, this pattern recognition task is fast compared to the time needed for extra segmentation. Moreover, as it is independent of the upstream segmentation network, the trained QC model could be reused flexibly for training different CMR datasets or other segmentation models.

We found that SSL without QC achieved some improvement in Dice scores with respect to half dataset training for SAX, but not for the aorta dataset. Including CRF postprocessing led to improvements in both datasets (Table 2). This is in keeping with earlier reports of SSL approaches [4]. However, the variability of the impact in aorta and SAX illustrates the uncertainty in uncontrolled data augmentation in SSL; the benefit or negative effect depends entirely on the quality of the output segmentations that are obtained for further training.

Figure 3 shows an example of segmentations obtained using the different approaches. This further illustrates the uncertain impact of SSL without QC; as the segmentations obtained (panel C) are worse than the ones obtained using half dataset training alone (panel B). Of all methods, SemiQCSeg (panel E) performed the best compared to the ground truth (panel F).

In this paper, we demonstrated our method in time-series CMR data. The success of our approach opens up the potential to use time-sensitive architectures, such as Long Short Term Memory models, without the need to provide fully labelled training examples that include each timeframe. Extension of our approach to this domain will be the subject of our future work.

Our approach could also be used in other segmentation tasks. For example, in computed tomography (multi-)organ segmentation, where segmenting only part of all available slices could be combined with semiQCSeg to provide whole body training sets, or brain imaging, where information obtained from multiple acquisitions can be exploited during QC.

In conclusion, we have shown that semiQCSeg is an efficient method for training DL segmentation networks for medical imaging tasks when labelled datasets are scarce.

Acknowledgements. Dr. Bram Ruijsink is supported by the NIHR Cardiovascular MedTech Co-operative awarded to the Guy's and St Thomas' NHS Foundation Trust. This work was further supported by the EPSRC (EP/R005516/1 and EP/P001009/1), the Wellcome EPSRC Centre for Medical Engineering at King's College London (WT 203148/Z/16/Z) and National Institute for Health Research (NIHR). This research has been conducted using the UK Biobank Resource under Application Number 17806.

References

1. Bai, W., et al.: Semi-supervised learning for network based cardiac MR image segmentation. In: Descoteaux, M., Maier-Hein, L., Franz, A., Jannin, P., Collins, D.L., Duchesne, S. (eds.) MICCAI 2017, Part II. LNCS, vol. 10434, pp. 253–260. Springer, Cham (2017). https://doi.org/10.1007/978-3-319-66185-8_29
2. Bai, W., et al.: Automated cardiovascular magnetic resonance image analysis with fully convolutional networks. J. Cardiovasc. Magn. Reson. **20**(1), 65 (2018)
3. Bozorgtabar, B., et al.: Informative sample generation using class aware generative adversarial networks for classification of chest X rays. Comput. Vis. Image Underst. **184**, 57–65 (2019)
4. Cheplygina, V., de Bruijne, M., Pluim, J.P.: Not-so-supervised: a survey of semi-supervised, multi-instance, and transfer learning in medical image analysis. Med. Image Anal. **54**, 280–296 (2019)
5. Hann, E., et al.: Quality control-driven image segmentation towards reliable automatic image analysis in large-scale cardiovascular magnetic resonance aortic cine imaging. In: Shen, D., et al. (eds.) MICCAI 2019, Part II. LNCS, vol. 11765, pp. 750–758. Springer, Cham (2019). https://doi.org/10.1007/978-3-030-32245-8_83
6. Hung, W.C., Tsai, Y.H., Liou, Y.T., Lin, Y.Y., Yang, M.H.: Adversarial learning for semi-supervised semantic segmentation. arXiv preprint arXiv:1802.07934 (2018)
7. Kerfoot, E., Clough, J., Oksuz, I., Lee, J., King, A.P., Schnabel, J.A.: Left-ventricle quantification using residual U-Net. In: Pop, M., Sermesant, M., Zhao, J., Li, S., McLeod, K., Young, A., Rhode, K., Mansi, T. (eds.) STACOM 2018. LNCS, vol. 11395, pp. 371–380. Springer, Cham (2019). https://doi.org/10.1007/978-3-030-12029-0_40
8. Kervadec, H., Dolz, J., Granger, É., Ben Ayed, I.: Curriculum semi-supervised segmentation. In: Shen, D., et al. (eds.) MICCAI 2019, Part II. LNCS, vol. 11765, pp. 568–576. Springer, Cham (2019). https://doi.org/10.1007/978-3-030-32245-8_63
9. Oksuz, I., et al.: Automatic CNN-based detection of cardiac MR motion artefacts using k-space data augmentation and curriculum learning. Med. Image Anal. **55**, 136–147 (2019)
10. Petersen, S.E., et al.: UK biobank's cardiovascular magnetic resonance protocol. J. Cardiovasc. Magn. Reson. **18**(1), 8 (2015). https://doi.org/10.1186/s12968-016-0227-4
11. Qiao, S., Shen, W., Zhang, Z., Wang, B., Yuille, A.: Deep co-training for semi-supervised image recognition. In: Ferrari, V., Hebert, M., Sminchisescu, C., Weiss, Y. (eds.) ECCV 2018, Part XV. LNCS, vol. 11219, pp. 142–159. Springer, Cham (2018). https://doi.org/10.1007/978-3-030-01267-0_9
12. Radosavovic, I., Dollár, P., Girshick, R., Gkioxari, G., He, K.: Data distillation: towards omni-supervised learning. In: Proceedings of the IEEE Conference on Computer Vision and Pattern Recognition, pp. 4119–4128 (2018)

13. Ronneberger, O., Fischer, P., Brox, T.: U-Net: convolutional networks for biomedical image segmentation. In: Navab, N., Hornegger, J., Wells, W.M., Frangi, A.F. (eds.) MICCAI 2015, Part III. LNCS, vol. 9351, pp. 234–241. Springer, Cham (2015). https://doi.org/10.1007/978-3-319-24574-4_28
14. Ruijsink, B., et al.: Fully automated, quality-controlled cardiac analysis from CMR: validation and large-scale application to characterize cardiac function. JACC Cardiovas. Imaging **13**(3), 684–695 (2020)
15. Sandfort, V., Yan, K., Pickhardt, P.J., Summers, R.M.: Data augmentation using generative adversarial networks (CycleGAN) to improve generalizability in CT segmentation tasks. Sci. Rep. **9**(1), 1–9 (2019)
16. Shorten, C., Khoshgoftaar, T.M.: A survey on image data augmentation for deep learning. J. Big Data **6**(1), 60 (2019). https://doi.org/10.1186/s40537-019-0197-0

Estimation of Imaging Biomarker's Progression in Post-infarct Patients Using Cross-sectional Data

Marta Nuñez-Garcia[1,2]($\boxtimes$), Nicolas Cedilnik[3], Shuman Jia[3], Hubert Cochet[1,4], Marco Lorenzi[3], and Maxime Sermesant[3]

[1] Electrophysiology and Heart Modeling Institute (IHU LIRYC), Pessac, France
marta.nunez-garcia@ihu-liryc.fr
[2] Université de Bordeaux, Bordeaux, France
[3] Inria, Université C'ôte d'Azur, Sophia Antipolis, France
[4] CHU Bordeaux, Department of Cardiovascular and Thoracic Imaging and Department of Cardiac Pacing and Electrophysiology, Pessac, France

Abstract. Many uncertainties remain about the relation between post-infarct scars and ventricular arrhythmia. Most post-infarct patients suffer scar-related arrhythmia several years after the infarct event suggesting that scar remodeling is a process that might require years until the affected tissue becomes arrhythmogenic. In clinical practice, a simple time-based rule is often used to assess risk and stratify patients. In other cases, left ventricular ejection fraction (LVEF) impairment is also taken into account but it is known to be suboptimal. More information is needed to better stratify patients and prescribe appropriate individualized treatments. In this paper we propose to use probabilistic disease progression modeling to obtain an image-based data-driven description of the infarct maturation process. Our approach includes monotonic constraints in order to impose a regular behaviour on the biomarkers' trajectories. 49 post-MI patients underwent Computed Tomography (CT) and Late Gadolinium Enhanced Cardiac Magnetic Resonance (LGE-CMR) scans. Image-derived biomarkers were computed such as LVEF, LGE-CMR scar volume, fat volume, and size of areas with a different degree of left ventricular wall narrowing, from moderate to severe. We show that the model is able to estimate a plausible progression of post-infarct scar maturation. According to our results there is a progressive thinning process observable only with CT imaging; intramural fat appears in a late stage; LGE-CMR scar volume almost does not change and LVEF slightly changes during the scar maturation process.

Keywords: Disease progression modeling · Cross-sectional data · Ventricular arrhythmia · Post-infarct cardiac remodeling

1 Introduction

Scar-related substrate after myocardial infarction (MI) induces most life-threatening ventricular arrhythmias [8]. Common recommendations for primary

© Springer Nature Switzerland AG 2021
E. Puyol Anton et al. (Eds.): STACOM 2020, LNCS 12592, pp. 108–116, 2021.
https://doi.org/10.1007/978-3-030-68107-4_11

prevention of ventricular arrhythmia and sudden cardiac death advise cardioverter defibrillator (ICD) implantation in post-MI patients with left ventricular ejection fraction (LVEF) lower than 35% and symptoms of heart failure [1]. However, a risk stratification rule solely based on LVEF lacks sensitivity and specificity [4]. Additionally, several studies have reported long periods between MIs and arrhythmia [9] suggesting scar remodeling or maturation is a dynamic process, and it may take many years until the affected tissue becomes arrhythmogenic. Therefore, a deeper investigation of the nature of the arrhythmogenic substrate and its potential evolution over time is required. New biomarkers and more information about the right time point to evaluate them are needed.

Non-invasive imaging techniques have the potential to improve our knowledge about the nature of scar-related arrhythmogenic substrate and its dynamic remodeling process. Late Gadolinium Enhancement Cardiovascular Magnetic Resonance (LGE-CMR) imaging is the reference method to classify myocardial tissue. Many methods have been proposed to detect and quantify scar in the left ventricular (LV) wall using LGE-CMR images, most of them relying on a threshold-based rule. After segmenting the LV wall, tissue is typically divided into healthy tissue (darkest voxels), dense scar (brightest voxels) and border zone (image intensities between the other 2 thresholds). Scar remodeling using longitudinal LGE-CMR data have been recently described by Jáuregui et al. [5]. After scanning post-MI patients at 7 days, 6 months and 4 years after the infarction, the authors showed that dense scar and border zone mass decreased over time suggesting the existence of a long-term scar healing process. Bayesi

Due to the relatively low spatial resolution of LGE-CMR images and several contraindications (patients carrying ICDs, claustrophia, etc.) Computed Tomography (CT) imaging has been recently introduced to accurately characterize scar. Furthermore, several studies have shown the existence of a progressive thinning of the LV wall in scarred areas [2]. Wall thickness and intramural fat mapping from CT are promising imaging biomarkers that could be useful to improve risk stratification in patients considered for ICD implantation and to better personalize ablation therapies.

Disease progression modeling (DPM) provides a data-driven description of the natural evolution of a given pathology and it aims at revealing long-term pathological trajectories from short-term clinical data [7]. Reformulating DPM within a probabilistic setting has the potential of predicting a plausible evolution of the different biomarkers considered.

In this paper, we used CT and LGE-CMR imaging biomarkers from cross-sectional (i.e. one single time point per patient) post-MI data, and a probabilistic DPM to characterize scar maturation. The model shows a plausible evolution of the different biomarkers taken into account and allows us to estimate a relative temporal timeline of disease progression.

2   Methods

Data Processing

Forty-nine post-MI patients underwent CT and LGE-CMR imaging at the Centre Hospitalier Universitaire (CHU) de Bordeaux (France). A schematic

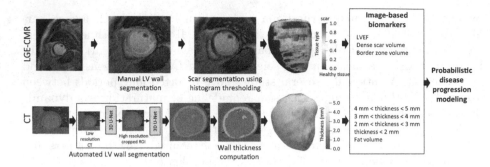

Fig. 1. Data processing pipeline. LGE-CMR images are manually segmented and the LV wall tissue is classified using a threshold-based approach. CT images are automatically segmented using 2 successive U-nets, and wall thickness is computed afterwards. Then, 8 imaging biomarkers are calculated and used to estimate the probabilistic disease progression model. LGE-CMR = Late Gadolinium Enhanced Cardiac Magnetic Resonance; CT = Computed Tomography; LVEF = Left Ventricular Ejection Fraction.

representation of the proposed method can be seen in Fig. 1. The LV wall was manually segmented from the LGE-CMR images by experts. As typically done in clinical practice, a threshold-based approach was used to classify voxels into healthy tissue (image intensity below the 40% of the maximum intensity in the LV wall), border zone (between 40% and 60% of the maximum intensity) and dense scar (above 60% of the maximum intensity).

The LV wall was automatically segmented from CT images using a deep learning approach based on two successive U-net networks [3,6,11]. The high spatial resolution of CT images requires high memory resource while at the same time the ventricles take only a fraction of the entire CT volume. For that reason, we input to the first network a low-resolution version of the CT data and the output (coarse segmentation) was used to locate the ventricles. Only the region around them was kept and a cropped high-resolution version of the image was fed to the second 3D U-net. The output masks were post-processed (including up-sampling to the original CT image resolution) to obtain clean and non-overlapping masks. The model was trained using 450 CT scans with available expert segmentations of the LV endocardium and epicardium, and the right ventricular epicardium. 50 CT scans were used for validation with a loss function defined as the opposite of the Dice score.

To estimate the LV wall thickness a previously described method based on solving a partial differential equation using the endocardium and epicardium masks was used [12]. This method assigns a thickness value to each voxel of the LV wall. Wall thickness measurements were then projected onto a mid-wall 3D surface mesh obtained from the LV wall segmentation where areas of different thickness bins were computed. We considered 4 wall thickness bins representing different degrees of wall thinning, from moderate to severe: thickness between 5 and 4 mm; thickness between 4 and 3 mm; thickness between 3 and 2 mm; and thickness below 2 mm.

Table 1. Population characteristics. Values are mean ± STD [Min, Max]. STD = Standard deviation. LAD = Left Anterior Descending; LCX = Left Circumflex; RCA = Right Coronary Artery.

Parameter	Mean ± STD [Min, Max]
Patient's age (years)	65.53 ± 11.62 [38, 91]
Scar age (years)	7.39 ± 9.06 [0.3, 40]
LV end-diastolic volume (mL)	137.47 ± 38.00 [90.61, 242.14]
LV end-diastolic area (cm^2)	181.10 ± 32.33 [123.40, 250.69]
LVEF (%)	44.04 ± 12.96 [23, 75]
Dense scar volume (mL)	8.24 ± 8.44 [0.72, 50.07]
Border zone volume (mL)	13.95 ± 8.68 [2.45, 34.69]
Fat volume (mL)	0.69 ± 1.66 [0, 9.13]
Area of wall thickness between 5 mm–4 mm (cm^2)	17.81 ± 11.07 [1.62, 45.44]
Area of wall thickness between 4 mm–3 mm (cm^2)	10.69 ± 7.44 [1.37, 27.99]
Area of wall thickness between 3 mm–2 mm (cm^2)	7.51 ± 6.37 [1.56, 25.44]
Area of wall thickness below 2 mm (cm^2)	7.16 ± 4.07 [3.59, 23.43]
Gender	5 (10.20%) female patients
LAD scar	20 (40.81%) patients
LCX scar	17 (34.69%) patients
RCA scar	23 (46.94%) patients

Eight biomarkers were included in the analysis. From LGE-CMR: LVEF; volume of dense scar; and volume of border zone. From CT: fat volume; and areas of wall thinning considering the 4 thickness bins mentioned before. Population characteristics are shown in Table 1 including scar age (i.e. lapse of time between MI and imaging studies).

Disease Progression Modeling

The statistical framework used in this study (detailed in [7]) formulates disease progression modeling based on Bayesian Gaussian Process (GP) regression [10] by modeling individual time transformations encoding the information of the associated pathological stage, and introducing monotonicity constraints to impose a plausible behaviour on the biomarkers' trajectories from normal to pathological stages. Briefly, biomarkers' evolution are modeled as monotonic GPs, while individuals are assigned to their specific time point relative to the regression time axis, according to the severity of the associated biomarker measurements. The time axis corresponds to the patient's scar age, i.e. delay between the infarction event and the imaging study. Monotonic constraints were defined as follows: increasing for fat volume and all degrees of wall thinning [2], and decreasing for LVEF, dense scar volume, and border zone volume [5].

3 Results

The predicted biomarkers temporal trajectories are shown in Fig. 2. It can be observed that the model predicts a plausible progressive thinning of the LV wall and an increase of fat volume in a late stage. On the contrary, the trajectories corresponding to border zone volume, dense scar volume, and LVEF remain almost constant.

All biomarkers trajectories can be seen together in Fig. 3 (left). Figure 4 and Table 2 show the estimated distributions of the maximum change time for all biomarkers suggesting when the biggest change for each biomarker occurs. These parameters allow us to estimate the timeline for remodeling changes by inferring which biomarkers may change before the others. The progressive wall thinning as seen by CT thickness seems clear showing, moreover, an initial fast thinning followed by a more softened, spaced in time, narrowing. According to the model, fat appears only in a late stage. The high variance of dense scar and border zone volumes indicates that no clear evolution pattern is seen for these features.

Importantly, due to the lack of longitudinal data (i.e. only one time point per patient was available), the time axis can only be interpreted as relative time and it does not represent real years. We believe however that this methodology can be useful to describe the dynamics across biomarkers, while distinguishing the most informative features to state the pathological stage of an individual during the course of the scar maturation process. Additionally, our method may enable to position a given patient among other patients at a similar state on the scar maturation process. It may be crucial to determine when the scarred tissue starts to become severely thin (thickness < 2 mm) because the total area of severe thinning has been shown to be a good predictor for ventricular arrhythmia. In this cohort the area of severe thinning was significantly greater in patients with arrhythmia (17 patients), $p = 0.0019$ in a two-sample t-test. Other significant features (at the 5% level) were: the area of thickness between 3 and 2 mm ($p = 0.0028$); area of thickness between 4 and 3 mm ($p = 0.037$); dense scar volume ($p = 0.010$); LVEF ($p = 0.028$); and fat volume ($p = 0.035$).

Table 2. Distributions of the maximum change time in ascending order. STD = Standard deviation.

Biomarker	Time (mean ± STD)
4 mm $<$ thickness $<$ 5 mm	2.1335 ± 4.4195
3 mm $<$ thickness $<$ 4 mm	4.06589 ± 1.7665
2 mm $<$ thickness $<$ 3 mm	16.3780 ± 3.5544
LVEF	18.06976 ± 15.06029
Dense scar volume	23.2351 ± 21.7677
thickness $<$ 2 mm	25.9893 ± 3.6027
Border zone volume	30.9844 ± 23.8233
Fat volume	35.7065 ± 5.5985

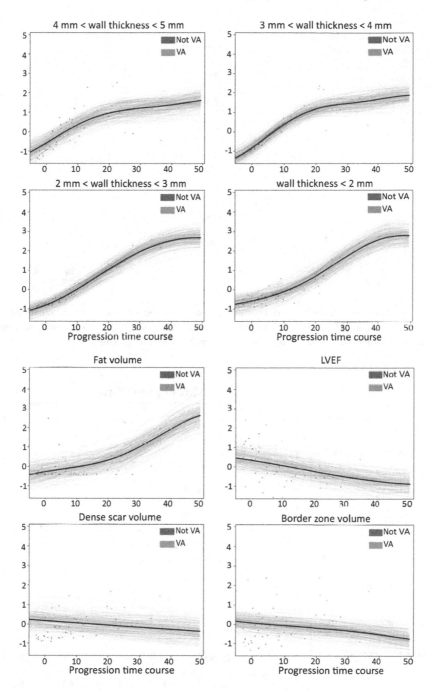

Fig. 2. Modeled biomarker progressions. Each red line displays the estimated trajectory for a given patient and a given biomarker. Dots correspond to individual samples colored according to disease status: not VA (blue) and VA (orange). Note that since only one time point per patient was available the time axis can only be interpreted as relative time and it does not represent real years. LVEF = Left Ventricular Ejection Fraction; VA = Ventricular arrhythmia. (Color figure online)

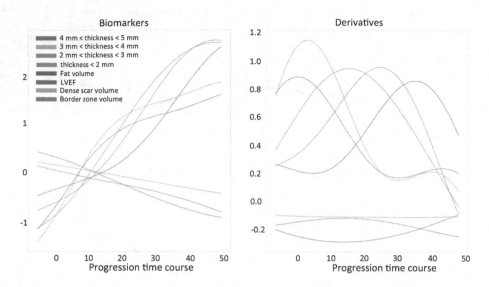

Fig. 3. On the left, all estimated biomarker trajectories; on the right, corresponding derivatives showing the temporal point of maximum change (i.e. point of maximum slope in the trajectories) and the relative magnitude of the change.

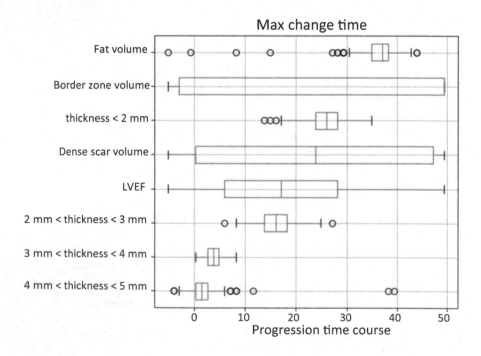

Fig. 4. Distributions of the biomarkers' maximum change time.

4   Conclusions

This work suggests that probabilistic disease progression modeling can be used to estimate the long-term progression of imaging biomarkers related to post-MI scar maturation using purely cross-sectional data. According to the model, it exists a progressive LV wall thinning process in post-MI patients that lasts for years. CT may outperform LGE-CMR imaging in characterizing post-infarct scar remodeling as well as in predicting disease stage. This prediction may be useful for example to assess individual cardiac risk to advise ICD insertion (currently determined by the time passed since the infarction, and suboptimal metrics such as LVEF). It can also be useful to adapt treatments (e.g. ablation therapy) to the patient's stage according to the model.

Acknowledgements. Part of this work was funded by the ERC starting grant EC-STATIC (715093), the IHU LIRYC (ANR-10-IAHU-04), the Equipex MUSIC (ANR-11-EQPX-0030) and the ANR ERACoSysMed SysAFib projects. This work was also supported by the French government, through the 3IA Côte d'Azur Investments in the Future project managed by the National Research Agency (ANR) with the reference number ANR-19-P3IA-0002, and by the ANR JCJC project Fed-BioMed 19-CE45-0006-01. We would like to thank all patients who agreed to make available their clinical data for research.

References

1. Al Khatib, S., et al.. AHA/ACC/HRS guideline for management of patients with ventricular arrhythmias and the prevention of sudden cardiac death: executive summary: a report of the American College of Cardiology/American Heart Association task force on clinical practice guidelines and the heart rhythm society. Heart Rhythm **35**, e-91–e-220 (2017)
2. Anversa, P., Olivetti, G., Capasso, J.M.: Cellular basis of ventricular remodeling after myocardial infarction. Am. J. Cardiol **68**(14), 7–16 (1991)
3. Cedilnik, N., Duchateau, J., Sacher, F., Jaïs, P., Cochet, H., Sermesant, M.: Fully automated electrophysiological model personalisation framework from CT imaging. In: Coudière, Y., Ozenne, V., Vigmond, E., Zemzemi, N. (eds.) FIMH 2019. LNCS, vol. 11504, pp. 325–333. Springer, Cham (2019). https://doi.org/10.1007/978-3-030-21949-9_35
4. Goldberger, J.J., Cain, M.E., Hohnloser, S.H., et al.: American Heart Association/American College of Cardiology Foundation/Heart Rhythm Society scientific statement on noninvasive risk stratification techniques for identifying patients at risk for sudden cardiac death: a scientific statement from the American Heart Association Council on clinical cardiology committee on electrocardiography and arrhythmias and council on epidemiology and prevention. J. Am. Coll. Cardiol. **52**(14), 1179–1199 (2008)
5. Jáuregui, B., Soto-Iglesias, D., Penela, D., Acosta, J., et al.: Follow-up after myocardial infarction to explore the stability of arrhythmogenic substrate: the FOOTPRINT study. JACC Clin. Electrophysiol. **6**(2), 207–218 (2020)

6. Jia, S., et al.: Automatically segmenting the left atrium from cardiac images using successive 3D U-nets and a contour loss. In: International Workshop on Statistical Atlases and Computational Models of the Heart, pp. 221–229 (2018)
7. Lorenzi, M., Filippone, M., Frisoni, G.B., Alexander, D.C., et al.: Probabilistic disease progression modeling to characterize diagnostic uncertainty: application to staging and prediction in Alzheimer's disease. NeuroImage **190**, 56–68 (2019)
8. Martin, C.A., Gajendragadkar, P.R.: scar tissue: never too old to remodel. JACC Clini. Electrophysiol. **6**(2), 219–220 (2020)
9. Martin, R., et al.: Characteristics of scar-related ventricular tachycardia circuits using ultra-high-density mapping: a multi-center study. Circ. Arrhythm. Electrophysiol **11**(10), e006569 (2018)
10. Rasmussen, C.E.: Gaussian processes in machine learning. In: Bousquet, O., von Luxburg, U., Rätsch, G. (eds.) ML -2003. LNCS (LNAI), vol. 3176, pp. 63–71. Springer, Heidelberg (2004). https://doi.org/10.1007/978-3-540-28650-9_4
11. Ronneberger, O., Fischer, P., Brox, T.: U-Net: convolutional networks for biomedical image segmentation. In: Navab, N., Hornegger, J., Wells, W.M., Frangi, A.F. (eds.) MICCAI 2015, Part III. LNCS, vol. 9351, pp. 234–241. Springer, Cham (2015). https://doi.org/10.1007/978-3-319-24574-4_28
12. Yezzi, A.J., Prince, J.L.: An Eulerian PDE approach for computing tissue thickness. IEEE Trans. Med. Imaging **22**(10), 1332–1339 (2003)

PC-U Net: Learning to Jointly Reconstruct and Segment the Cardiac Walls in 3D from CT Data

Meng Ye[1]([✉]), Qiaoying Huang[1], Dong Yang[2], Pengxiang Wu[1], Jingru Yi[1], Leon Axel[3], and Dimitris Metaxas[1]

[1] Department of Computer Science, Rutgers University, Piscataway, NJ 08854, USA
my389@cs.rutgers.edu
[2] Nvidia Corporation, Bethesda, MD 20814, USA
[3] Department of Radiology, New York University, New York, NY 10016, USA

Abstract. The 3D volumetric shape of the heart's left ventricle (LV) myocardium (MYO) wall provides important information for diagnosis of cardiac disease and invasive procedure navigation. Many cardiac image segmentation methods have relied on detection of region-of-interest as a pre-requisite for shape segmentation and modeling. With segmentation results, a 3D surface mesh and a corresponding point cloud of the segmented cardiac volume can be reconstructed for further analyses. Although state-of-the-art methods (e.g., U-Net) have achieved decent performance on cardiac image segmentation in terms of accuracy, these segmentation results can still suffer from imaging artifacts and noise, which will lead to inaccurate shape modeling results. In this paper, we propose a PC-U net that jointly reconstructs the point cloud of the LV MYO wall directly from volumes of 2D CT slices and generates its segmentation masks from the predicted 3D point cloud. Extensive experimental results show that by incorporating a shape prior from the point cloud, the segmentation masks are more accurate than the state-of-the-art U-Net results in terms of Dice's coefficient and Hausdorff distance. The proposed joint learning framework of our PC-U net is beneficial for automatic cardiac image analysis tasks because it can obtain simultaneously the 3D shape and segmentation of the LV MYO walls.

Keywords: Point cloud · Segmentation · LV shape modeling

1 Introduction

Cardiovascular disease is one of the major causes of human death worldwide. The shape of the left ventricle (LV) myocardium (MYO) wall plays an important role in diagnosis of cardiac disease, and in surgical and other invasive cardiac procedures [6]. It allows the subsequent cardiac LV motion function analysis to determine the presence and location of possible cardiac disease [4]. In order to automatically evaluate cardiac function measures such as the ejection fraction

© Springer Nature Switzerland AG 2021
E. Puyol Anton et al. (Eds.): STACOM 2020, LNCS 12592, pp. 117–126, 2021.
https://doi.org/10.1007/978-3-030-68107-4_12

(EF), most previous methods first segment the LV blood pool or LV MYO from images [5] then generate a shape model from the segmentation results. However, this sequential pipeline is not ideal since the shape reconstruction strongly depends on the precision of the segmentation algorithms. As the example shown in Fig. 1, due to the imperfect segmentation result, the generated 3D shape is erroneous and not suitable in a practical clinical scenario. It is therefore imperative to develop a more robust approach for LV MYO segmentation jointly with accurate and clinically useful myocardium 3D surface reconstruction.

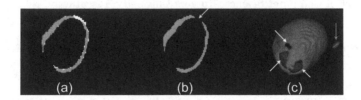

Fig. 1. The shape generated from U-Net based segmentation results can be erroneous (holes indicated by yellow arrows) due to the thin wall structure of the myocardium apex (green arrow). (a) is the ground truth (in white color) of segmentation overlaid by the U-Net segmentation result shown in (b). (c) is the LV MYO wall shape reconstructed from (b). (Color figure online)

The 3D shape of an object can be described as a 3D mesh or point cloud (PC). A 3D mesh contains a set of vertices, edges and faces, which can be represented as a graph $M = (V, E)$, where M is the mesh, $V = \{v_i\}_{i=1}^{N}$ is the set of N vertices and $E = \{e_j\}_{j=1}^{K}$ is a set of K edges with each connecting two vertices. Each vertex v has a 3D coordinate (x, y, z). The set of the vertices V forms a point cloud. Point cloud-based shape representations have received increasing attention in medical image analysis. In [1], a convolutional neural network (CNN)-based classifier was first trained to estimate a point cloud from 3D volumetric images. In particular, a point cloud network was trained to classify each point into foreground or background in order to refine the segmentation mask of the peripheral nerve. In [2], an organ point-network was introduced to learn the shape of abdomen organs which was described as points located on the organ surface. This point-network was trained together with a CNN-based segmentation model to help improving organ segmentation results in a multi-task learning manner. Recently, another method, named PointOutNet was proposed to predict a 3D point cloud from a single RGB image [3]. In [14], the same model was applied to conduct one-stage shape instantiation for the right ventricle (RV) from a single 2D long-axis cardiac MR image. However, their method required an optimum imaging plane, which was decided by a predefined 3D imaging process to acquire a synchronized 2D image as the input to the PointOutNet.

In this paper, we propose a PC-U net to reconstruct a 3D point cloud of the LV MYO shape from a volume of CT images. In order to segment the

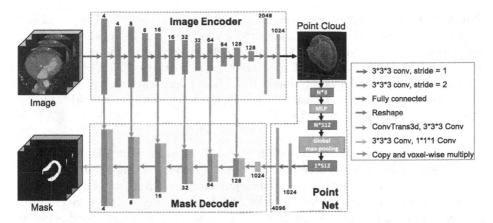

Fig. 2. An overview of the proposed PC-U net for 3D point cloud reconstruction and segmentation of the LV MYO wall simultaneously (best view in color). The numbers on the top or bottom of a bar depict the number of output channels for convolutional layers or fully connected layers. N is the point number of a point cloud. (Color figure online)

myocardium wall for other applications such as global heart function evaluation, we design this PC-U net to reconstruct the slice segmentation masks from the point cloud directly. The benefits of our PC-U net are twofold. First, the reconstructed LV MYO wall 3D shape is no longer dependent on the precision of the segmentation results. Second, the LV MYO wall segmentation results are reconstructed from the point cloud, therefore it combines the reconstructed shape prior into the segmentation branch of the network, which, in turn, makes the segmentation results more accurate [8,13]. To the best of our knowledge, this is the first work that explicitly utilizes features from an estimated shape prior of an organ in the form of a point cloud to reconstruct segmentation masks.

2   Proposed Method

2.1   PC-U Net

Figure 2 shows the architecture of our proposed PC-U net for 3D point cloud reconstruction and segmentation of the LV MYO wall. The PC-U net consists of three parts: image encoder, point net, and mask decoder.

Image Encoder. As shown in Fig. 2, the image encoder takes 3D volumes of CT images as input and extracts image features $IF(x', y', z')$ at each voxel (x', y', z'). This branch is used for 3D point cloud reconstruction, which is similar to the PointOutNet [14], except that the inputs are 3D volumetric images and convolution kernels are 3D.

Point Net. The point net consists of a multi-layer perceptron (MLP), a global max-pooling layer and two fully connected layers. The MLP is consisted of three convolutional layers (with 32, 128, 512 channels, respectively), followed by ReLU activation. This branch is used to extract point features $PF(x, y, z)$ at each 3D point (x, y, z) of the point cloud [10]. These point features will be used to reconstruct segmentation masks. In this way, the point cloud can provide a shape prior for the segmentation task explicitly.

Mask Decoder. The lower branch is the mask decoder, which is used for LV MYO wall segmentation. We refer the process of reconstructing dense segmentation masks from a sparse point cloud as segmentation. It consists of up-sampling layers and convolutional layers. In order to take advantage of the more detailed contextual image features in the shallow layers, we use skip connections between the upper branch and the lower branch. This fashion makes the upper and lower branches similar to the encoder and decoder of the U-Net [11]. In particular, the features from the encoder are multiplied by the features in the decoder voxel-wisely. We empirically found that voxel-wise multiplication operation could make the output masks have less outliers. The significant difference between the proposed PC-U net and a U-Net is that PC-U net al.lows the intermediate output of the image encoder be the point cloud, which serves as an effective regularizer and provides a shape prior for segmentation. Given the point features $PF(x, y, z)$ and image features $IF(x', y', z')$, the mask decoder branch will learn an implicit function f represented by the up-sampling layers and convolutional layers:

$$f(PF(x, y, z), IF(x', y', z')) = m : m \in \{0, 1\}, \tag{1}$$

where m is the binary segmentation mask.

2.2   Loss Function

We use the Chamfer loss [12] to measure the distance between the predicted point cloud and the ground-truth point cloud. It can be formulated as:

$$l_p = \frac{1}{|Y_P|} \sum_{y_p \in Y_P} min_{y_g \in Y_G} \|y_p - y_g\|_2 + \frac{1}{|Y_G|} \sum_{y_g \in Y_G} min_{y_p \in Y_P} \|y_g - y_p\|_2, \tag{2}$$

where the first term $min_{y_g \in Y_G} \|y_p - y_g\|_2$ enforces that any 3D point y_p in the predicted point cloud Y_P has a matching 3D point y_g in the ground-truth point cloud Y_G, and vice versa for the second term $min_{y_p \in Y_P} \|y_g - y_p\|_2$.

For segmentation, we adopt the soft Dice loss [7] to measure the error between the predicted mask and the ground truth:

$$l_s = 1 - \frac{2 \sum_i^N p_i g_i}{\sum_i^N (p_i^2 + g_i^2)}, \tag{3}$$

where N is the total number of voxels, p_i is the ith voxel of the predicted mask and g_i is the ith voxel of the ground-truth mask.

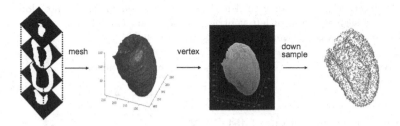

Fig. 3. The process of how to get the ground-truth point cloud: (1) A mesh is first built from the 3D volumetric masks; (2) The vertices of the mesh form a dense point cloud; (3) We down-sample a sparse point cloud from the dense point cloud.

The PC-U net is trained end-to-end and optimized by the following weighted loss of l_p and l_s.

$$Loss = l_p + \lambda \cdot l_s, \tag{4}$$

where λ is a balancing weight. We empirically set $\lambda = 0.001$ via the grid search.

3 Experiments

We validated the proposed PC-U net on a publicly available 3D cardiac CT dataset [15], which contains 20 subjects. In order to deal with the rather small number of available training samples, we used elastic deformation to augment the training set by 200 times. After augmentation, there are 4,000 samples in total. We first cropped out the region-of-interest of LV and sampled all volumes and their labels to 1.0 mm isotropically. Then the image volume was resized to a fixed size as $(128, 128, 64)$. The intensity values of the CT volumes were first divided by 2048 and then clamped between $[-1, 1]$. The ground-truth point cloud generation process is shown in Fig. 3. All point clouds were centered at $(0, 0, 0)$ by subtracting their own center coordinates. In our experiments, the total number of points was set to be $N = 4096$. To show the robustness and effectiveness of our PC-U net, we performed 4-fold cross validation experiments.

Since there is no other work that jointly reconstructs a point cloud and segments the LV MYO wall, we will compare the two tasks separately. For point cloud reconstruction, PointOutNet [3] is the baseline model, which takes as input a single image slice [14] and outputs a 3D point cloud, as shown in the upper encoder branch of Fig. 2. Depending on the input data (a single slice of long axis image or the whole 3D volumetric images) and the convolutional kernel (2D or 3D), we create three variants: *PointOutNet-single-slice*, *PointOutNet-volume-2DConv* and *PointOutNet-volume-3DConv*. For segmentation, the state-of-the-art model U-Net [9] is compared to our method. We consider 3D volumetric images as input, giving two models: *UNet-volume-2DConv* and *UNet-volume-3DConv*. Our *PC-Unet-2DConv* and *PC-Unet-3DConv* also take as input 3D volumetric images, and output the 3D point clouds and segmentation masks simultaneously. The goal of this experiment is to validate the significance of

Table 1. Average performance of 4-fold cross validation on the point cloud reconstruction and segmentation of left ventricle myocardium (MYO) wall. The mean and standard deviation values over all the folds are reported.

Method	PC reconstruction	Segmentation	
	CD (mm)	Dice	HD (mm)
PointOutNet-single-slice	1.489 ± 0.547	–	–
PointOutNet-volume-2DConv	1.454 ± 0.422	–	–
PointOutNet-volume-3DConv	1.330 ± 0.330	–	–
UNet-volume-2DConv	–	0.843 ± 0.024	16.477 ± 8.311
UNet-volume-3DConv	–	0.877 ± 0.012	10.446 ± 4.489
PC-Unet-2DConv	1.278 ± 0.249	0.838 ± 0.026	10.894 ± 2.176
PC-Unet-3DConv (Ours)	**1.276 ± 0.168**	**0.885 ± 0.011**	**7.050 ± 1.103**

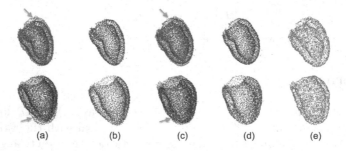

Fig. 4. LV MYO wall point cloud reconstruction results. (a) PointOutNet (blue) overlaid with ground-truth point cloud (red). (b) PointOutNet. (c) PC-U net (blue) overlaid with ground-truth point cloud (red). (d) PC-U net (Ours). (e) ground-truth point cloud. (Color figure online)

using 3D input and solving two tasks together as well as show how they benefit each other.

All models were trained on a single NVIDIA Tesla K40 GPU with Adam optimizer and the learning rate was set at $1e^{-4}$. The Chamfer distance (CD) error between prediction and ground truth was used to measure the point cloud reconstruction performance. Dice's coefficient and Hausdorff distance (HD) error were used to evaluate the segmentation performance.

Experimental Results. Table 1 shows the average performance on LV MYO wall point cloud reconstruction and segmentation of the 4-fold cross validation. As we can see from the table, with more information contained in a volume input (PointOutNet-single-slice *v.s.* PointOutNet-volume-2DConv), the reconstructed point cloud gets better in terms of CD error. In addition, 3D convolution outperforms 2D convolution (PointOutNet-volume-2DConv *v.s.* PointOutNet-volume-3DConv) on the point cloud reconstruction task. PC-Unet-3DConv achieves the

Fig. 5. LV MYO wall shape modeling from U-Net segmentation (first row) and point cloud predicted by the proposed PC-U net (Ours, second row). Case one: first to third column. Case two: fourth to sixth column. Each column for each case represents a specific view.

Table 2. Average performance of the ablation study.

Method	PC reconstruction	Segmentation	
	CD (mm)	Dice	HD (mm)
PC-Mask-Decoder	-	0.725	8.682
PC-Unet-no-skip	1.182	0.699	7.896
PC-Unet (Ours)	**1.149**	**0.884**	**5.862**

best point cloud reconstruction result among all the models, which could be attributed to the joint learning of point cloud reconstruction and segmentation, apart from using a volume input and 3D convolution.

Figure 4 shows two cases (the first and the second rows) of visual comparisons between the baseline (PointOutNet-single-slice) and the proposed model (PC-Unet-3DConv). It is clear that baseline model fails to reconstruct the points on the thin areas of myocardium including the atrioventricular ring and the apex (indicated by arrows). This limitation has been pointed out by the authors in [14]. When combining with more 3D contextual information in contrast to a single input 2D image slice, the proposed PC-U net successfully reconstructs the 3D points on these thin regions of myocardium.

For segmentation of the LV MYO, PC-Unet-3DConv achieves a slightly better result in terms of Dice's coefficient compared to UNet-volume-3DConv. Note that it achieves a significant smaller Hausdorff distance error. Again, 3D convolution outperforms 2D convolution (UNet-volume-2DConv $v.s.$ UNet-volume-3DConv, PC-Unet-2DConv $v.s.$ PC-Unet-3DConv) on the segmentation task. Although PC-Unet-2DConv performs slightly worse than UNet-volume-2DConv in terms of Dice's coefficient, its Hausdorff distance error is much smaller than UNet-volume-2DConv, suggesting the significant benefit of using the shape prior provided by the point cloud on the segmentation task. We qualitatively show two cases of the shape reconstructed from the segmentation results of UNet-volume-3DConv and the predicted point cloud of the proposed PC-Unet-3DConv in Fig. 5, respectively. It is clear that due to error segmentation, there can be holes

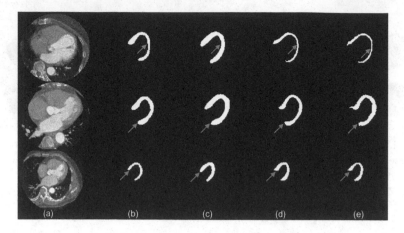

Fig. 6. LV MYO segmentation results. (a) the CT image slice. (b) PC-Mask-Decoder. (c) PC-Unet-no-skip. (d) PC-U net (Ours). (e) ground-truth mask.

on the surface of the LV MYO wall shape (indicated by blue arrows). When our PC-U net takes as input the 3D volumetric images it successfully predicts a complete shape of the LV MYO wall without any holes. The large Hausdorff distance error of UNet-volume-3DConv indicates that there are some isolated outliers in the segmentation results, shown as green arrows in Fig. 5.

Ablation Study. In this experiment, we demonstrate how the PC-U net can recover segmentation details using skip connections between encoder and decoder. We test 3 models here with a 1-fold experiment. The first model is composed of the Point Net and Mask Decoder with the ground-truth point cloud as input (PC-Mask-Decoder). With this model, we plan to demonstrate that CNNs can reconstruct dense masks from a sparse point cloud. The second model is the PC-U net without skip connections (PC-Unet-no-skip). The third model is the proposed PC-U net model. All the models use 3D convolution. The point cloud reconstruction and segmentation results are shown in Table 2.

From Table 2, although the point cloud is a sparse representation of the 3D shape of LV MYO wall, we can use the Point Net to extract point features efficiently with which the Mask Decoder can reconstruct a dense segmentation mask. To our knowledge, this is the first work that uses CNNs to reversely reconstruct a dense mask from a sparse point cloud. However, as shown in Fig. 6, the PC-Mask-Decoder alone can only reconstruct a coarse mask. As indicated by the arrows, some details of the mask cannot be recovered very well. The PC-Unet-no-skip model takes input as the volumetric images but it does not consider the detailed contextual image features extracted by the Image Encoder, again resulting in coarse segmentation. Another reason for its lower Dice's coefficient compared to PC-Mask-Decoder is that the point cloud reconstructed by the PC-Unet-no-skip model still has some errors, which will introduce bias in point features. The proposed PC-U net achieves the best point cloud reconstruction and recovers most details of the segmentation by using skip connections and joint learning.

4 Conclusion

In this work, we proposed a PC-U net to reconstruct an LV MYO wall point cloud from 3D volumetric CT images. By combining with the 3D context information in the 3D volumetric images, the PC-U net could successfully reconstruct the 3D point cloud of the LV MYO wall. Our PC-U net models the shape of the LV MYO wall directly from the 3D volumetric images, which is accurate and avoids errors induced by potentially inaccurate segmentation results. We demonstrated that the PC-U net could not only find the shape of LV MYO wall, but also obtain its segmentation results simultaneously. With a shape prior provided by the point cloud, the segmentation results are more accurate than state-of-the-art U-Net results in terms of Dice's coefficient and Hausdorff distance.

References

1. Balsiger, F., Soom, Y., Scheidegger, O., Reyes, M.: Learning shape representation on sparse point clouds for volumetric image segmentation. In: Shen, D., et al. (eds.) MICCAI 2019, Part II. LNCS, vol. 11765, pp. 273–281. Springer, Cham (2019). https://doi.org/10.1007/978-3-030-32245-8_31
2. Cai, J., Xia, Y., Yang, D., Xu, D., Yang, L., Roth, H.: End-to-end adversarial shape learning for abdomen organ deep segmentation. In: Suk, H.-I., Liu, M., Yan, P., Lian, C. (eds.) MLMI 2019. LNCS, vol. 11861, pp. 124–132. Springer, Cham (2019). https://doi.org/10.1007/978-3-030-32692-0_15
3. Fan, H., Su, H., Guibas, L.J.: A point set generation network for 3d object reconstruction from a single image. In: Proceedings of the IEEE Conference on Computer Vision and Pattern Recognition, pp. 605–613 (2017)
4. Juergens, K.U., et al.: Multi-detector row CT of left ventricular function with dedicated analysis software versus MR imaging: initial experience. Radiology 230(2), 403–410 (2004)
5. Lopez-Perez, A., Sebastian, R., Ferrero, J.M.: Three-dimensional cardiac computational modelling: methods, features and applications. Biomed. Eng. Online 14(1), 35 (2015)
6. Medrano-Gracia, P., Zhang, X., Suinesiaputra, A., Cowan, B., Young, A.: Statistical shape modeling of the left ventricle: myocardial infarct classification challenge. In: Proceedings of MICCAI-STACOM (2015)
7. Milletari, F., Navab, N., Ahmadi, S.A.: V-net: Fully convolutional neural networks for volumetric medical image segmentation. In: 2016 Fourth International Conference on 3D Vision (3DV), pp. 565–571. IEEE (2016)
8. Painchaud, N., Skandarani, Y., Judge, T., Bernard, O., Lalande, A., Jodoin, P.-M.: Cardiac MRI segmentation with strong anatomical guarantees. In: Shen, D., Liu, T., Peters, T.M., Staib, L.H., Essert, C., Zhou, S., Yap, P.-T., Khan, A. (eds.) MICCAI 2019, Part II. LNCS, vol. 11765, pp. 632–640. Springer, Cham (2019). https://doi.org/10.1007/978-3-030-32245-8_70

9. Payer, C., Štern, D., Bischof, H., Urschler, M.: Multi-label whole heart segmentation using CNNs and anatomical label configurations. In: Pop, M., Sermesant, M., Jodoin, P.-M., Lalande, A., Zhuang, X., Yang, G., Young, A., Bernard, O. (eds.) STACOM 2017. LNCS, vol. 10663, pp. 190–198. Springer, Cham (2018). https://doi.org/10.1007/978-3-319-75541-0_20

10. Qi, C.R., Su, H., Mo, K., Guibas, L.J.: Pointnet: deep learning on point sets for 3D classification and segmentation. In: Proceedings of the IEEE Conference on Computer Vision and Pattern Recognition, pp. 652–660 (2017)

11. Ronneberger, O., Fischer, P., Brox, T.: U-Net: convolutional networks for biomedical image segmentation. In: Navab, N., Hornegger, J., Wells, W.M., Frangi, A.F. (eds.) MICCAI 2015, Part III. LNCS, vol. 9351, pp. 234–241. Springer, Cham (2015). https://doi.org/10.1007/978-3-319-24574-4_28

12. Yang, Y., Feng, C., Shen, Y., Tian, D.: Foldingnet: oint cloud auto-encoder via deep grid deformation. In: Proceedings of the IEEE Conference on Computer Vision and Pattern Recognition, pp. 206–215 (2018)

13. Yue, Q., Luo, X., Ye, Q., Xu, L., Zhuang, X.: Cardiac segmentation from LGE MRI using deep neural network incorporating shape and spatial priors. In: Shen, D., et al. (eds.) MICCAI 2019, Part II. LNCS, vol. 11765, pp. 559–567. Springer, Cham (2019). https://doi.org/10.1007/978-3-030-32245-8_62

14. Zhou, X.-Y., Wang, Z.-Y., Li, P., Zheng, J.-Q., Yang, G.-Z.: One-stage shape instantiation from a single 2D image to 3D point cloud. In: Shen, D., et al. (eds.) MICCAI 2019. LNCS, vol. 11767, pp. 30–38. Springer, Cham (2019). https://doi.org/10.1007/978-3-030-32251-9_4

15. Zhuang, X., Shen, J.: Multi-scale patch and multi-modality atlases for whole heart segmentation of mri. Medical image analysis **31**, 77–87 (2016)

Shape Constrained CNN for Cardiac MR Segmentation with Simultaneous Prediction of Shape and Pose Parameters

Sofie Tilborghs[1,4]([✉]), Tom Dresselaers[2,4], Piet Claus[3,4], Jan Bogaert[2,4], and Frederik Maes[1,4]

[1] Department of Electrical Engineering, ESAT/PSI, KU Leuven, Leuven, Belgium
sofie.tilborghs@kuleuven.be
[2] Department of Imaging and Pathology, Radiology, KU Leuven, Leuven, Belgium
[3] Department of Cardiovascular Sciences, KU Leuven, Leuven, Belgium
[4] Medical Imaging Research Center, UZ Leuven, Leuven, Belgium

Abstract. Semantic segmentation using convolutional neural networks (CNNs) is the state-of-the-art for many medical segmentation tasks including left ventricle (LV) segmentation in cardiac MR images. However, a drawback is that these CNNs lack explicit shape constraints, occasionally resulting in unrealistic segmentations. In this paper, we perform LV and myocardial segmentation by regression of pose and shape parameters derived from a statistical shape model. The integrated shape model regularizes predicted segmentations and guarantees realistic shapes. Furthermore, in contrast to semantic segmentation, it allows direct calculation of regional measures such as myocardial thickness. We enforce robustness of shape and pose prediction by simultaneously constructing a segmentation distance map during training. We evaluated the proposed method in a fivefold cross validation on a in-house clinical dataset with 75 subjects containing a total of 1539 delineated short-axis slices covering LV from apex to base, and achieved a correlation of 99% for LV area, 94% for myocardial area, 98% for LV dimensions and 88% for regional wall thicknesses. The method was additionally validated on the LVQuan18 and LVQuan19 public datasets and achieved state-of-the-art results.

Keywords: Cardiac MRI segmentation · Convolutional neural network · Statistical shape model

1 Introduction

Cardiac magnetic resonance (CMR) imaging provides high quality images of the heart and is therefore frequently used to assess cardiac condition. Clinical measures of interest include left ventricular (LV) volume and myocardial thickness, which can be calculated from a prior segmentation of LV and myocardium. In the last years, convolutional neural networks (CNNs) have shown to outperform traditional model-based segmentation techniques and quickly became the

© Springer Nature Switzerland AG 2021
E. Puyol Anton et al. (Eds.): STACOM 2020, LNCS 12592, pp. 127–136, 2021.
https://doi.org/10.1007/978-3-030-68107-4_13

method of choice for this task [1]. However, since CNNs are trained to predict a class probability (i.e. LV or background) for each voxel, they are missing explicit shape constraints, occasionally resulting in unrealistic segmentations with missing or disconnected regions and hence requiring postprocessing. In this respect, several authors have proposed to integrate a shape prior in their CNN. Examples are atlases [2,3] or hidden representations of anatomy [4–6]. In contrast to CNNs, Active Shape Models (ASM) [7] construct a landmark-based statistical shape model from a training dataset and fit this model to a new image using learned local intensity models for each landmark, yielding patient-specific global shape coefficients. In this paper, we combine the advantages of both methods: (1) a CNN is used to extract complex appearance features from the images and (2) shape constraints are imposed by regressing the shape coefficients of the statistical model. Compared to Attar et al. [8], who used both CMR images and patient metadata to directly predict the coefficients of a 3D cardiac shape, we enforce robustness of coefficient prediction by simultaneously performing semantic segmentation. A similar approach combining segmentation with regression was used by Vigneault et al. [9] to perform pose estimation of LV, by Gessert and Schlaefer [10] and by Tilborghs and Maes [11] to perform direct quantification of LV parameters and by Cao et al. [12] for simultaneous hippocampus segmentation and clinical score regeression from brain MR images. In our approach, the semantic segmentation is performed by regression of signed distance maps, trained using a loss function incorporating both distance and overlap measures. Previous methods to incorporate distance losses include the boundary loss of Kervadec et al. [13], the Hausdorff distance loss of Karimi and Salcudean [14] and the method of Dangi et al. [15] who used separate decoders for the prediction of distance maps and segmentation maps. Different to Dangi et al. our CNN only generates a distance map, while the segmentation map is directly calculated from this distance map, guaranteeing full correspondence between the two representations.

2   Methods

2.1   Shape Model

The myocardium in a short-axis (SA) cross-section is approximated by a set of N endo- and epicardial landmarks radially sampled over uniform angular offsets of $2\pi/N$ rad, relative to an anatomical reference orientation θ. From a training set of images, a statistical shape model representing the mean shape and the modes of variation is calculated using principal component analysis. For each image i, the myocardial shape $\mathbf{p}_i$ is first normalized by subtracting the LV center position $\mathbf{c}_i$ and by rotating around θ_i, resulting in the pose-normalized shape $\mathbf{s}_i$:

$$\begin{bmatrix} \mathbf{s}_{i,x} \\ \mathbf{s}_{i,y} \end{bmatrix} = \begin{bmatrix} \cos(\theta_i) & \sin(\theta_i) \\ -\sin(\theta_i) & \cos(\theta_i) \end{bmatrix} \begin{bmatrix} \mathbf{p}_{i,x} - \mathbf{c}_{i,x} \\ \mathbf{p}_{i,y} - \mathbf{c}_{i,y} \end{bmatrix} \tag{1}$$

Representing the shapes as vectors $\mathbf{s}_i = (x_1, ..., x_{2N}, y_1, ..., y_{2N})$, the mean shape $\bar{\mathbf{s}}$ is calculated as $\bar{\mathbf{s}} = \frac{1}{I} \sum_{i=1}^{I} \mathbf{s}_i$ with I the number of training images.

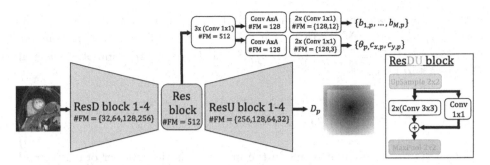

Fig. 1. Proposed CNN architecture with three outputs: shape $\{b_{1,p}, ..., b_{M,p}\}$, pose $\{\theta_p, c_{x,p}, c_{y,p}\}$ and distance maps (D_p). The details of residual (Res), downsampling Res (ResD) and upsampling Res (ResU) blocks are given on the right. Every convolutional (Conv) layer is followed by batch normalization and a parameterized rectified linear unit, except for the final layer in every output. The number of feature maps ($\#FM$) is the same for all Conv layers in one Res block. The filter size A in a Conv layer is equal to the dimensions of that layer's input. Same padding is used.

The normalized eigenvectors $\mathbf{V} = \{\mathbf{v}_1, ..., \mathbf{v}_m, ..., \mathbf{v}_{4N}\}$ and corresponding eigenvalues λ_m are obtained from the singular value decomposition of the centered shapes $\mathbf{s}_i - \bar{\mathbf{s}}$. The shape of the myocardium is approximated by the M first eigenmodes:

$$\mathbf{s}_i \approx \bar{\mathbf{s}} + \sum_{m-1}^{M} b_{i,m} \cdot \sqrt{\lambda_m} \cdot \mathbf{v}_m \tag{2}$$

Using this definition, the variance of the distribution of shape coefficients b_m is the same for every mode m.

2.2 CNN

A schematic representation of the CNN architecture is shown in Fig. 1. It has three outputs: (1) M predicted shape coefficients $\{b_{1,p}, ..., b_{M,p}\}$, (2) pose parameters $\{\theta_p, c_{x,p}, c_{y,p}\}$ and (3) segmentation map D_p. Semantic segmentation is performed by the regression of distance maps D. D is an image representing the Euclidean distance d between pixel position and contour. The sign is negative for pixels inside structure S:

$$D(x) = \begin{cases} -d(x), & \text{if } x \in S \\ d(x), & \text{if } x \notin S \end{cases} \tag{3}$$

For both endo- and epicardium, separate distance maps D_{endo} and D_{epi} are created. The loss function is a weighted sum of the shape loss L_1, pose loss L_2 and segmentation loss L_3:

$$L = \gamma_1 L_1 + \gamma_2 L_2 + \gamma_3 L_3 \tag{4}$$

with L_1 the mean squared error (MSE) between true and predicted coefficients b_m: $L_1 = \frac{1}{M}\sum_{m=1}^{M}(b_{m,t} - b_{m,p})^2$, L_2 the MSE for pose parameters $O = \{\theta, c_x, c_y\}$: $L_2 = \frac{1}{3}\sum_{j=1}^{3}(o_{j,t} - o_{j,p})^2$, and L_3 a weighted sum of categorical Dice loss and MSE:

$$L_3 = \left(1 - \frac{1}{K}\sum_k \frac{2\cdot\sum_x S_{k,t}(x)\cdot S_{k,p}(x)}{\sum_x S_{k,t}(x) + \sum_x S_{k,p}(x)}\right) + \mu\frac{1}{K\cdot X}\sum_{k,x}(D_{k,t}(x) - D_{k,p}(x))^2 \tag{5}$$

where X is the number of pixels in the image, K is the number of classes and S_k is the binarized distance map using a sigmoid as conversion function:

$$S_k = \frac{e^{-\alpha\cdot D_k}}{1 + e^{-\alpha\cdot D_k}} \tag{6}$$

where α affects the steepness of the sigmoid function.

2.3   Implementation Details

Endo- and epicardium are both represented by $N = 18$ landmarks and θ is defined as the orientation of the line connecting the center of LV with the middle of the septum. The network predicts the first $M = 12$ shape coefficients, representing over 99% of shape variation. Pose parameters θ, c_x and c_y are normalized to the range $[-1,1]$. Given the notable difference in magnitude of the different losses, they are weighted with $\gamma_1 = 1$, $\gamma_2 = 10$, $\gamma_3 = 100$ and $\mu = 0.1$. These weights were heuristically defined and assure significant contribution of each of the losses. Parameter α in Eq. 6 is set to 5 to approximate a binary map with an error of only $6.7e^{-3}$ for a distance of one pixel from the contour. The network is trained end-to-end over 5000 epochs with Adam optimizer, learning rate 2e-3 and batch size 32.

Online data augmentation is applied by adapting pose and shape parameters. Position and orientation offsets are sampled from uniform distributions between $[-40,40]$ mm and $[-\pi, \pi]$ rad, respectively. Additionally, shape coefficients were adapted as $b_{m,aug} = b_m + a$, where a is sampled from a uniform distribution between -1 and 1. The input images and distance maps are modified accordingly. For the input image, a thin-plate-spline point-to-point registration is performed using the $2N$ original and augmented landmarks while the distance maps are recreated from the augmented landmarks, connected using cubic spline interpolation, according to Eq. 3. Furthermore, Gaussian noise with standard deviation between 0 and 0.1 is online added to the MR images during training.

3   Experiments

The models were constructed and validated in a fivefold cross validation on a clinical dataset ('D1') containing images of 75 subjects (M = 51, age = 48.2 ± 15.6 years) suffering from a wide range of pathologies including hypertrophic cardiomyopathy, dilated cardiomyopathy, myocardial infarction, myocarditis, pericarditis, LV aneurysm... The subjects were scanned on a 1.5T MR scanner (Ingenia, Philips Healthcare, Best, The Netherlands), with a 32-channel phased array receiver coil setup. The endo- and epicardium in end-diastole and end-systole in the SA cine images were manually delineated by a clinical expert. To allow calculation of θ, the RV attachment points were additionally indicated. This resulted in a total of 1539 delineated SA images, covering LV from apex to base. All images of a patient were assigned to the same fold. For each fold, a separate shape model was constructed using the four remaining folds. The images were resampled to a pixel size of 2 mm $\times$ 2 mm and image size of 128 $\times$ 128, which results in a value of 8 for parameter A in Fig. 1.

We validated the performance of our method and the added value of each choice with five different setups: (1) semantic segmentation using categorical Dice loss ('$S_{\mu=0}$'), (2) semantic segmentation using combined loss ('S'): $L = \gamma_3 L_3$, (3) regression of shape and pose parameters ('R'): $L = \gamma_1 L_1 + \gamma_2 L_2$, (4) regression and segmentation losses ('RS') as in Eq. 4, (5) loss as in Eq. 4 and with pose and shape data augmentation ('RS-A$_{ps}$'). For setups 1–4, data augmentation only consisted of the addition of Gaussian noise. Due to faster convergence of training without pose and shape data augmentation, setups 1–4 were only trained for 1000 epochs. For each setup, Dice similarity coefficient (DSC), mean boundary error (MBE) and Hausdorff distance (HD) were calculated from the binarized distance maps ('Map'), as well as from the predicted shape and pose parameters by converting the parameters to landmarks using Eq. 1 and 2 and interpolating with cubic splines ('Contour'). The position and orientation errors were respectively defined as $\Delta d = \sqrt{(c_{x,t} - c_{x,p})^2 + (c_{y,t} - c_{y,p})^2}$ and $\Delta\theta = |\theta_t - \theta_p|$. The influence of every shape coefficient was validated by calculating the Euclidean distance between ground truth landmarks and landmarks reconstructed using an increasing number of predicted coefficients. To only capture the impact of shape coefficients, ground truth pose parameters were used for reconstruction. Furthermore, LV area, myocardial area, LV dimensions in three different orientations and regional wall thickness (RWT) for six cardiac segments were calculated from the predicted landmarks. LV dimensions and RWT were directly obtained by calculating the distance between two corresponding landmarks and averaging the different values in one segment. For these four physical measures, mean absolute error (MAE) and Pearson correlation coefficient (ρ) were calculated. Statistical significant improvement of every choice was assessed by the two-sided Wilcoxon signed rank test with a significance level of 5%.

Additionally, we applied the proposed method to two different public datasets: LVQuan18 [16] and LVQuan19 [17]. Both datasets contain mid-cavity SA slices for 20 time frames spanning the complete cardiac cycle and provide ground truth values for LV and myocardial area, three LV dimensions and six RWT. In LVQuan18 (145 patients, 2879 images), the 80 $\times$ 80 images were normalized for pose and size while in LVQuan19 (56 patients, 1120 images), no preprocessing was applied. LVQuan19 was identically processed as D1, including prior resampling. Since LVQuan18 contained small, centered images, these

Table 1. Results for D1 obtained from the binarized distance maps ('Map') or shape and pose parameters ('Contour'). Mean and standard deviation for DSC, MBE, HD, position error (Δd) and orientation error ($\Delta\theta$) are reported. Best values are indicated in bold. Statistical significant improvement with respect to the previous row is indicated with *.

	DSC LV [%]	DSC myo [%]	MBE [mm]	HD [mm]	Δd [mm]	$\Delta\theta$ [°]
Map						
$S_{\mu=0}$	90.5±13.9	81.2±14.0	1.99±3.47	18.38±42.39	/	/
S	91.7±12.3*	83.1±12.6*	1.34±0.90*	4.32±6.19*	/	/
RS	91.8 ± 11.6	83.1 ± 12.4	1.35±0.92	4.23±4.29	/	/
RS-A$_{ps}$	**92.8±10.1***	**85.3±10.6***	**1.18±0.69***	**3.64±3.00***	/	/
Contour						
R	65.1±25.5	38.1±21.9	7.15±5.29	15.41±10.70	10.1±9.1	10.4±10.9
RS	82.6±18.9*	64.3±21.5*	3.29±3.29*	7.70±7.39*	4.1±5.4*	11.7±12.4
RS-A$_{ps}$	**88.1±11.9***	**72.7±14.1***	**2.16±1.03***	**5.37±3.49***	**2.5±1.8***	**9.5±7.5***

images were not resampled, no pose regression was applied, the number of epochs was decreased to 1000 and parameter A in Fig. 1 equals 5. For both datasets, a fivefold cross validation was performed and LV area, myocardial area, LV dimensions and RWT were calculated.

4   Results

Table 1 shows the results of DSC, MBE, HD, Δd and $\Delta\theta$ for the different setups. The combined MSE and Dice loss (S) significantly improved DSC, MBE and HD compared to the setup with only Dice loss ($S_{\mu=0}$), most notably for HD. $S_{\mu=0}$ resulted in 10.2% unrealistic shapes and S in 0%. While adding L_1 and L_2 (RS) did not alter the performance of distance map regression, shape and pose data augmentation (RS-A$_{ps}$) did significantly improve all metrics. For the 'Contour' experiments, the addition of semantic segmentation and data augmentation both significantly improved the results, except for $\Delta\theta$. However, DSC, MBE and HD remain worse compared to the 'Map' experiments. The distance errors on the landmarks are visualized in Fig. 2, which indicates again that both modifications to a standard regression CNN contribute to significant improvement. Furthermore, whereas the first coefficients, accounting for the largest variation, are relatively well predicted, the latter coefficients were not accurately estimated. The average landmark error for setup RS-A$_{ps}$ using 12 shape coefficients is 1.44 mm, which is lower than the MBE, indicating that the inferior segmentation results are partially due to pose estimation.

Table 2 reports MAE and ρ of LV area, myocardial area, LV dimensions and RWT, averaged over all segments. The results on D1 show that these metrics can be more accurately estimated by simultaneous semantic segmentation and by addition of data augmentation. For LV and myocardial area and LV dimensions, RS-A$_{ps}$ obtained better results compared to the winner of the LVQuan18

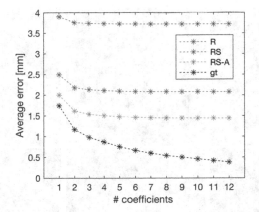

Fig. 2. Average distance between ground truth landmarks and landmarks reconstructed using a limited number of coefficients. The results are given for predicted as well as ground truth (gt) coefficients.

Table 2. MAE and ρ for LV area, myocardial area, LV dimensions and RWT. Best values are indicated in bold. For D1, statistical significant improvement with respect to the previous column is indicated with *. [1]In [19], a threefold cross validation was used. [2]In [10], the average MAE of LV and myocardial area was reported to be 122 mm^2.

		D1 [%]			LVQuan18		LVQuan19			
		R	RS	RS-A$_{ps}$	[18]	RS-A$_{ps}$	[19][1]	[10]	[11]	RS-A$_{ps}$
MAE	Area LV [mm^2]	472	256*	**139***	135	**117**	**92**	122^2	186	134
	Area Myo [mm^2]	299	192*	**145***	177	**162**	**121**	122^2	222	201
	Dim [mm]	7.06	3.58*	**2.37***	2.03	**1.50**	**1.52**	1.84	3.03	2.10
	RWT [mm]	1.86	1.38*	**1.18***	1.38	1.52	**1.01**	1.22	1.67	1.78
ρ [%]	Area LV	81	95	**99**	/	99	/	/	97	**98**
	Area Myo	77	90	**94**	/	93	/	/	88	**93**
	Dim	84	96	**98**	/	98	/	/	95	**97**
	RWT	60	83	**88**	/	84	/	/	73	**83**

challenge [18], who used a parameter regression approach, while the estimation of RWT was slightly worse. For LVQuan19, the results of RS-A$_{ps}$ are compared to the top three entries of the challenge. While the results of [19] and [10] are superior, our error on LV and myocardial area and LV dimensions is lower compared to the errors reported in [11], and the correlation is higher for all metrics. Figure 3 depicts representative segmentation examples.

5 Discussion

In contrast to semantic segmentation, the predicted shape coefficients are directly linked to an oriented landmark-based representation and as such allow straightforward calculation of regional metrics including myocardial thickness or strain.

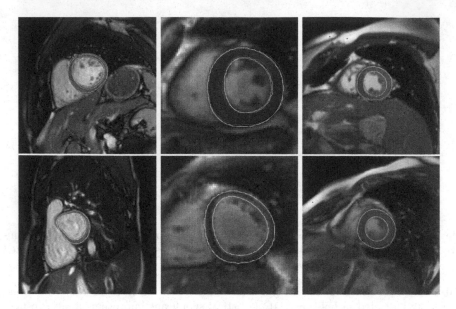

Fig. 3. Representative segmentation examples for datasets D1, LVQuan18 and LVQuan19 (left to right). Ground truth (red), semantic segmentation output (yellow) and segmentation reconstructed from shape and pose output (cyan) are shown. (Color figure online)

Furthermore, contrary to conventional semantic segmentation using Dice loss ($S_{\mu=0}$), our approach did not result in any missing or disconnected regions since the shape model is inherently unable to predict such unrealistic shapes. While some initial experiments showed that pose and shape data augmentation was able to significantly improve the segmentation for setup $S_{\mu=0}$, the results remained significantly worse compared to the proposed approach RS-A$_{ps}$.

For the LVQuan19 challenge data, we obtained higher MAE compared to the leading results of [19]. There are multiple possible explanations for this. First, the two methods use significantly different approaches: Acero et al. [19] calculated the LV parameters from a segmentation obtained with a semantic segmentation CNN while we calculated the LV parameters from the 12 predicted shape coefficients. When calculating the LV parameters from the predicted distance maps and position instead, slightly lower MAE of 109 mm^2 for LV area, 188 mm^2 for myocardial area, 1.69 mm for LV dimensions and 1.74 mm for RWT were achieved. This is in accordance with the lower performance of the 'Contour' experiments compared to the 'Map' experiments in Table 1. Second, preprocessing steps such as resampling strategy and intensity windowing, data augmentation and training approach all have an impact on CNN performance. In the LVQuan18 challenge, the images were preprocessed by the challenge organizers, eliminating some of these sources of variability. Third, contrary to the challenge entries [10,11,19], our method was not specifically developed and tuned for this task. It should be noted that all three challenge entries reported substantially worse results on LVQuan19 test set, which is not publicly available.

We found that the regression of shape coefficients is a more difficult task compared to semantic segmentation. In semantic segmentation using distance maps, 128×128 correlated values should be estimated for every image while shape coefficient regression required the estimation of 12 uncorrelated values from relatively little training data. The combination with semantic segmentation and addition of data augmentation was however able to significantly improve the shape coefficient regression. In future work, we want to investigate if an extra loss term enforcing consistency between semantic segmentation and pose and shape parameters can further improve these results.

6   Conclusion

In this paper, we presented a proof-of-concept of our shape constrained CNN on 2D cardiac MR images for segmentation of LV cavity and myocardium. In the future, this can be expanded to 3D segmentation and to other applications.

Acknowledgement. Sofie Tilborghs is supported by a Ph.D fellowship of the Research Foundation - Flanders (FWO). The computational resources and services used in this work were provided in part by the VSC (Flemisch Supercomputer Center), funded by the Research Foundation - Flanders (FWO) and the Flemisch Government - department EWI. This research also received funding from the Flemish Government under the "Onderzoeksprogramma Artificiële intelligentie (AI) Vlaanderen" programme and is also partially funded by KU Leuven Internal Funds C24/19/047 (promotor F. Maes).

References

1. Bernard, O., et al.: Deep learning techniques for automatic MRI cardiac multi-structures segmentation and diagnosis: is the problem solved? IEEE Trans. Med. Imaging **37**(11), 2514–2525 (2018)
2. Duan, J., et al.: Automatic 3D bi-ventricular segmentation of cardiac images by a shape-refined multi-task deep Learning approach. IEEE Trans. Med. Imaging **38**(9), 2151–5164 (2019)
3. Zotti, C., et al.: Convolutional neural network with shape prior applied to cardiac MRI segmentation. IEEE J. Biomed. Health Inform. **23**(3), 1119–1128 (2019)
4. Oktay, O., et al.: Anatomically Constrained Neural Networks (ACNNs): application to cardiac image enhancement and segmentation. IEEE Trans. Med. Imaging **37**(2), 384–395 (2018)
5. Painchaud, N., Skandarani, Y., Judge, T., Bernard, O., Lalande, A., Jodoin, P.-M.: Cardiac MRI segmentation with strong anatomical guarantees. In: Shen, D., et al. (eds.) MICCAI 2019, Part II. LNCS, vol. 11765, pp. 632–640. Springer, Cham (2019). https://doi.org/10.1007/978-3-030-32245-8_70
6. Yue, Q., Luo, X., Ye, Q., Xu, L., Zhuang, X.: Cardiac segmentation from LGE MRI using deep neural network incorporating shape and spatial priors. In: Shen, D., et al. (eds.) MICCAI 2019, Part II. LNCS, vol. 11765, pp. 559–567. Springer, Cham (2019). https://doi.org/10.1007/978-3-030-32245-8_62

7. Cootes, T.F., et al.: Active shape models - their training and application. Comput. Vis. Image Underst **61**(1), 38–59 (1995)

8. Attar, R., et al.: 3D cardiac shape prediction with deep neural networks: simultaneous use of images and patient metadata. In: Shen, D., et al. (eds.) MICCAI 2019, Part II. LNCS, vol. 11765, pp. 586–594. Springer, Cham (2019). https://doi.org/10.1007/978-3-030-32245-8_65

9. Vigneault, D., et al.: Ω-Net (omega-net): fully automatic, multi-view cardiac mr detection, orientation, and segmentation with deep neural networks. Med. Image Anal. **48**, 95–106 (2018)

10. Gessert, N., Schlaefer, A.: Left ventricle quantification using direct regression with segmentation regularization and ensembles of pretrained 2D and 3D CNNs. In: Pop, M., et al. (eds.) STACOM 2019. LNCS, vol. 12009, pp. 375–383. Springer, Cham (2020). https://doi.org/10.1007/978-3-030-39074-7_39

11. Tilborghs, S., Maes, F.: Left ventricular parameter regression from deep feature maps of a jointly trained segmentation CNN. In: Pop, M., et al. (eds.) STACOM 2019. LNCS, vol. 12009, pp. 395–404. Springer, Cham (2020). https://doi.org/10.1007/978-3-030-39074-7_41

12. Cao, L., et al.: Multi-task neural networks for joint hippocampus segmentation and clinical score regression. Multimed. Tools Appl. **77**, 29669–29686 (2018)

13. Kervadec, H., et al.: Boundary loss for highly unbalanced segmentation. In: MIDL 2019, JMLR, vol. 102, pp. 285–296 (2019)

14. Karimi, D., Salcudean, S.E.: Reducing the Hausdorff distance in medical image segmentation with convolutional neural networks. IEEE Trans. Med. Imaging **39**(2), 499–513 (2020)

15. Dangi, S., et al.: A distance map regularized CNN for cardiac cine MR image segmentation. Med. Phys. **46**(12), 5637–5651 (2019)

16. Li, S., Xue, W.: Left ventricle full quantification hallenge. In: MICCAI 2018. (2018) https://lvquan18.github.io/

17. Yang, G., et al.: Left ventricle full quantification challenge. In: MICCAI 2019 (2019). https://lvquan19.github.io/

18. Li, J., Hu, Z.: Left ventricle full quantification using deep layer aggregation based multitask relationship learning. In: Pop, M., et al. (eds.) STACOM 2018. LNCS, vol. 11395, pp. 381–388. Springer, Cham (2019). https://doi.org/10.1007/978-3-030-12029-0_41

19. Corral Acero, J., et al.: Left ventricle quantification with cardiac MRI: deep learning meets statistical models of deformation. In: Pop, M., et al. (eds.) STACOM 2019. LNCS, vol. 12009, pp. 384–394. Springer, Cham (2020). https://doi.org/10.1007/978-3-030-39074-7_40

Left Atrial Ejection Fraction Estimation Using SEGANet for Fully Automated Segmentation of CINE MRI

Ana Lourenço[1,2]([✉]), Eric Kerfoot[1], Connor Dibblin[1], Ebraham Alskaf[1],
Mustafa Anjari[3], Anil A. Bharath[4], Andrew P. King[1], Henry Chubb[1],
Teresa M. Correia[1], and Marta Varela[1,5]

[1] School of Biomedical Engineering and Imaging Sciences, King's College London,
London, UK
{ana.lourenco,teresa.correia}@kcl.ac.uk
[2] Faculty of Sciences, University of Lisbon, Lisbon, Portugal
[3] Department of Neuroradiology, National Hospital for Neurology and Neurosurgery,
University College London Hospitals NHS Trust, London, UK
[4] Bioengineering Department, Imperial College London, London, UK
[5] National Heart and Lung Institute, Imperial College London, London, UK
marta.varela@imperial.ac.uk

Abstract. Atrial fibrillation (AF) is the most common sustained cardiac arrhythmia, characterised by a rapid and irregular electrical activation of the atria. Treatments for AF are often ineffective and few atrial biomarkers exist to automatically characterise atrial function and aid in treatment selection for AF. Clinical metrics of left atrial (LA) function, such as ejection fraction (EF) and active atrial contraction ejection fraction (aEF), are promising, but have until now typically relied on volume estimations extrapolated from single-slice images.

In this work, we study volumetric functional biomarkers of the LA using a fully automatic SEGmentation of the left Atrium based on a convolutional neural Network (SEGANet). SEGANet was trained using a dedicated data augmentation scheme to segment the LA, across all cardiac phases, in short axis dynamic (CINE) Magnetic Resonance Images (MRI) acquired with full cardiac coverage. Using the automatic segmentations, we plotted volumetric time curves for the LA and estimated LA EF and aEF automatically.

The proposed method yields high quality segmentations that compare well with manual segmentations (Dice scores [0.93 ± 0.04], median contour [0.75 ± 0.31] mm and Hausdorff distances [4.59 ± 2.06] mm). LA EF and aEF are also in agreement with literature values and are significantly higher in AF patients than in healthy volunteers. Our work opens up the possibility of automatically estimating LA volumes and functional biomarkers from multi-slice CINE MRI, bypassing the limitations of current single-slice methods and improving the characterisation of atrial function in AF patients.

T. M. Correia and M. Varela—Contributed equally.

© Springer Nature Switzerland AG 2021
E. Puyol Anton et al. (Eds.): STACOM 2020, LNCS 12592, pp. 137–145, 2021.
https://doi.org/10.1007/978-3-030-68107-4_14

Keywords: Atrial fibrillation · Ejection fraction · Left atrium · Convolutional neural network · Segmentation · CINE MRI

1   Introduction

Atrial fibrillation (AF) is the most common sustained cardiac arrhythmia, characterized by a rapid and irregular contraction of the upper chambers of the heart, the atria. AF affects more than 33 million people worldwide, is associated with great increases in morbidity and mortality, and typically has a negative impact on patients' quality of life. Although AF is mainly managed with medical therapy, catheter ablation is arguably the gold standard treatment to directly terminate AF in selected patients. In their current form, ablations are only first-time effective in approximately 50% of AF patients [5].

Treatment efficacy for AF could be greatly increased if biomarkers capable of characterising atrial function and predicting treatment outcome were available. Left atrial (LA) dimensions and shape have been shown to be reasonable predictors of post-ablation patient outcome [13]. It is expected that dynamic characterisations of atrial dimensions and volumes across the cardiac cycle may be even more informative [7]. In particular, LA ejection fraction (EF) and active atrial contraction ejection fraction (aEF) contain important clinical information that can contribute to the management of AF. The LA EF characterises the LA global function across the cardiac cycle, whereas the aEF assesses the LA pump function, which is likely to provide important additional clues about atrial tissue health. Both these biomarkers rely on ratios of LA volumes across the cardiac cycle. LA volumes are usually estimated by applying standard formulae derived from assumptions on the LA shape obtained from 2D atrial echocardiography or dynamic (CINE) Magnetic Resonance Images (MRI) [4,12]. Volume estimation from 3D images is expected to be more accurate and precise. Furthermore, aEF relies on the identification of the onset of active atrial contraction (preA), which is difficult to perform visually in a single 2D view.

CINE MRI combines good spatial and temporal resolution with large cardiac coverage and is thus ideally suited for assessing LA function. Nevertheless, detailed volumetric temporal analyses of atrial volumes are not currently available for clinical use for two key reasons. First, atrial CINE MRI images are typically acquired on single-slice 2-chamber and 4-chamber views, which do not sample the full atrial volume [9]. Second, reliable automatic techniques for segmenting the atria in short axis views across the entire cardiac cycle are not yet available. This is in contrast with the segmentation of ventricular structures from short axis CINE MRI and the LA in 2- and 4-chamber views [1]. We expect that automatic processing tools for short axis atrial images may lead to an increased interest in imaging this chamber in this orientation.

In this work, we propose a dedicated neural network for fully automatic SEGmentation of the left Atrium based on a convolutional neural Network (SEGANet) to address this important clinical gap. We acquired short-axis CINE images with full ventricular and atrial coverage and used SEGANet to segment the LA.

The quality of the SEGANet LA segmentation is assessed using the Dice coefficient (DC), median contour distances (MCD) and Hausdorff distances (HD). We additionally compare the SEGANet results with the inter-observer variability of manual segmentations. To demonstrate the clinical utility of our method, we use SEGANet to estimate LA volumes across all cardiac phases and automatically calculate LA EFs and aEFs in both healthy subjects and AF patients.

2   Methods

2.1   Data Acquisition

Short axis CINE MR image stacks with full cardiac coverage were acquired in a 1.5T Philips Ingenia scanner with a 32-channel cardiac coil. Imaging was performed in 60 AF patients (31–72 years old, 75% male) and 12 healthy volunteers (24–36 years old, 50% female) under ethical approval and following informed written consent.

A 2D bSSFP protocol (flip angle: 60°, TE/TR: 1.5/2.9 ms, SENSE factor 2) was used. Images were captured with ECG-based retrospective gating with a typical field of view of $385 \times 310 \times 150 \, mm^3$, acquisition matrix of 172×140 and slice thickness of 10 mm. Images were reconstructed to a resolution of $1.25 \times 1.25 \times 10 \, mm^3$ and 30–50 cardiac phases with 60–70% view sharing. These acquisitions provide full coverage of the ventricles and atria, with typically 4–6 slices covering the atria. Each slice was acquired in a separate breath hold. Due to the large slice thickness in all datasets, it can be difficult, even for an expert, to identify the separation between the atria and ventricles. In order to solve this challenge, we used a pre-trained left ventricular (LV) segmentation network [8] to identify the slices containing ventricular tissue. The most basal atrial slice was automatically identified as the slice contiguous to the top ventricular slice segmented by this LV network.

Ground truth (GT or M1) segmentations were obtained by a consensus of two medical experts, who manually segmented the LA on a slice by slice basis at 3 phases of the cardiac cycle. The LA appendage was included in the segmentation, but the pulmonary vein insertions were not.

2.2   SEGANet

The proposed neural network is based on the U-Net architecture [11] with the following modifications: 1) the addition of residual units [6] throughout the encode and decode paths, which prevent vanishing gradients by allowing some measure of the original input to be included in each layer's output; 2) instance layer-normalization, which minimises the influence outlier images may have in the whole batch; 3) Parametric Rectified Linear Unit (PReLU) activation, which allows the network to learn from negative values. The network is built as a stack of layers which combine the encode and decode paths as illustrated in Fig. 1. Four layers are stacked together using convolutions with a stride of 2 to downsample data in the encode path, and with output tensors to the layers below with channel dimensions of 16, 32, 64, 128, and 256.

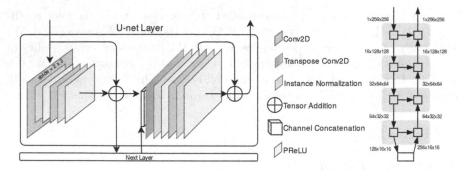

Fig. 1. The proposed LA segmentation network SEGANet is based on a U-Net architecture. The segmentation network is built as a stack of layers (shown on the right), which combine the encode path of the network on the left and the decode path on the right. Subsequent layers, or the bottom connection, are denoted by "Next Layer". The bottom connection is composed of a single residual unit including two sets of convolution/normalisation/regularisation sequences.

The network was trained for LA segmentation on short axis CINE MRI in $50,000$ iterations with mini-batches of 300 slices drawn randomly from the 715 slices of the training dataset. We used a binary formulation of the Dice loss function and the Adam optimiser with a learning rate of 0.0001.

To extract more generalised information from our relatively small dataset and to regularise the network to prevent overfitting, we used a number of randomly-chosen data augmentation functions to apply to each slice. These data augmentations were chosen to replicate some of the expected variation in real images without modifying the cardiac features of interest. They include: 1) rotations, translations and flips; 2) croppings and rotations; 3) additive stochastic noise; 4) k-space corruption to simulate acquisition artefacts; 5) smooth non-rigid deformations encoded using free-form deformations and 6) intensity scalings by a random factor.

2.3 Data Analysis

In order to assess the quality of the automatic segmentations, an additional medical expert was asked to provide manual LA segmentations (M2) in 13 AF patients. Segmentation quality metrics (DC, HD and MCD) were estimated relative to the ground truth for both the SEGANet segmentations and the manual segmentations M2. We also estimated the agreement between the two sets of manual segmentations.

For a complete visualisation of the heart volume variations throughout the cardiac cycle, LA volumetric time curves were plotted. Using these, maximal (V_{max}) and minimal (V_{min}) LA volumes, as well as the LA volume at the onset of atrial contraction (V_{preA}) were automatically computed. EF and aEF were also estimated using $EF = \frac{V_{max} - V_{min}}{V_{max}} \times 100$ and $aEF = \frac{V_{preA} - V_{min}}{V_{preA}} \times 100$.

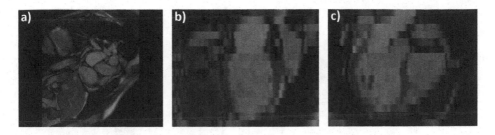

Fig. 2. Segmentations obtained with the proposed method for one representative subject during atrial diastole, overlaid on the CINE **a)** short axis, **b)** pseudo 2-chamber, and **c)** pseudo 4-chamber views. The acquired short axis CINE image stacks include both ventricles and atria. (Color figure online)

LA EFs and aEFS obtained from SEGANet segmentations were calculated for both AF patients in sinus rhythm and healthy subjects. We used paired Student t-tests to assess whether we could detect a significant difference in these metrics between the two groups.

3 Results

SEGANet generated very good quality LA segmentations. This can be qualitatively observed in Fig. 2, which depicts, in three orthogonal views, the CINE MRI for one representative patient overlaid with the automatic LA segmentations in red. Figure 3 shows that automatic segmentations are on par with manual segmentations, with DC: 0.93 ± 0.04, HD: 4.59 ± 2.06 mm; MCD: 0.75 ± 0.31 mm for the SEGANet segmentations when compared to the GT. These values are comparable to those of the additional manual segmentations M2, further highlighting the good performance of the network (Fig. 3).

In previous studies, the network for LV segmentation had demonstrated accurate results when it was tested using other datasets [8] and the method allowed us to automatically select the slices that covered the atria.

The obtained LA volumes across the cardiac cycle for AF patients (V_{min} : 79.40 ± 25.34 mL; V_{max} : 111 ± 24.00 mL; V_{preA} : 103.40 ± 25.34 mL) and for healthy subjects (V_{min} : 22.43 ± 25.34 mL; V_{max} : 44.23 ± 24.73 mL; V_{preA} : 35.47 ± 19.48 mL) show high inter-subject variability and agree well with the literature [10]. The LA volume varies smoothly across the cardiac cycle, further suggesting the good quality of segmentations throughout the different cardiac phases (Fig. 4).

Regarding the LA EF, we obtained $31.1\% \pm 9.9\%$ in AF patients and $49.8\% \pm 7.6\%$ in healthy volunteers, which is in good agreement with published echocardiographic and 2D CINE MRI values [3,10,14]. The obtained values for aEF (AF patients: $24.3\% \pm 9.0\%$, healthy volunteers: $37.9\% \pm 10.1\%$) are also in accordance with the literature [9,10,14]. As expected, both LA EF and aEF are significantly higher ($p < 1e^{-4}$ and $p < 1e^{-7}$, respectively) in healthy subjects than in AF patients (Fig. 4).

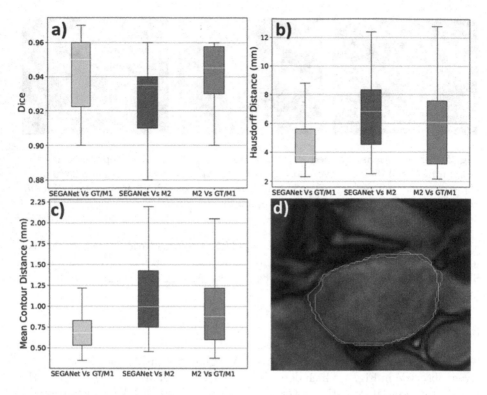

Fig. 3. a) Dice coefficient, **b)** Hausdorff distance and **c)** mean contour distance comparing automatic (SEGANet) and manual (M2) LA segmentations with the ground truth (GT/M1) in 13 subjects. **d)** Contours indicating the results of the manual and automatic LA segmentations in a representative slice: yellow is the SEGANet automatic segmentation, red is the manual M2 segmentation, and green is the GT/M1.

4   Discussion

We present SEGANet, a convolutional neural network that allows automatic LA segmentation in short-axis CINE MRI across the whole cardiac cycle. To the best of our knowledge, there are currently no fully automatic segmentation methods designed for this important clinical application. We demonstrate the clinical potential of our technique by automatically estimating atrial biomarkers of interest, such as atrial ejection fractions.

The proposed method shows a very good segmentation performance for the LA (Figs. 2 and 3), on par with the performance of other neural networks: for the segmentation of ventricular structures in CINE-MRI [1]; and for LA segmentations in static MRI with a different resolution and contrast to ours. In particular, our results for CINE MRI compare well with the metrics for the segmentation of delayed enhancement MRI in the STACOM 2018 Left Atrial Segmentation Challenge (DC: 0.90 ± 0.93, HD: 14.23 ± 4.83 mm) [2] .

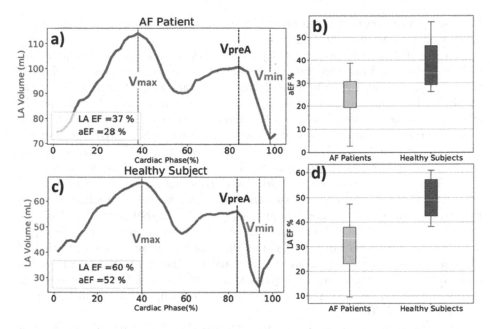

Fig. 4. Left atrium (LA) volume throughout the cardiac cycle for two representative subjects: **a)** an AF patient and **c)** an healthy subject . Three LA volumes are indicated: the maximal volume (V_{max}), the minimal volume (V_{min}) and the second peak volume, corresponding to the volume at the onset of the P wave, just before active atrial contraction (V_{preA}). LA EF and aEF values are shown in each subfigure. Moreover, box plots of **b)** aEF and **d)** LA EF for AF patients and healthy subjects show that values are significantly higher in healthy subjects than in patients (LA EF p-value: $< 1e^{-4}$; aEF p-value $< 1e^{-7}$).

In this work, we rely on a modified short axis CINE MRI acquisition that provides volumetric and functional information for the LA, as well as for the ventricles. To keep our analysis pipeline fully automatic, we employed a pre-trained LV segmentation network [8] to identify the most basal atrial slice. This approach was found to be extremely reliable for all images and carried a negligible time penalty.

We chose to segment multi-slice CINE MRI data on a slice-by-slice basis. Although this setup does not make direct use of the 3D spatial and temporal correlations of this dataset, this approach has other advantages, as highlighted previously [1]. One of the advantages is the comparative simplicity and low parameter numbers of the SEGANet compared to its 3D or 3D + time counterparts and its immunity to misregistrations across slices caused by variations in breath hold positions. Additionally, treating each slice independently allows us to train the network with comparatively fewer number of patient datasets than for 3D (or 3D + time) networks. Furthermore, due to the relatively small size of the training dataset (715 slices), we applied an extensive data augmentation strategy, which was crucial for the good performance shown by the network.

We add that our segmentation also uniquely includes the LA appendage, which is relevant for assessing stroke risk in AF.

To demonstrate the clinical utility of our work, we used SEGANet to obtain fully automatic estimates of clinical biomarkers such as LA volumes, LA EFs and aEFs. Up until now, LA volumes and functional markers have been estimated in CINE MRI by applying empirical formulae to single-slice long axis views [4,12], after manual annotation. These shortcomings have contributed to a comparative lack of interest in imaging left atrial function using CINE MRI. Our method allows instead for biomarker estimates from volumetric CINE MRI, which are expected to be more accurate, reproducible and thus more clinically useful.

Estimates of LA volumes from long axis 2D slices typically rely on manual delineations of left atrial apex-base dimensions and assume, unrealistically (see Fig. 2), that the LA shape is as geometrically regular as that of the LV. Furthermore, given that atrial deformations across the cardiac cycle are not isotropic, it is not clear whether single 2D views allow for the accurate identification of the cardiac phases corresponding to maximal LA and minimal LA volume, as required for EF estimation. It is particularly difficult to identify the onset of atrial contraction (V_{preA}), required for aEF estimation, which our method performs automatically.

We thus expect that SEGANet will improve the clinical usefulness of atrial CINE MRI imaging and ultimately make an important contribution to the stratification and treatment of AF patients.

5   Conclusion

We have presented SEGANet for fully automatic segmentation of the LA from CINE MRI and showed it can accurately segment the LA across the cardiac cycle. SEGANet provides reliable automatic estimates of clinical biomarkers such as LA volumes, LA EF and aEF.

Acknowledgments. This research was supported by the Wellcome EPSRC Centre for Medical Engineering at the School of Biomedical Engineering and Imaging Sciences, King's College London [WT203148/Z/16/Z] and the National Institute for Health Research (NIHR) Biomedical Research Centre at Guy's and St Thomas' NHS Foundation Trust and King's College London. We also acknowledge funding from the Engineering and Physical Sciences Research Council [EP/N026993/1] and the British Heart Foundation [RE/18/4/34215].).

References

1. Chen, C., et al.: Deep learning for cardiac image segmentation: A review. arXiv (2019)
2. Chen, C., Bai, W., Rueckert, D.: Multi-task learning for left atrial segmentation on GE-MRI. In: Pop, N., et al. (eds.) STACOM 2018. LNCS, vol. 11395, pp. 292–301. Springer, Cham (2019). https://doi.org/10.1007/978-3-030-12029-0_32

3. Chubb, H., et al.: The reproducibility of late gadolinium enhancement cardiovascular magnetic resonance imaging of post-ablation atrial scar: a cross-over study. J. Cardiovasc. Magn. Reson. **20**(1), 21 (2018). https://doi.org/10.1186/s12968-018-0438-y

4. Erbel, R., et al.: Comparison of single-plane and biplane volume determination by two-dimensional echocardiography 1. Asymmetric model hearts*. Eur. Heart J. **3**(5), 469–480 (1982)

5. Ganesan, A.N., et al.: Long-term outcomes of catheter ablation of atrial fibrillation: a systematic review and meta-analysis. J. Am. Heart Assoc. **2**(2), e004549 (2013)

6. He, K., Zhang, X., Ren, S., Sun, J.: Identity mappings in deep residual networks. In: Leibe, B., Matas, J., Sebe, N., Welling, M. (eds.) ECCV 2016 Part IV. LNCS, vol. 9908, pp. 630–645. Springer, Cham (2016). https://doi.org/10.1007/978-3-319-46493-0_38

7. Hoit, B.D.: Evaluation of left atrial function: current status. Struct. Heart **8706**, 112 (2017)

8. Kerfoot, E., Puyol Anton, E., Ruijsink, B., Clough, J., King, A.P., Schnabel, J.A.: Automated CNN-based reconstruction of short-axis cardiac mr sequence from real-time image data automated CNN-based reconstruction of short-axis cardiac mr sequence from real-time image data. In: Stoyanov, D., et al. (eds.) RAMBO/BIA/TIA -2018. LNCS, vol. 11040, pp. 32–41. Springer, Cham (2018). https://doi.org/10.1007/978-3-030-00946-5_4

9. Kowallick, J.T., et al.: Quantification of atrial dynamics using cardiovascular magnetic resonance: inter-study reproducibility. J. Cardiovasc. Magn. Reson. **17**(1), 36 (2015). https://doi.org/10.1186/s12968-015-0140-2

10. Mika, M., et al.: Impact of reduced left atrial functions on diagnosis of paroxysmal atrial fibrillation: Results from analysis of time-left atrial volume curve determined by two-dimensional speckle tracking. J. Cardiol. **57**(1), 89–94 (2011)

11. Ronneberger, O., Fischer, P., Brox, T.: U-Net: convolutional networks for biomedical image segmentation. In: Navab, N., Hornegger, J., Wells, W.M., Frangi, A.F. (eds.) MICCAI 2015, Part III. LNCS, vol. 9351, pp. 234–241. Springer, Cham (2015). https://doi.org/10.1007/978-3-319-24574-4_28

12. Ujino, K., et al.: Two-dimensional echocardiographic methods for assessment of left atrial volume. Am. J. Cardiol. **98**(9), 1185–1188 (2006)

13. Varela, M., et al.: Novel computational analysis of left atrial anatomy improves prediction of atrial fibrillation recurrence after ablation. Front. Physio. **8**, 68 (2017)

14. Victoria, D., et al.: Fate of left atrial function as determined by real-time three-dimensional echocardiography study after radiofrequency catheter ablation for the treatment of atrial fibrillation. Am. J. Cardio. **101**(9), 1285–1290 (2008)

Estimation of Cardiac Valve Annuli Motion with Deep Learning

Eric Kerfoot[1], Carlos Escudero King[1], Tefvik Ismail[1], David Nordsletten[1,2], and Renee Miller[1(✉)]

[1] King's College London, London, UK
{eric.kerfoot,carlos.escudero,tefvik.ismail,david.nordsletten,
renee.miller}@kcl.co.uk
[2] University of Michigan, Ann Arbor, USA

Abstract. Valve annuli motion and morphology, measured from non-invasive imaging, can be used to gain a better understanding of healthy and pathological heart function. Measurements such as long-axis strain as well as peak strain rates provide markers of systolic function. Likewise, early and late-diastolic filling velocities are used as indicators of diastolic function. Quantifying global strains, however, requires a fast and precise method of tracking long-axis motion throughout the cardiac cycle. Valve landmarks such as the insertion of leaflets into the myocardial wall provide features that can be tracked to measure global long-axis motion. Feature tracking methods require initialisation, which can be time-consuming in studies with large cohorts. Therefore, this study developed and trained a neural network to identify ten features from unlabeled long-axis MR images: six mitral valve points from three long-axis views, two aortic valve points and two tricuspid valve points. This study used manual annotations of valve landmarks in standard 2-, 3- and 4-chamber long-axis images collected in clinical scans to train the network. The accuracy in the identification of these ten features, in pixel distance, was compared with the accuracy of two commonly used feature tracking methods as well as the inter-observer variability of manual annotations. Clinical measures, such as valve landmark strain and motion between end-diastole and end-systole, are also presented to illustrate the utility and robustness of the method.

Keywords: Cardiac valve motion · Landmark detection · Regression network · Deep learning

1 Introduction

Cardiovascular disease in patients can result in reduced quality of life and premature mortality, becoming increasingly prevalent with age [3]. The integrity of all four cardiac valves, mitral, tricuspid, aortic and pulmonary, is crucial for appropriate haemodynamic function. The annulus is a ring-shaped structure

© Springer Nature Switzerland AG 2021
E. Puyol Anton et al. (Eds.): STACOM 2020, LNCS 12592, pp. 146–155, 2021.
https://doi.org/10.1007/978-3-030-68107-4_15

that forms the perimeter of the valve opening, and its shape changes dynamically during the cardiac cycle [13]. Characterising the physiological deformation of the valve annuli and how their shapes change in cardiac pathologies has become a topic of increased research. For example, excessive dilation of the aortic valve annulus has been identified as a predictor of recurrent aortic regurgitation following surgical intervention in aortic valve disease [11]. Further, the mitral valve annulus has been shown to be "saddle-shaped" in healthy individuals, but becomes increasingly dilated and flattened during systole in mitral regurgitation [7]. Many previous studies have focused on the role of a single valve annulus in cardiac function. However, simultaneous characterisation of the motion and morphology of all four valves could provide a more comprehensive view of their interactions throughout the cardiac cycle. Long-axis motion and strain is used as an indicator of systolic function. For example, mitral/tricuspid annular plane systolic excursion (MAPSE/TAPSE), measured as the distance that the lateral valve landmark moves between end-diastole and end-systole, can provide a quantitative marker of systolic function, particularly in patients who have impaired left ventricular function despite a normal ejection fraction (e.g. patients with hypertrophic cardiomyopathy, HCM) [5]. Long-axis strain, quantified using annotations of valve landmarks, provides a normalized metric for assessing cardiac function [4]. One recent study showed that long-axis strain measurements obtained by a semi-automated method of tracking valve positions was able to differentiate patients with heart failure from controls [14].

Previous approaches to measuring the motion of valve annuli from cardiac magnetic resonance (MR) have utilised semi-automated feature tracking methods [13]. In principle, most cardiac MR tracking tools follow the features of the blood-myocardium boundary using a form of a maximum likelihood estimation [17]. Tracking methods require manual initialisation of the feature or boundary to be tracked and have been shown to have low reproducibility [16]. However, methods exist which integrate machine learning methods, such as marginal space learning and trajectory spectrum learning, to accurately capture the shape and deformation of valves through the cardiac cycle [10].

The aim of this study is to train a single deep learning neural network to robustly identify 10 valve landmarks, describing the positions of three out of the four valves, from three different long-axis MR image orientations which are acquired as part of a standard clinical routine. The valve landmarks, identified through the entire cardiac cycle, are then used to characterise motion in a diverse cohort of patients and healthy controls. We anticipate that this tool can provide a method for rapidly assessing valve annulus motion and morphology, which does not require manual initialisation or input. Ultimately, applied to a large cohort, this tool can provide further understanding of the interaction and role of the valve annuli in healthy and pathological cardiac function.

2   Methods

2.1   Training Data and Manual Annotation

The training dataset consisted of 8574 long-axis images collected from hypertrophic cardiomyopathy (HCM, n = 3069) and myocardial infarction (MI, n = 5505) patients. All HCM patients were part of an ongoing imaging study. MI patients were a subset from the DETERMINE study obtained through the Cardiac Atlas Project (http://stacom.cardiacatlas.org/) [6]. All images were acquired in the 2-chamber (2CH, n = 2952), 3-chamber (3CH, n = 2712) or 4-chamber (4CH, n = 2910) view on either Philips (n = 4463), Siemens (n = 3195) or GE scanners (n = 916). The validation set consisted of 930 images. Variations in the positioning of these views occurred due to natural and pathological variations in patient heart shape and valve positioning. Generally, the 4CH image (Fig. 1C) shows the mitral and tricuspid valves, which control flow of blood into the left and right ventricles, respectively. The 3CH image (Fig. 1B) provides views of the mitral and aortic valves. Finally, the 2CH image (Fig. 1A) contains a view solely of the mitral valve. Each long-axis view was composed of approximately 30 frames through the entire cardiac cycle, with slight variation based on patient heart rate and breath-hold duration required during acquisition. Ten valve landmarks, including the mitral (1–6), aortic (7–8) and tricuspid (9–10) valves, were manually annotated at every frame by an expert reviewer to provide a ground-truth for training (Fig. 1). Image annotation was conducted using ITK-SNAP [19]. Ten cases (960 images) were annotated by two additional expert observers to calculate the inter-observer variability.

2.2   Feature Tracking

As a comparison with the neural network prediction, valve landmarks were tracked using two common feature tracking algorithms available in Matlab 2019b (Kanade-Lucas-Tomasi algorithm) [18] and MIRTK [15].

2.3   Neural Network

The network architecture, shown in Fig. 2, is composed of a sequence of dilated dense blocks [8] represented by each square with input and output volume sizes given. Each dense block applies 2D convolutions with different dilation rates to their inputs and concatenates the results from each together as its output. A dense block produces an output volume containing feature information derived from multiple sizes of convolutional fields-of-view, subsequent convolutions can thus combine information at different scales to recognise small and large features.

The dimensions of the output volume from each dense block is half as large as the input in the spatial dimensions. After five such blocks, the volume is passed to a series of small networks, each composed of a convolution followed by a fully-connected layer, which compute one landmark coordinate each from the volume. These small networks can learn a specific reduction for each landmark, and are able to correctly learn to produce a result of $(0, 0)$ for landmarks not present in the image. The network was trained for 120,000 iterations using a straightforward supervised approach with 8,574 image-landmark pairs. The Adam optimizer [12] was used with a learning rate of 0.0001, and the loss computed as the mean absolute error between ground truth and predicted landmarks.

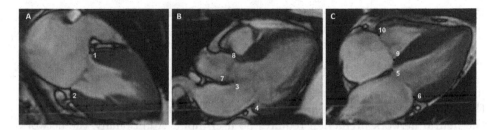

Fig. 1. Manually annotated valve landmarks in the A) 2-chamber, B) 3-chamber, and C) 4-chamber long-axis images labelled from 1–10. Red: mitral valve, blue: aortic valve, green: tricuspid valve. (Color figure online)

Minibatches were collected together by choosing random images from the training set and applying a randomly-selected series of flip, rotation, shift and transpose operations to the image and landmark pairs. The images were further augmented by adding stochastic noise, applying a dropout function in k-space to partially corrupt the image, and varying the image intensity with a smooth shift function. Lastly, free-form deformation was applied to the images and the landmarks to further vary the data. The result of these augmentations was to minimize over-fitting to the training data set and vastly increase the variation in features the network observed over the course of training.

During training, after every epoch of 400 iterations, a per-landmark accuracy value was calculated by applying the network to a validation data set of 930 image-landmark pairs. For each landmark, the mean Cartesian distance between the set of ground truth and predicted coordinates was computed as the per-landmark prediction error. The minibatch selection procedure was then adjusted such that each image's probability of being selected was computed to be proportional to the distance error of the landmarks defined for it. This step ensured that landmarks exhibiting lower accuracy would, at later stages in the training, become more common in the selected minibatches. This curriculum learning approach [1,2] forced the network to concentrate on those landmarks which were more difficult to accurately place compared to others, enforcing their importance to the overall loss calculation.

The trained neural network was then used to predict valve landmarks in long-axis images from patients with HCM, MI, and dilated cardiomyopathy (DCM), as well as healthy volunteers (Vol). Long-axis strain was calculated as the normalised distance from a stationary apical point to the valve landmarks.

Mitral/tricuspid annular plane systolic excursion (MAPSE/TAPSE) values, calculated as the distance between lateral valve landmarks (6 and 10) in the 4CH view at end-diastole and end-systole, were also computed for each group.

3   Results and Discussion

Boxplots in Fig. 3 illustrate the differences between inter-observer, tracking and predicted point errors for each valve landmark. All mean pixel errors (± one standard deviation) are shown in Table 1. Mean inter-observer error was lower than mean predicted point errors in all landmarks except 7 and 8 (aortic valve).

Although mean errors for the network-predicted landmarks were higher than inter-observer errors, the mean errors were less than or comparable to errors from the tracking methods. Tracking methods require manual initialisation whereas the neural network prediction requires no user input. Additionally, the network is able to not only provide accurate point locations but also accurate labels (e.g. 1–10), meaning that not only the long-axis view but also the image orientation is correctly identified by the neural network. For these reasons, the neural network could be used as a stand-alone tool or in conjunction with a more robust tracking method as a tool to automatically initialise landmark locations.

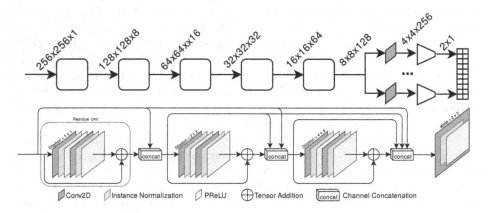

Fig. 2. (Top) Regressor network topology. (Bottom) Dilated dense block of convolutions used to construct the regressor network. This dense block is composed of three series of residual units whose convolutions have a dilation of 1, 2 and 4, respectively. The outputs from each unit is concatenated with that from the previous, such that the final volume is composed of features derived from viewing areas of different size. A final convolution is used to down-sample the output volume by a factor of two. All convolution kernels have 3×3 shape.

The current network predicts the presence and location of 10 valve landmarks in individual long-axis images, not incorporating any temporal consistency. Therefore, time-series images are not needed and valve landmarks can be obtained from a single frame. However, incorporating temporal consistency may prove beneficial in cases with poor image quality in individual frames, for example, due to flow artefacts during the ejection phase. For the moment, post-processing steps can be used to interpolate landmark positions in these frames. The addition of temporal consistency will be evaluated in future work.

3.1 Clinical Measurements

To illustrate the clinical utility of this prediction tool for rapid identification of valve landmarks, an illustrative plot of long-axis strain is shown in Fig. 4a. Global longitudinal strain, an accepted measurement for assessing cardiac function, is plotted for each valve landmark. The strain plots clearly show peak strain at end-systole and enable further quantification of clinical measurements such as peak strain and peak strain rates. Peak strain predicted for each valve view was plotted versus the same peak strain measured from manually annotated landmarks in Fig. 4b (red). Similarly, peak strain measured from two different observers are plotted for the cases used to measure inter-observer variability (Fig. 4b, blue). A perfect correspondence (y = x) is shown as a black dashed line. The 2CH mitral valve, 3CH mitral valve and 4CH tricuspid valve show good agreement for the predicted versus manual peak strain. The predicted versus manual peak strain showed poor agreement for the aortic valve (as illustrated by the discrepancy between red and black dashed lines). This is unsurprising since, at end-systole, rapid ejection causes flow artifacts in the aorta, making the leaflet landmarks hard to detect. In the small sample size used in this plot, the predicted versus manual peak strain also showed poor agreement for the mitral valve 4CH view. However, in general, this shows that, for certain long axis views, the network was

Table 1. Mean (± one standard deviation) pixel error for the interobserver error (O), Kanade-Lucas-Tomasi (KLT) tracking (T), MIRTK tracking (T2) and neural network predicted landmarks (P).

Landmark	Inter-observer	KLT tracking	MIRTK tracking	Neural network
1	2.013 ± 1.571	2.311 ± 3.017	1.979 ± 1.764	2.334 ± 1.398
2	2.422 ± 1.791	2.640 ± 2.307	1.927 ± 1.650	2.560 ± 2.154
3	1.463 ± 1.089	2.375 ± 1.745	2.861 ± 2.297	1.773 ± 0.905
4	2.024 ± 1.070	2.458 ± 2.167	2.367 ± 1.945	2.401 ± 1.704
5	1.684 ± 1.470	2.169 ± 1.915	1.851 ± 1.193	2.270 ± 1.571
6	2.717 ± 1.808	3.204 ± 2.210	1.619 ± 1.251	2.758 ± 1.763
7	3.042 ± 2.948	2.309 ± 2.149	2.683 ± 2.357	1.993 ± 1.172
8	4.311 ± 3.086	3.051 ± 2.692	2.422 ± 1.789	3.765 ± 2.691
9	1.949 ± 1.505	5.411 ± 3.661	2.666 ± 2.371	3.545 ± 2.213
10	3.440 ± 2.832	3.724 ± 2.636	2.140 ± 1.632	3.522 ± 2.235

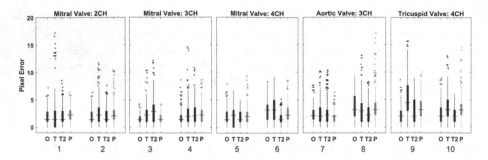

Fig. 3. Boxplots show the differences between inter-observer (O - red), Kanade-Lucas-Tomasi tracking (T - blue), MIRTK tracking (T2 - green) and neural network predicted landmarks (P - grey) errors for all valve landmarks (1–10). Lines represent the median, boxes the interquartile ranges and circles are outliers. (Color figure online)

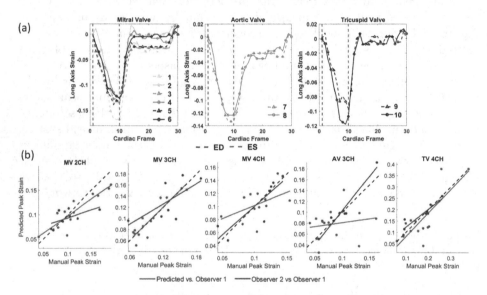

Fig. 4. a) Long axis strain measured from predicted points for each valve landmark (1–10), grouping together landmarks for the mitral valve (left), aortic valve (centre) and tricuspid valve (right). This is an example case for a set of images previously unseen by the network. b) Peak long axis strain measured from the midpoint of the annulus landmarks plotted for predicted versus manual (red), two different observers (blue) and a perfect correspondence (black dashed line). (Color figure online)

capable of extracting not only valve landmarks but also valuable clinical metrics with similar reproducibility as that obtained between multiple observers.

MAPSE/TAPSE values (measured from neural network predicted landmarks) plotted in Fig. 5 show variations in valve plane motion between patient groups and healthy volunteers. Patients with MI exhibit reduced long-axis strain due to areas of non-contracting myocardium [9], which can be seen in these

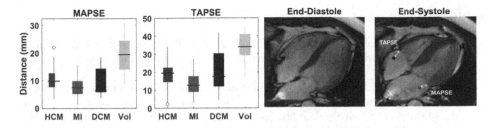

Fig. 5. Boxplots of MAPSE/TAPSE measured from predicted valve landmarks for each group: HCM, MI, DCM and volunteers.

results. MAPSE and TAPSE show large variability within the DCM group, illustrating large variations in phenotype within this patient group. These measurements also reveal that HCM and DCM groups exhibit reduced long-axis motion when compared to healthy volunteers.

4   Conclusions

This study presents a deep learning regression network which utilised standard data augmentation techniques such as flip, rotation, shift and transpose operations in addition to an adaptive minibatch selection process, for the identification of 10 valve landmark points from cardiac long-axis MR images. The network was capable of predicting both the presence and location of these valve landmarks throughout the cardiac cycle in images from a diverse cohort of pathologies. The network robustly predicted valve landmarks with similar or improved accuracy when compared with two easily applied tracking algorithms.

The network was then applied to predict landmarks in a diverse cohort, including patients with hypertrophic cardiomyopathy, myocardial infarctions, and dilated cardiomypathy, as well as healthy volunteers. These predicted points were used to illustrate differences in clinical measurements such as MAPSE/TAPSE, common measures of systolic function. The landmarks from the network were also used to estimate long-axis strain. Differences between peak long axis strain measured from predicted landmarks versus those measured from manually annotated landmarks were similar to differences seen between multiple observers. This tool provides a rapid method for studying, not only valve shape and motion, but also long axis motion in large cohorts with varying pathologies, providing an advance over existing techniques by bypassing the need for manual supervision.

Acknowledgements. This research was supported by the Wellcome EPSRC Centre for Medical Engineering at the School of Biomedical Engineering and Imaging Sciences, King's College London (WT 203148/Z/16/Z).

References

1. Bengio, Y., Louradour, J., Collobert, R., Weston, J.: Curriculum learning. In: International Conference on Machine Learning, pp. 41–48. Association for Computing Machinery, New York (2009)
2. Berger, L., Eoin, H., Cardoso, M.J., Ourselin, S.: An adaptive sampling scheme to efficiently train fully convolutional networks for semantic segmentation. In: Nixon, M., Mahmoodi, S., Zwiggelaar, R. (eds.) MIUA 2018. CCIS, vol. 894, pp. 277–286. Springer, Cham (2018). https://doi.org/10.1007/978-3-319-95921-4_26
3. Bhatnagar, P., Wickramasinghe, K., Wilkins, E., Townsend, N.: Trends in the epidemiology of cardiovascular disease in the UK. Heart 102(24), 1945–1952 (2016)
4. Brecker, S.J.D.: The importance of long axis ventricular function. Heart 84, 577–578 (2000)
5. Bulluck, H., et al.: A simple technique to measure TAPSE and MAPSE on CMR and normal values. JCMR 16(S1), 1–2 (2014)
6. Fonseca, C.G., et al.: The cardiac atlas project-an imaging database for computational modeling and statistical atlases of the heart. Bioinformatics 27(16), 2288–2295 (2011)
7. Grewal, J., et al.: Mitral annular dynamics in myxomatous valve disease: new insights with real-time 3-dimensional echocardiography. Circulation 121(12), 1423–1431 (2010)
8. Huang, G., Liu, Z., Weinberger, K.Q.: Densely connected convolutional networks. In: Proceedings of the IEEE Conference on Computer Vision and Pattern Recognition (2017)
9. Hung, C.L., et al.: Longitudinal and circumferential strain rate, LV remodeling, and prognosis after myocardial infarction. J. Am. Coll. Cardiol. 56(22), 1812–1822 (2010)
10. Ionasec, R.I., et al.: Patient-specific modeling and quantification of the aortic and mitral valves from 4D cardiac CT and TEE. IEEE Trans. Med. Imaging 29(9), 1636–1651 (2010)
11. de Kerchove, L., et al.: The role of annular dimension and annuloplasty in tricuspid aortic valve repair. Eur. J. Cardiothorac. Surg. 49(2), 428–437 (2016). https://doi.org/10.1093/ejcts/ezv050. discussion 437–8
12. Kingma, D.P., Ba, J.: Adam: a method for stochastic optimization. In: 3rd International Conference on Learning Representations, pp. 1–15 (2014)
13. Leng, S., et al.: Imaging 4D morphology and dynamics of mitral annulus in humans using cardiac cine MR feature tracking. Sci. Rep. 8(1), 81 (2018)
14. Leng, S., Tan, R.-S., Zhao, X., Allen, J.C., Koh, A.S., Zhong, L.: Fast long-axis strain: a simple, automatic approach for assessing left ventricular longitudinal function with cine cardiovascular magnetic resonance. Eur. Radiol. 30(7), 3672–3683 (2020). https://doi.org/10.1007/s00330-020-06744-6
15. Schuh, A., et al.: Construction of a 4D brain atlas and growth model using diffeomorphic registration. In: Durrleman, S., Fletcher, T., Gerig, G., Niethammer, M., Pennec, X. (eds.) STIA 2014. LNCS, vol. 8682, pp. 27–37. Springer, Cham (2015). https://doi.org/10.1007/978-3-319-14905-9_3
16. Schuster, A., et al.: Cardiovascular magnetic resonance feature-tracking assessment of myocardial mechanics: intervendor agreement and considerations regarding reproducibility. Clin. Radiol. 70(9), 989–998 (2015)
17. Schuster, A., Hor, K.N., Kowallick, J.T., Beerbaum, P., Kutty, S.: Cardiovascular magnetic resonance myocardial feature tracking: concepts and clinical applications. Circ. Cardiovasc. Imaging 9(4), 1–9 (2016)

18. Tomasi, C., Kanade, T.: Detection and tracking of point features. Technical report (1991)
19. Yushkevich, P.A., et al.: User-guided 3D active contour segmentation of anatomical structures: significantly improved efficiency and reliability. Neuroimage **31**(3), 1116–1128 (2006)

4D Flow Magnetic Resonance Imaging for Left Atrial Haemodynamic Characterization and Model Calibration

Xabier Morales[1]([✉]), Jordi Mill[1], Gaspar Delso[2], Filip Loncaric[3,4,5],
Ada Doltra[3,4,5], Xavier Freixa[3,4,5], Marta Sitges[3,4,5], Bart Bijnens[5,6],
and Oscar Camara[1]

[1] Physense, BCN Medtech, Department of Information
and Communications Technologies, Universitat Pompeu Fabra, Barcelona, Spain
xabier.morales@upf.edu
[2] GE Healthcare Spain, Barcelona, Spain
[3] Cardiovascular Institute, Hospital Clínic, Universitat de Barcelona,
Barcelona, Spain
[4] CIBERCV, Instituto de Salud Carlos III (CB16/11/00354), CERCA
Programme/Generalitat de Catalunya, Madrid, Spain
[5] Institut d'investigacions biomèdiques august pi i sunyer (IDIBAPS),
Barcelona, Spain
[6] Institució Catalana de Recerca i Estudis Avançats (ICREA), Barcelona, Spain

Abstract. 4D flow magnetic resonance imaging (MRI) and in-silico simulations have seen widespread use in the characterization of blood flow patterns in the aorta and subsequent calibration of haemodynamic computational models. Computational Fluid Dynamics (CFD) simulations offer a complete overview on local haemodynamics but require patient-specific boundary conditions to provide realistic simulations. Despite the inherent low spatial resolution of 4D flow MRI near the boundaries, it can provide rich haemodynamic details to improve existing simulations. Unfortunately, very few works exist imaging the left atria (LA) with 4D flow MRI due to the acquisition and processing challenges associated to the low magnitude of velocities, the small size of the structure and the complexity of blood flow patterns, especially in pathologies such as atrial fibrillation (AF). The main goal of this study was to develop a computational pipeline to extract qualitative and quantitative indices of LA haemodynamics from 4D flow MRI to: assess differences between normal and AF left atria; and calibrate existing fluid models with improved boundary conditions. The preliminary results obtained in two cases demonstrate the potential of 4D flow MRI data to identify haemodynamic differences between healthy and AF left atria. Furthermore, it can help to bring flow computational simulations to a new level of realism, allowing more degrees of freedom to better capture the complexity of LA blood flow patterns.

Keywords: 4D flow magnetic resonance imaging · Computational Fluid Dynamics · Haemodynamics · Left atrium · Model calibration

© Springer Nature Switzerland AG 2021
E. Puyol Anton et al. (Eds.): STACOM 2020, LNCS 12592, pp. 156–165, 2021.
https://doi.org/10.1007/978-3-030-68107-4_16

1   Introduction

3D time-resolved phase-contrast magnetic resonance imaging (MRI), more colloquially known as 4D flow MRI, has become an invaluable tool for the comprehensive study of in-vivo haemodynamics, allowing for the calculation of a multitude of complex fluid mechanics parameters [12]. However, while the aorta has been the primary focus of such studies, structures such as the left atrium (LA) have been largely left aside [5], with some exceptions ([6] among others). In this regard, characterizing abnormal haemodynamics in the LA has become increasingly relevant due to its role in the development of pathological processes such as LA remodelling or thrombogenesis in patients who suffer from atrial fibrillation (AF) [3].

The low magnitude of blood flow velocities and the high spatio-temporal complexity of LA haemodynamics make 4D flow MRI acquisition and processing of the LA difficult, especially in pathological cases (e.g. with AF). Moreover, the reliability of 4D flow MRI velocities near the vessel wall is hugely dependant on the spatial resolution and velocity encoding range (VENC), as pointed out by several studies [11], which may limit the computation of indices such as wall shear stress (WSS) that could give great insight in atrial pathophysiology.

Computational Fluid Dynamics (CFD) simulations have been extensively used in biomedical applications to characterize blood flow patterns in different organs and systems, providing accurate in-silico indices even at the boundary layer of vessels and structures. Although some CFD-based studies of LA hemodynamics have been published, only a handful of them [7,8,10] use patient-specific functional data, while most of them set the boundary conditions based on literature hampering the realism of the simulations. Some more advanced models [7,10] included atrial wall movement obtained from Computerized Tomography (CT) scans, but at the moment only Mill et al. [8] included flow information that can be reliably extracted from echocardiographic doppler data, while it still may be insufficient to constrain and guide certain CFD simulations. Researchers have employed a large variety of boundary conditions to compensate the lack of patient-specific data, interchanging velocity and pressure as inlets or outlets along with different approaches to mimic wall behaviour, but no consensus has yet been reached on the optimal set up [9]. In this regard, 4D flow MRI offers several advantages when including more accurate hemodynamic data into the computational simulations, as recently demonstrated by Koltukluoğlu [4] in the aorta.

In this study we aimed at processing 4D flow MRI data to extract advanced indices to identify differences in blood flow patterns of healthy and AF left atria. Velocities at pulmonary veins (PV) and the mitral valve (MV), wall shear stress, oscillatory index and vorticity indices were estimated. In addition, 4D flow MRI data was used to re-calibrate CFD models previously personalized with simpler ultrasound data, helping to introduce realistic pressure profiles at different pulmonary veins at choice.

2   Methods

The general pipeline of the study is shown in Fig. 1. First, the 3D + t velocity field and the magnitude image from the 4D flow acquisition were combined to generate a 4D flow magnetic resonance angiogram (MRA), where the LA segmentation was performed. Afterwards, 4D flow data was masked with the LA segmentation to isolate the LA flow. The mean flow in the PV and the MV was then extracted to serve as boundary conditions for the CFD simulations. Finally, the advanced fluid dynamics parameters were computed for both 4D flow acquisitions and their corresponding simulations and compared to each other.

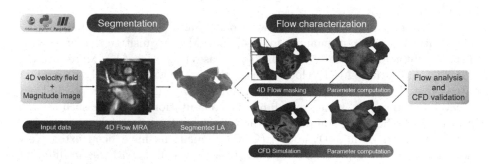

Fig. 1. General pipeline of the project. MRA: Magnetic Resonance Angiography, LA: Left Atrium, CFD: Computational Fluid Dynamics.

2.1   Data

The 4D flow data was provided by the Department of Cardiology at Hospital Clínic de Barcelona, Spain, as part of an ongoing study seeking early prediction of atrial fibrillation in patients at risk but that have not yet developed the disease. The data was acquired through a SIGNA Architect 3.0T MRI scanner (General Electric Medical Systems) resulting in an image of dimension $276 \times 276 \times 176$ with a slice thickness of 2.2 mm with 1.1 mm increment and an in-plane resolution of 1.4063×1.4063 mm. The acquisition was performed in free breathing with retrospective gating. The rest of relevant scan parameters went as follow: Flip angle = 15°, Echo time = 2.216 ms, Repetition time = 4.172 ms. We analyzed a healthy subject and an AF patient in sinus rhythm. Due to different protocols at the time of acquisition two distinct velocity encodings (VENC) were utilized; the healthy subject had a VENC = 80 m/s and the AF patient a VENC = 120 m/s.

Blood flow velocity components and the magnitude or anatomic image were acquired covering the whole cardiac cycle in 30 frames with an approximate temporal resolution of 33 ms. The post-processing of the raw velocity data was performed through Arterys[1] that allowed to save the velocity data in numpy arrays.

---

[1] https://arterys.com/.

2.2 Segmentation

Seeking to segment the LA in the most precise way a 4D flow MR angiography (MRA) was generated allowing the segmentation of the region of interest directly over the 4D flow data. The 4D flow MRA was generated as in Bustamante et al. [1]:

$$4D \text{ MRA} = \frac{1}{N} \sum_{t=1}^{N} M(t) \cdot (V_x{}^2(t) + V_y{}^2(t) + V_x{}^2(t))^\gamma, \tag{1}$$

where $\gamma = 0.2$ is used as a correction factor to enhance the velocity values, ensuring that lower velocities such as those found on the heart chambers also contribute to the image. $M(t)$ is the signal magnitude, V_x, V_y and V_z are the three components of the velocity and N the total number of frames.

We performed the segmentation manually with 3D Slicer[2]. The bulk of the LA could be easily segmented utilizing the MRA image, but the LA appendage (LAA) was hardly visible in 4D flow data due to the low velocities involved. Hence, we registered the 4D flow data to the available contrast-enhanced MRA images with the landmark registration module available in 3D Slicer and segmented the LAA in the registered contrast-enhanced image. Once completed, the segmentation was smoothed and exported as a binary mask. In addition, the segmentation was also exported as a triangular surface mesh for later volumetric mesh generation required by CFD simulations and WSS computation.

2.3 4D Flow Processing

In order to isolate the LA haemodynamics we masked the 4D velocity matrix with the segmented binary mask by cancelling out the velocity in all the voxels outside the mask. Afterwards, the cropped velocity matrix was saved as a series of unstructured VTK files for each time frame, enabling visualization and further processing in Paraview[3]. The processed 4D-flow data can be seen as a set of voxels loaded into Paraview on Fig. 1.

Before proceeding with the computation of the advanced fluid dynamic parameters we interpolated the voxel data to a volumetric tetrahedral mesh generated in gmsh[4] from the aforementioned segmented surface mesh. By doing so, we could directly obtain the normal vectors to the LA surface from the mesh surface elements during the WSS computation. In this regard we made use of a slightly altered version of an Open-Source code[5] that is a Python implementation of the study by Petersson et al. [11]. We calculated the WSS employing the parabolic fitting velocity-based method with an inward normal distance equivalent to the in-plane resolution.

Having calculated the WSS, we also derived parameters such as the Endothelial Cell Activation Potential (ECAP), defined by Di Achille et al. [2], since it

---

[2] https://www.slicer.org/.
[3] https://www.paraview.org/.
[4] https://gmsh.info/.
[5] https://github.com/EdwardFerdian/wss_mri_calculator.

provides useful insight on the risk of thrombus formation (i.e. for higher ECAP values due to low Time-Averaged WSS and high Oscilatory Shear Index, meaning low velocities and high flow complexity).

2.4 CFD Simulations

The velocity profiles for the boundary conditions were directly extracted from the 4D flow data. A set of planes perpendicular to the vessels of interest were placed to calculate the mean of the normal projection of the velocity throughout the cardiac cycle.

The 3D volumetric meshes were generated using gmsh. The meshes of the two analyzed cases had around 700 k elements. All simulations were computed using Ansys Fluent 19 R3[6] (ANSYS Inc, USA) over two heart beats. A dynamic mesh boundary condition was applied based on the MV ring movement defined by Veronesi et al. [13], being diffused with the spring-based method implemented in Ansys. Two different boundary conditions configurations were applied in the simulations: In the first case, which we will refer to as 2PI(pressure inlet), had two pulmonary veins (PV) defined as velocity inlets according to the flow patterns extracted from the 4D data and the 2 PVs as pressure inlets with a generic profile, as in [8]. In the second, named as 4PI, all 4 PVs were set up as pressure inlets. Both configurations had the MV as a velocity outlet extracted from the 4D flow.

3 Results

A quick glance at the pulmonary and mitral 4D flow velocity profiles (the red curves in Fig. 2) immediately reveals significant differences in blood flow patterns between the healthy and the AF case: e.g. the AF patient completely misses the A wave in the MV profile; the large difference in velocity magnitudes (maximum of 0.9 m/s and 0.15 m/s respectively); and bimodal vs monomodal PV velocity patterns. The healthy subject also showcases, independently of the BC configuration, a significant reverse flow at the beginning of the beat, in the PV defined as pressure inlets (dotted lines in Fig. 2) being greater in the 2PI than in the 4PI simulation. Furthermore, while the simulation PV velocity profiles in the healthy subject seem to mimic the pattern imposed by the MV, in the AF patient the flow patterns more closely resemble the 4D Flow MRI data. In addition, differences between the 2PI and 4PI scenarios in the AF patient are also noticeable. Unsurprisingly, those PVs defined as velocity inlet in the 2 PI configuration closely match the 4D inflow. But this is at the expense of the degrees of freedom of the system producing more unstable velocity curves at the rest of PVs. On the contrary, the 4 PI configuration resulted in more balanced velocity curves among all PVs, since there is no velocity imposed at the inlets.

Focusing on Fig. 3, differences can observed in the internal LA haemodynamics of the healthy subject, as the re-circulation seen in 4D flow data near the

---

[6] https://www.ansys.com/products/fluids/ansys-fluent.

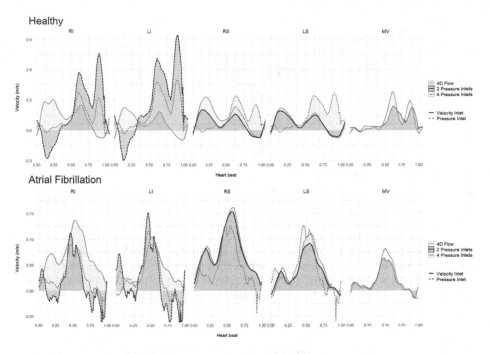

Fig. 2. Mean velocity profile across the 4 pulmonary veins and the mitral valve. The 4D Flow extracted profiles can be seen in red, while the 2 and 4 pressure inlet simulations can be found in green and blue respectively. The pressure inlets are marked as dotted lines. Note: different axis scale is used for each subject to facilitate the visual analysis. (Color figure online)

left superior PV during the end diastole, is almost unnoticeable in Scenario 2PI, and it is even weaker in 4PI. Regarding the AF patient, both simulations seem to be more faithful to the internal 4D Flow LA haemodynamics. In fact, a big re-circulation is generated at the center of the LA due to the crash of the four inflows of the PVs at the end of the ventricular diastole. This can be appreciated in both simulations, although attained speeds are not as high as in 4D data and the vortex in the 4PI case is smaller. Additionally, the superiority of CFD simulations when it comes to the analysis of local haemodynamics it is rather evident due to the spatial resolution constraints of 4D Flow data.

With respect to the ECAP maps in Fig. 4, the simulations seem to be a long way off from the 4D reference, while no significant differences are observed between different configurations. In the healthy subject an ECAP value close to zero was obtained in basically the whole LA except in the LAA, an structure known to be prone to thrombi generation Regarding the AF patient, ECAP values were extremely high, specially in the LAA. Although one might think that the results are hugely overestimated, it has to be noted that the patient represents a very severe case of AF as demonstrated by the abysmal 4D flow velocity curves. Whenever velocities are low, the chance of flow stagnation at the LAA raises and so does the risk of thrombogenesis.

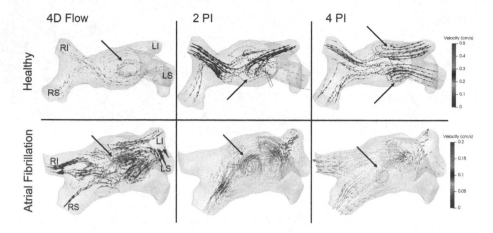

Fig. 3. Velocity streamlines are shown for the end diastole of both the healthy subject and the AF patient. The black arrows signal the main recirculations generated at the center of the left atrium in each case. RS: Right Superior, RI: Right Inferior, LS: Left Superior, LI: Inferior. Different velocity scales are used for visualization purposes.

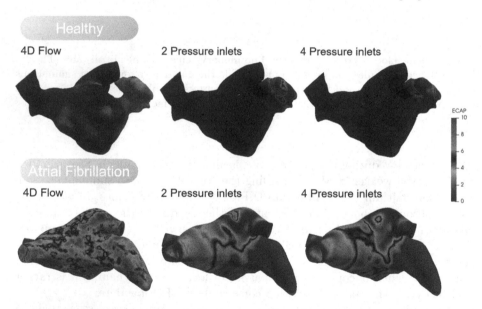

Fig. 4. Endothelial Cell Activation Potential (ECAP) maps of both patients for each of the configurations.

4   Discussion

As described in the previous section, the different PV and MV velocity patterns are easily distinguished in Fig. 2, also allowing the study of temporal dyssynchrony between inlets and outlets, which is not straightforward with echocardiography. On the other hand, ECAP maps obtained from 4D flow MRI are not as conclusive, with high values in LA regions where thrombus are rarely formed in AF, even if the LAA is clearly depicted as a critical area. As the ECAP index is computed from a surrogate of the WSS, it is not surprising the lack of robustness of 4D flow MRI data to estimate wall-related indices due to its issues to capture local haemodynamics.

A quick visual analysis of the PV and MV velocity profiles of the 4D-flow data (red curves in Fig. 2) allows to easily distinguish different blood flow patterns in healthy and AF cases: e.g. the missing A wave for the AF patient in the MV profile; the large difference in velocity magnitudes (maximum of 0.9 m/s and 0.15 m/s for healthy and AF, respectively); and bimodal vs monomodal PV velocity patterns for healthy and AF cases, respectively.

The available 4D flow MRI data has allowed us to test different inlet patient-specific boundary conditions in the fluid simulations that are difficult to obtain with echocardiography. Comparing the two tested scenarios, 4 vs 2 pressure inlets (4PI and 2PI, respectively), the 2PI configuration provided two reliable velocity inlets (see RS and LS veins in Fig. 2) but at the expense of an imbalance in the remaining PV, leading to abnormal haemodynamic patterns in the LA. On the contrary, the 4PI configuration shows a more balanced flow among the different PV, having all the same generic pressure profile. Although it can be paradoxical, in this particular case, more patient-specific boundary conditions do not necessarily lead to better results. It is important to note that the velocities in the velocity inlets do not match 100% with the 4D flow because the planes chosen to measure the average velocity were not located exactly in the inlet in order to see how the flow developed.

The obtained flow simulations were clearly more realistic for the AF patient than for the healthy case, being closer to 4D flow MRI data. The chosen atrial wall movement boundary conditions, i.e. dynamic mesh diffusing from MV movement, has been proven sufficient in the AF case, as demonstrated in [9], since it is the main movement in an LA left without active contraction. However, more sophisticated atrial deformation is required for the healthy case; although there was an atrial volume increase due to MV longitudinal movement, it was not large enough to absorb all flow coming from the PVs. A clear example of this unrealistic situation can be seen in Fig. 2 for the 2PI scenario, where a severe reversal flow is produced in RI and LI (pressure inlets) to fulfill mass conservation due to the imposed high velocities. The 4PI scenario was slightly better since all inlets were pressure-based, but the PV velocity patterns were fully determined by the MV profile. These observations could not be found in the simulated ECAP maps, which were as expected: mostly null ECAP values in the healthy case except some points in the LAA; and large high risk areas in the AF patient, especially in the LAA, which was a severe case with very low velocities and a high probability of flow stagnation.

In the near future we would like to implement a more complete atrial motion when simulating healthy subjects. Using data from dynamic CT images for this purpose could be an option but doesn't properly scale if working towards patient specific simulations. On the other hand, continuous feedback with the clinicians in charge of the 4D equipment is required to optimize the acquisition parameters, which could substantially improve the quality of the images and the resulting hemodynamic indexes. Lastly, we aim at developing a set of data assimilation processes to optimize the parameters of the models and make a better fitting with the data.

5   Conclusion

In conclusion, while there is a lot of room for improvement in regards to the CFD simulations the employment of 4D flow MRI data has been an invaluable tool in setting us on the path to greater realism and could vastly increase our current understanding of blood flow patterns in the whole cardiovascular system. The presented proof-of-concept shows the potential of this technology to explore the complex spatio-temporal nature of haemodynamics in structures such as the left atria in a way unavailable for conventional technologies such as echocardiography. Future work will be devoted to applying the developed pipeline to a more extensive database and the inclusion of a more sophisticated atrial wall motion.

Funding. This work was supported by the Agency for Management of University and Research Grants of the Generalitat de Catalunya under the Grants for the Contracting of New Research Staff Programme - FI (2020 FI_B 00608) and the Spanish Ministry of Economy and Competitiveness under the Programme for the Formation of Doctors (PRE2018-084062), the Maria de Maeztu Units of Excellence Programme (MDM-2015-0502) and the Retos investigación project (RTI2018-101193-B-I00).

Conflict of Interest Statement. The authors declare that the research was conducted in the absence of any commercial or financial relationships that could be construed as a potential conflict of interest.

References

1. Bustamante, M., et al.: Atlas-based analysis of 4D flow CMR: automated vessel segmentation and flow quantification. J. Cardiovasc. Magn. Reson. **17**(1), 87 (2015). https://doi.org/10.1186/s12968-015-0190-5
2. Di Achille, P., Tellides, G., Figueroa, C.A., Humphrey, J.D.: A haemodynamic predictor of intraluminal thrombus formation in abdominal aortic aneurysms. Proc. R. Soc. A Math. Phys. Eng. Sci. **470**(2172), 20140163–20140163 (2014)
3. Glikson, M., et al.: EHRA/EAPCI expert consensus statement on catheter-based left atrial appendage occlusion - an update. EP Europace **22**(2), 184–184 (2019)
4. Koltukluoğlu, T.S.: Fourier spectral dynamic data assimilation: interlacing CFD with 4D flow MRI. In: Shen, D., et al. (eds.) MICCAI 2019. LNCS, vol. 11765, pp. 741–749. Springer, Cham (2019). https://doi.org/10.1007/978-3-030-32245-8_82

5. Lantz, J., et al.: Impact of pulmonary venous inflow on cardiac flow simulations: comparison with in vivo 4D flow MRI. Ann. Biomed. Eng. **47**(2), 413–424 (2019)
6. Markl, M., et al.: Left atrial and left atrial appendage 4D blood flow dynamics in atrial fibrillation. Circ. Cardiovasc. Imaging **9**(9), e004984–e004984 (2016)
7. Masci, A., et al.: The impact of left atrium appendage morphology on stroke risk assessment in atrial fibrillation: a computational fluid dynamics study. Front. Physiol. **9**, 1938–1938 (2019)
8. Mill, J., et al.: Impact of flow dynamics on device-related thrombosis after left atrial appendage occlusion. Can. J. Cardiol. **36**(6), 968.e13–968.e14 (2020)
9. Mill, J., Olivares, A.L., Noailly, J., Bijnens, B., Freixa, X., Camara, O.: Optimal boundary conditions in fluid simulations fo predicting occluder-related thrombus formation in the left atria. In: Computational and Mathematical Biomedical Engineering -CMBE 2019, pp. 256–259 (2019)
10. Otani, T., Al-Issa, A., Pourmorteza, A., McVeigh, E.R., Wada, S., Ashikaga, H.: A computational framework for personalized blood flow analysis in the human left atrium. Ann. Biomed. Eng. **44**(11), 3284–3294 (2016). https://doi.org/10.1007/s10439-016-1590-x
11. Petersson, S., Dyverfeldt, P., Ebbers, T.: Assessment of the accuracy of MRI wall shear stress estimation using numerical simulations. J. Magn. Reson. Imaging **36**(1), 128–138 (2012)
12. Soulat, G., McCarthy, P., Markl, M.: 4D flow with MRI. Annu. Rev. Biomed. Eng. **22**(1), 103–126 (2020)
13. Veronesi, F., et al.: Quantification of mitral apparatus dynamics in functional and ischemic mitral regurgitation using real-time 3-dimensional echocardiography. J. Am. Soc. Echocardiogr. **21**(4), 347–354 (2008)

Segmentation-Free Estimation of Aortic Diameters from MRI Using Deep Learning

Axel Aguerreberry[1]([✉]), Ezequiel de la Rosa[2], Alain Lalande[3,4], and Elmer Fernández[1,5]

[1] Facultad de Ciencias Exactas, Fisicas y Naturales, Universidad Nacional de Córdoba, Cordoba, Argentina
axel.aguerreberry@gmail.com
[2] Department of Computer Science, Technical University of Munich, Munich, Germany
[3] ImViA Laboratory, University of Burgundy, Dijon, France
[4] Medical Imaging Department, University Hospital of Dijon, Dijon, France
[5] CIDIE-CONICET, Universidad Católica de Córdoba, Cordoba, Argentina

Abstract. Accurate and reproducible measurements of the aortic diameters are crucial for the diagnosis of cardiovascular diseases and for therapeutic decision making. Currently, these measurements are manually performed by healthcare professionals, being time consuming, highly variable, and suffering from lack of reproducibility. In this work we propose a supervised deep-learning method for the direct estimation of aortic diameters. The approach is devised and tested over 100 magnetic resonance angiography scans without contrast agent. All data was expert-annotated at six aortic locations typically used in clinical practice. Our approach makes use of a 3D+2D convolutional neural network (CNN) that takes as input a 3D scan and outputs the aortic diameter at a given location. In a 5-fold cross-validation comparison against a fully 3D CNN and against a 3D multiresolution CNN, our approach was consistently superior in predicting the aortic diameters. Overall, the 3D+2D CNN achieved a mean absolute error between 2.2–2.4 mm depending on the considered aortic location. These errors are less than 1 mm higher than the inter-observer variability. Thus, suggesting that our method makes predictions almost reaching the expert's performance. We conclude that the work allows to further explore automatic algorithms for direct estimation of anatomical structures without the necessity of a segmentation step. It also opens possibilities for the automation of cardiovascular measurements in clinical settings.

Keywords: Aortic diameter · Magnetic resonance imaging · Convolutional neural networks

A. Lalande and E. Fernández—These authors contributed equally.

E. Puyol Anton et al. (Eds.): STACOM 2020, LNCS 12592, pp. 166–174, 2021.
https://doi.org/10.1007/978-3-030-68107-4_17

1   Introduction

The determination of the maximum aortic diameters using imaging techniques is crucial for cardiovascular diseases (such as aortic dissection, aortic hematoma, aortic aneurysm and coarctation of the aorta) assessment. The diagnosis and monitoring of these diseases require precise and reproducible measurements of the aortic diameters for optimal treatment decision making [4]. Maximum aortic diameters are crucial not only for deciding surgical procedures, but are also crucial at postoperative stages, where an increase in the aortic diameter can eventually lead to complications.

Currently, aortic markers are manually delineated by physicians, being time demanding and suffering from observers variability and from poor reproducibility. The inter-observer measurement variability of proximal aortic diameters range from 1.6 to 5 mm [4]. Annotations are usually done using 3D imaging software, generally integrated in the scanner's workstations. Given the complexity of the aortic anatomy, a consensus has been established for defining where these annotations should be conducted by defining a set of reference points.

Traditional approaches for automatically estimating the aortic diameters require, as the first step, the segmentation of the vessel, such as the work of Suzuki et al. [7]. Segmentation is typically performed following supervised learning approaches [3], which require full expert delineation of the aorta. The process is not only time demanding, but also observer-biased. As a consequence, the segmentation performance affects the diameters' measurement. Different segmentation-free approaches have been proposed to estimate cardiac parameters. For instance, Xu et al. directly estimate the cardiac cavity, the myocardial area, and the regional wall thickness in cine-MRI sequences [10,11]. Debus and Ferrante make use of a 3D spatio-temporal CNN for regressing left ventricle volumes from MRI scans [2]. More recently, Zhang et al. proposed a model for the direct estimation of coronary artery diameter in X-ray angiograhpy images [12]. It consists of a multi-view module with attention mechanisms for learning spatio-temporal coronary features. Furthermore, machine learning approaches have also been explored for the direct estimation of cardiovascular markers. For instance, Zhen et al. conducts direct and joint bi-ventricular volume estimation by extracting multiple image features and fitting a random forest regression algorithm [13].

In this work, an alternative strategy is proposed for directly estimating aortic diameters bypassing segmentation. We propose a 3D+2D CNN that inputs MRI angiography scans and estimates the maximum aortic diameter at different locations.

2   Materials and Methods

2.1   Dataset

The dataset consists of 100 magnetic resonance angiography studies without contrast agent. A free-breathing, ECG-triggered steady-state free precession sequence was employed. The studies have been acquired with a 1.5T scanner (Siemens Healthineers, Erlangen, Germany) at Dijon University Hospital (France). The voxel resolution varies among scans in between $1.22 \times 1.22 \times 1.50$ mm^3 and $1.56 \times 1.56 \times 1.50$ mm^3.

We follow the conventional clinical process to evaluate and diagnose the aorta. The database annotations have been performed by several different physicians of the hospital, where each scan has been annotated a single time by one of the experts. For each scan, six aortic diameters typically used in clinical practice [5] have been measured: the diameter at the i) Valsalva sinuses, ii) sinotubular junction, iii) ascending aorta, iv) horizontal aorta, v) aortic isthmus and vi) thoracic (descending) aorta. The annotations were done using the 3D data by means of the software *syngo.via* (Siemens Healthineers, Erlangen, Germany). We directly retrieved the MRI scans with their corresponding diameter measurements. It is worth to mention that only the scalar of these measurements was available. Thus, the ground-truth does not include coordinate points or location information of the measurements. Our database included pathological cases with varying severity, with not only normal cases (89%, $n = 89$), but also cases with aortic dissection (4%, $n = 4$) and presence of prosthesis (7%, $n = 7$). Moreover, with the aim of assessing the inter-observer variability for each aortic diameter, a randomly chosen 20% of the database has been re-annotated by an expert in the field (AL), which was blinded to the first observer results.

2.2 Proposed Approach

We propose an end-to-end framework for automatically regressing the aortic diameters from MR angiography data. The approach receives as input the 3D angiography scans and outputs the measured aortic diameter at a specific location. It is worth pointing out that no aortic segmentation mask is required. The same framework and training strategy is used for each of the different aortic diameters.

2.3 Preprocessing

All images undergo a reslicing and affine registration step for homogenizing the scans resolution and for aligning and/or orienting all scans respectively. Images are resliced to a voxel resolution of a $1 \times 1 \times 1$ mm^3. This step, which is done automatically, is crucial for allowing the neural network to properly learn the voxel resolution scale and, hence, properly predict the aortic diameters. The registration is performed by aligning each scan to a single reference image, accounting the reference one with i) high voxel resolution and ii) good image quality (qualitatively assessed). The usage of a single reference scan is enough for our purposes, since we only aim to compensate for acquisition orientation differences across scans.

After image registration, with the aim of reducing GPU memory usage, we automatically select the 16 central slices of the scan which include the whole aorta anatomy. After that, the images are normalized following a z-score normalization.

2.4 CNN Architecture

An overview of the proposed CNN architecture is presented in Fig. 1. The network takes sequences of 16 MRI slices as input and outputs the corresponding aortic diameter at a given landmark. The architecture is fairly straightforward: it consists on 3D convolutional operations, followed by 2D convolutional operations with fully connected layers at the end. The last neuron represents the aortic diameter estimated (there is one specific network for each measurement). *LeakyReLU* activation functions were used across all layers, except for the last fully-connected layer, where linear functions are preferred, in order to map the neural network output to the expected diameter values. Besides, for faster convergence of the model, we include batch-normalization layers after each convolutional operation.

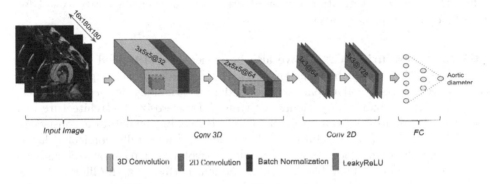

Fig. 1. Proposed CNN architecture. It consists on a 3D + 2D network. Inside each convolution block are displayed the amount and size of the kernels. Conv: convolutional layer. FC: Fully-connected layer.

2.5 Training Procedure

We train the proposed network by minimizing a mean-squared-error loss function. Given an MRI scan x and a ground-truth diameter y (for the sake of simplicity, we consider y to represent any of the six aortic diameters), the loss function is defined as:

$$\mathcal{L}(y, \hat{y}) = \frac{1}{N} \sum_{i=1}^{N} (y_i - \hat{y}_i)^2 \tag{1}$$

where y_i and $\hat{y}_i$ represent the predicted and ground-truth aortic diameters respectively. Optimization of Eq. 1 was conducted using Adam optimizer with mini-batches of size equal to eight.

For regularizing the model we make use of traditional data augmentation transformations (such as Gaussian noise addition and slight image rotations to compensate for inaccuracies during registration). Besides, we introduce a L2 regularization term to Eq. 1. The L2 weighting factor (known as λ in literature) was empirically set to 0.1. Dropout layers [6] were only included after the fully connected layers (dropout factor $p = 0.5$).

2.6 Comparison Against Other Methods

To the best of our knowledge this is the first approach addressing direct aortic diameters estimation. As such, comparing our proposal against reference approaches is not straightforward. We propose two baseline models for comparing our method with: a 3D regression CNN and a 3D multi-resolution CNN.

3D Regression CNN. This network is a fully 3D version of our 3D+2D proposal. All convolutional operations are 3D. We try to keep the network as close as possible to our proposal. Thus, the amount of parameters was kept similar to the 3D+2D CNN. Activation functions are *LeakyRelu* for all layers except for the last fully connected layer.

3D Multiresolution CNN. We also implemented and trained the multiresolution CNN proposed by Tousignant et al. [8]. The approach inputs 3D MRI data and outputs a scalar. Though the model was originally devised for predicting disability outcome in patients with multiple sclerosis, its architecture can be used for addressing our problem. Our adapted implementation consists of three consecutive 3D convolutional blocks, followed by a fully connected layer. Each convolutional block contains parallel convolutional pathways. Thus, the input to the convolutional block is split into four pathways with different kernel sizes. Finally, the pathways are concatenated and fed as the new input to the consecutive block.

2.7 Evaluation Criteria

In order to evaluate the performance of the proposed method, we conduct 5-fold cross-validation experiments for each aortic diameter. As performance metrics we use the mean squared error (Eq. 1), the mean absolute error ($MAE = \frac{1}{N}\sum_{i=1}^{N} | y_i - \widehat{y_i} |$), Pearson correlation coefficient (PCC) and Bland-Altman analysis [1].

3 Results and Discussion

All experiments have been performed in a 12 GB GPU with processor Intel Core i7-7700. Training of the model took on average ~3.9 h. While predicting one aortic diameter took ~8.34 μs, performing its manual annotation took ~50 s per diameter per scan.

In this work we assume that the same model and training strategy could be shared across the different aortic diameters. So, for selecting the optimal model and for conducting hyper-parameter tuning, we first experiment taking into account the annotation at the level of the Valsalva sinuses. This diameter was chosen given the fact that is generally a challenging measurement in clinical practice and, typically, suffers from higher inter-observer variability according to the experts. Once the optimal model and training strategy have been defined, the analysis was extended by training and validating the model over all the remaining aortic landmarks (i.e. sinotubular junction, ascending aorta, etc.).

3.1   Effect of the Registration Approach

We have compared three different preprocessing pipelines using: i) No registration, ii) *rigid* registration and iii) *affine* registration. The results are summarized in Table 1, where it can be appreciated the advantage of using affine registrations in the pipeline.

Table 1. Prediction performance for different registration approaches trained at the Valsalva sinuses' diameter using the model 3D+2D. MAE: Mean absolute error; MSE: mean squared error; PCC: Pearson correlation coefficient.

Registration	MAE (mm)	MSE (mm^2)	PCC
Affine	**2.341**	**8.721**	**0.914**
Rigid	2.839	13.951	0.886
No registration	4.549	32.825	0.778

3.2   Effect of the Chosen Architecture

In Table 2 the performance obtained for each architecture trained at the level of the Valsalva sinuses are shown.

Table 2. Prediction performance for the three compared architectures trained at the Valsalva sinuses' diameter. MAE: Mean absolute error; MSE: mean squared error; PCC: Pearson correlation coefficient.

Model	MAE (mm)	MSE (mm^2)	PCC
3D	3.493	18.151	0.859
3D+2D	**2.341**	**8.721**	**0.914**
Multiresolution	4.269	30.342	0.785

The CNN made up of 3D+2D convolutional layers was consistently better than the other models, by reaching the lowest MAE, the lowest MSE and the highest PCC. These results show that incorporating 2D convolutions after the 3D model improves the overall performance.

3.3   Performance at Each Aortic Location

Table 3 summarizes the inter-observer errors and the performance obtained at each aortic location when using the model and training strategy defined above. The prediction errors for each landmark are very similar, with tiny differences that can be attributed to the complexity of the aortic anatomy at the considered location. Overall, our model predicted with a MAE in between 2.2–2.4 mm. This is less than 1 mm of MAE difference when compared against the experts, which obtained a MAE in between 1.3–1.7 mm. The lowest prediction and inter-observer errors correspond to the diameters at the aortic isthmus location. The highest MAE at prediction time were obtained for the thoracic descending aorta. The Bland Altman plots (Fig. 2) show a low mean bias at all locations without a clear trend towards under/over-estimation of the diameters.

Table 3. Results at each aortic location.

Landmark	Predicted			Inter-observer		
	MAE (mm)	MSE (mm^2)	PCC	MAE (mm)	MSE (mm^2)	PCC
Valsalva sinuses	2.341	8.721	0.914	1.65	3.95	0.961
Sinotubular junction	2.316	8.652	0.923	1.5	3.26	0.980
Ascending aorta	2.283	8.245	0.927	1.75	4.26	0.989
Horizontal aorta	2.270	7.832	0.913	1.55	7.05	0.913
Aortic isthmus	2.251	7.364	0.912	1.35	2.52	0.956
Thoracic aorta	2.435	9.162	0.893	1.45	3.00	0.940

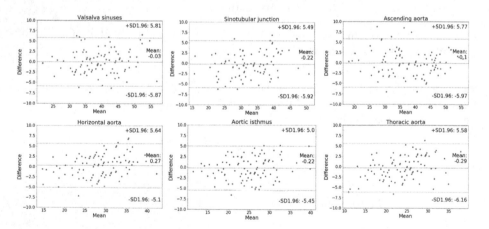

Fig. 2. Bland-Altman plots for each given landmark.

4   Conclusions

In this work we propose an end-to-end, segmentation-free 3D+2D CNN for the direct estimation of aortic diameters. To our knowledge, this is the first work attempting to regress aortic biomarkers bypassing the anatomical aorta segmentation. The method outperforms a 3D CNN and a multiresolution 3D CNN in terms of MAE, MSE and PCC. The inclusion of a 2D CNN in the model showed some sort of information gain when compared against fully 3D CNNs. The proposed model achieved performances almost at the inter-observer variability error. For all the considered aortic diameters, the CNN obtained a MAE less than 1 mm compared to the experts variability (i.e., errors on average up to one voxel of difference).

This work allows to continue researching intelligent algorithms devised to directly estimate anatomical structures without image delineation. The method is general and could be tested over different image modalities, different anatomical structures and markers. Particularly, for the problem addressed in this work, there is much more effort to consider towards its translation to clinical settings. We aim to improve the model performance by making use of problem-specific data augmentation. Random scaling at image and ground truth levels could help to simulate anatomical size variations. Moreover, devising deep-learning interpretable models is a big milestone for this project. We aim to allow anatomical localization into the CNN, such that clinicians could get some knowledge regarding where the measurements have been performed. Another approach could be to compare our method with a landmark detection model. For instance, Vloznt-zos et al. [9] proposed the use of multi-agent reinforcement learning for landmark localization. They proposed a detection approach for multiple landmarks based on a Deep Q-Network architecture. Currently, since we do not have landmark annotations in our database, but only the measurements retrieved from the clinical reports, these approaches are not applicable to our problem. However, landmark localization through reinforcement learning are valid approaches for future research. Finally, another future perspective is exploring multi-task learning models for performing multiple aortic location estimations with a single model.

References

1. Bland, J.M., Altman, D.: Statistical methods for assessing agreement between two methods of clinical measurement. Lancet **327**(8476), 307–310 (1986)
2. Debus, A., Ferrante, E.: Left ventricle quantification through spatio-temporal CNNs. In: Pop, M., et al. (eds.) STACOM 2018. LNCS, vol. 11395, pp. 466–475. Springer, Cham (2019). https://doi.org/10.1007/978-3-030-12029-0_50
3. Duquette, A.A., Jodoin, P.M., Bouchot, O., Lalande, A.: 3D segmentation of abdominal aorta from CT-scan and MR images. Comput. Med. Imaging Graph. **36**(4), 294–303 (2012)

4. Goldstein, S.A., et al.: Multimodality imaging of diseases of the thoracic aorta in adults: from the American society of echocardiography and the European association of cardiovascular imaging: endorsed by the society of cardiovascular computed tomography and society for cardiovascular magnetic resonance. J. Am. Soc. Echocardiogr. **28**(2), 119–182 (2015)
5. Mongeon, F.P., Marcotte, F., Terrone, D.G.: Multimodality noninvasive imaging of thoracic aortic aneurysms: time to standardize? Can. J. Cardiol. **32**(1), 48–59 (2016)
6. Srivastava, N., Hinton, G., Krizhevsky, A., Sutskever, I., Salakhutdinov, R.: Dropout: a simple way to prevent neural networks from overfitting. J. Mach. Learn. Res. **15**(1), 1929–1958 (2014)
7. Suzuki, H., et al.: Automated assessment of aortic and main pulmonary arterial diameters using model-based blood vessel segmentation for predicting chronic thromboembolic pulmonary hypertension in low-dose CT lung screening. In: Medical Imaging 2018: Computer-Aided Diagnosis. International Society for Optics and Photonics, vol. 10575, p. 105750X (2018)
8. Tousignant, A., Lemaître, P., Precup, D., Arnold, D.L., Arbel, T.: Prediction of disease progression in multiple sclerosis patients using deep learning analysis of MRI data. In: International Conference on Medical Imaging with Deep Learning, pp. 483–492 (2019)
9. Vlontzos, A., Alansary, A., Kamnitsas, K., Rueckert, D., Kainz, B.: Multiple landmark detection using multi-agent reinforcement learning. In: Shen, D., et al. (eds.) MICCAI 2019. LNCS, vol. 11767, pp. 262–270. Springer, Cham (2019). https://doi.org/10.1007/978-3-030-32251-9_29
10. Xue, W., Brahm, G., Pandey, S., Leung, S., Li, S.: Full left ventricle quantification via deep multitask relationships learning. Med. Image Anal. **43**, 54–65 (2018)
11. Xue, W., Nachum, I.B., Pandey, S., Warrington, J., Leung, S., Li, S.: Direct estimation of regional wall thicknesses via residual recurrent neural network. In: Niethammer, M., et al. (eds.) IPMI 2017. LNCS, vol. 10265, pp. 505–516. Springer, Cham (2017). https://doi.org/10.1007/978-3-319-59050-9_40
12. Zhang, D., Yang, G., Zhao, S., Zhang, Y., Zhang, H., Li, S.: Direct quantification for coronary artery stenosis using multiview learning. In: Shen, D., et al. (eds.) MICCAI 2019. LNCS, vol. 11765, pp. 449–457. Springer, Cham (2019). https://doi.org/10.1007/978-3-030-32245-8_50
13. Zhen, X., Wang, Z., Islam, A., Bhaduri, M., Chan, I., Li, S.: Direct estimation of cardiac Bi-ventricular volumes with regression forests. In: Golland, P., Hata, N., Barillot, C., Hornegger, J., Howe, R. (eds.) MICCAI 2014. LNCS, vol. 8674, pp. 586–593. Springer, Cham (2014). https://doi.org/10.1007/978-3-319-10470-6_73

Multi-centre, Multi-vendor, Multi-disease Cardiac Image Segmentation Challenge (M&Ms)

Histogram Matching Augmentation for Domain Adaptation with Application to Multi-centre, Multi-vendor and Multi-disease Cardiac Image Segmentation

Jun Ma[✉][iD]

Department of Mathematics, Nanjing University of Science and Technology,
Nanjing, China
junma@njust.edu.cn

Abstract. Convolutional Neural Networks (CNNs) have achieved high accuracy for cardiac structure segmentation if training cases and testing cases are from the same distribution. However, the performance would be degraded if the testing cases are from a distinct domain (e.g., new MRI scanners, clinical centers). In this paper, we propose a histogram matching (HM) data augmentation method to eliminate the domain gap. Specifically, our method generates new training cases by using HM to transfer the intensity distribution of testing cases to existing training cases. The proposed method is quite simple and can be used in a plug-and-play way in many segmentation tasks. The method is evaluated on MICCAI 2020 M&Ms challenge, and achieves average Dice scores of 0.9051, 0.8405, and 0.8749, and Hausdorff Distances of 9.996, 12.49, and 12.68 for the left ventricular, myocardium, and right ventricular, respectively. Our results rank third place in the M&Ms challenge. The code and trained models are publicly available at https://github.com/JunMa11/HM_DataAug.

Keywords: Cardiac segmentation · Deep learning · Domain adaptation · Histogram matching · Generalization

1 Introduction

Accurate segmentation of the left ventricular cavity, myocardium and right ventricle from cardiac magnetic resonance images plays an import role for quantitative analysis of cardiac function, which can be used in clinical cardiology for patient management, disease diagnosis, risk evaluation, and therapy decision [11]. In the recent years, many deep learning-based methods have achieved unprecedented performance [1,3], especially when testing cases have the same distribution as training cases. However, segmentation accuracy can be greatly

E. Puyol Anton et al. (Eds.): STACOM 2020, LNCS 12592, pp. 177–186, 2021.
https://doi.org/10.1007/978-3-030-68107-4_18

degraded when these methods are tested on unseen datasets acquired from distinct MRI scanners or clinical centres [12]. This problem makes it difficult for these methods to be applied consistently across multiple clinical centres, especially when subjects are scanned using different MRI protocols or machines.

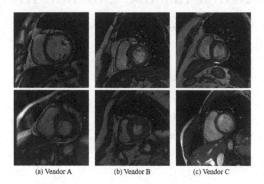

(a) Vendor A   (b) Vendor B   (c) Vendor C

Fig. 1. Visual examples from different vendors. Images from different vendors have significant appearance variations.

The M&Ms challenge is the first international competition to date on cardiac image segmentation combining data from different centres, vendors, diseases and countries at the same time. It evaluates the generalisation ability of machine/deep learning and cross-domain transfer learning techniques for cardiac image segmentation [2]. Figure 1 shows six cardiac MR cases from three vendors. It can be observed that the appearance varies remarkably among different vendors. Thus, how to develop a robust segmentation model that it can generalize to different centers, vendors, and diseases is an important but challenging problem.

Recently, many studies have been proposed to tackle this issue, such as domain adaptation [4,6] and domain generalization [5]. Basically, domain adaptation aims to learn to align source and target domain in a domain-invariant high level feature space, which usually needs few annotated or unlabelled cases from the target domain during training. Domain generalization aims to train a model that it can directly generalize to new domains without need of retraining, which does not use data from the target domain. In practice, both the two popular methods need to modify network architectures or loss functions.

Motivated by a recent study [7] where CNNs are more sensitive to texture and intensity features. We aim to improve the generalization ability of CNNs by transferring the intensity distribution of the target dataset to the source dataset. Specifically, we use histogram matching to bring the intensity appearance of the target dataset to the source dataset. Instead of modifying the network architecture or loss function, our method only augments the training dataset, which is very simple and can be a plug-and-play method to any segmentation tasks.

2   Proposed Method

Histogram matching has been a widely used method, which generates a processed image with a specified histogram [8]. We give a formal introduction of histogram matching as follows. Let S and T denote continuous intensities (considered as random variables) of the source image and the target image, respectively. P_S and P_T denote their corresponding continuous probability density functions (PDF). We can estimate P_S from the source image, and P_T is the target probability density function. Let r be a random variable with the property

$$r = M(S) = (L - 1) \int_0^S P_S(x)dx, \tag{1}$$

where L is the number of intensity levels and x is a dummy variable of integration. Suppose that a random variable w has the property

$$G(T) = (L - 1) \int_0^T P_T(x)dx = r, \tag{2}$$

it than follows from these two equations that $M(S) = G(T)$ and therefore, that T must satisfy the condition

$$T = G^{-1}[M(S)] = G^{-1}(r) \tag{3}$$

Equations (1)–(3) show that an image whose intensity levels have a specified probability density function can be obtained from a given image by using the following four steps:

- Step 1. Obtain P_S from the source image and use Eq. (1) to obtain the value of r.
- Step 2. Use the specified PDF in Eq. (2) to obtain the transformation function $G(T)$.
- Step 3. Obtain the inverse transformation $T = G^{-1}(r)$.
- Step 4. Obtain the output image by first equalizing the input image using Eq. (1); the pixel values in this image are the r values. For each pixel with value r in the equalized images, perform the inverse mapping $T = G^{-1}(r)$ to obtain the corresponding pixel in the output image. When all pixels have been thus processed, the PDF of the output image will be equal to the specified PDF.

Scikit-image [13] has a build-in function *match_histograms*[1] for histogram matching. In this paper, we use histogram matching to augment the training dataset so as to introduce the intensity distribution of the testing set. Specifically, we randomly select image pairs from labelled cases and unlabelled cases, and then transform the intensity distribution of the unlabelled case to labelled case. In this way, we can obtain many new training cases where its intensity distribution is similar to the unlabelled cases. Figure 2 presents some examples of the source images, target images and augmented images.

---

[1] https://scikit-image.org/docs/stable/.

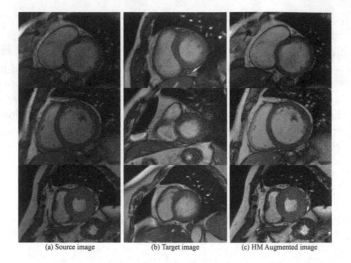

(a) Source image   (b) Target image   (c) HM Augmented image

Fig. 2. Visual examples of the augmented images by histogram matching.

3   Experiments and Results

3.1   Dataset and Training Protocols

Dataset. The M&Ms challenge cohort is composed of 350 patients with hypertrophic and dilated cardiomyopathies as well as healthy subjects. All subjects were scanned in clinical centres in three different countries (Spain, Germany and Canada) using four different magnetic resonance scanner vendors (Siemens, General Electric, Philips and Canon). The training set will contain 150 annotated images from two different MRI vendors (75 each) and 25 unlabelled images from a third vendor. The CMR images have been segmented by experienced clinicians from the respective institutions, including contours for the left (LV) and right ventricle (RV) blood pools, as well as for the left ventricular myocardium (MYO). The 200 test cases correspond to 50 new studies from each of the vendors provided in the training set and 50 additional studies from a fourth unseen vendor, that will be tested for model generalization ability. 20% of these datasets will be used for validation and the rest will be reserved for testing and ranking participants.

During preprocessing, we resample all the images to $1.25 \times 1.25 \times 8$ mm^3 and apply Z-score (mean subtraction and division by standard deviation) to normalize each image. We employ nnU-Net [9] as the default network. The patch size is $288 \times 288 \times 14$, and batch size is 8. We train 2D U-Net and 3D U-Net with five-fold cross validation. Each fold is trained on a TITAN V100 GPU with 1000 epochs. For each fold, we save the best-epoch model[2] and final-epoch model. The code and trained models will be publicly available for research community

---

[2] Best-epoch model stands for the model that can achieves the best Dice on the validation set.

after anonymous review. We declare that the segmentation method has not used any pre-trained models nor additional MRI datasets other than those provided by the organizers.

3.2 Five-Fold Cross Validation Results

Table 1 presents the five-fold cross validation results of the final-epoch models of 2D U-Net and 3D U-Net[3]. Basically, we found that the performance of 3D U-Net is slightly better than 2D U-Net. Figure 3 presents some visual segmentation results of different models. The methods achieve the best Dice scores for left ventricular, while the performance of myocardium is inferior than the left and right ventricular, indicating that the myocardium is more challenging to obtain accurate segmentation results.

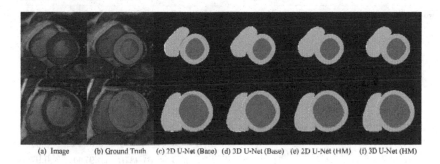

(a) Image   (b) Ground Truth   (c) 2D U-Net (Base) (d) 3D U-Net (Base) (e) 2D U-Net (HM) (f) 3D U-Net (HM)

Fig. 3. Visual examples of segmentation results. Base and HM stand for the baseline dataset and the histogram matching augmented dataset.

3.3 Validation Set Results

The validation set is hidden by challenge organizers. We package our code and model in a singularity container and submit it to the organizers. Due to the limited number of submission tries, we do not submit the baseline models because the models trained on augmented datasets would be better than baseline models. In summary, we submit the following five solutions on the validation set.

- Solution 1. 3D U-Net best-epoch model;
- Solution 2. 3D U-Net final-epoch model;
- Solution 3. Ensemble of 3D U-Net best-epoch model and 2D U-Net best-epoch model;
- Solution 4. Ensemble of 3D U-Net final-epoch model and 2D U-Net final-epoch model;
- Solution 5. Ensemble of the above four solutions;

---

[3] The results corresponding to the best-epoch model are not reported because it over-fits the validation set, where the corresponding Dice score is meaningless.

Table 1. Five-fold cross validation results. It should be noted that The models trained on the default and the augmented dataset are not comparable, because the five-fold splits are different between the two datasets.

Dataset	Model	Fold	LV_Dice	Myo_Dice	RV_Dice
Default Dataset (Baseline)	2D U-Net	0	0.9124	0.8695	0.8842
		1	0.9254	0.8756	0.9074
		2	0.9218	0.8753	0.8945
		3	0.9265	0.8684	0.8845
		4	0.9291	0.8645	0.8874
	3D U-Net	0	0.9256	0.8873	0.8981
		1	0.9344	0.8879	0.9141
		2	0.9399	0.8828	0.9010
		3	0.9372	0.8760	0.8930
		4	0.9308	0.8753	0.8944
Histogram Matching Augmented Dataset	2D U-Net	0	0.9832	0.9639	0.9762
		1	0.9871	0.9729	0.9825
		2	0.9871	0.9747	0.9803
		3	0.9831	0.9683	0.9784
		4	0.9835	0.9735	0.9772
	3D U-Net	0	0.9895	0.9633	0.9760
		1	0.9887	0.9690	0.9826
		2	0.9824	0.9632	0.9719
		3	0.9871	0.9654	0.9796
		4	0.9870	0.9741	0.9762

Table 2 shows the quantitative results of the five solutions on validation set. It can be observed that assembling multiple models may improve Dice, but would degrade the HD and ASSD. We also apply paired T-test between the solution 1 and the other four solutions to show whether their performances are statistically significant different. Surprisingly, the statistical significance level $p > 0.05$ for all comparisons. In other words, the performances of solution 2–4 do not have statistically significant difference compared with the solution 1. It can be found that ensemble more models can obtain sightly better Dice scores, but could degrade the Hausdorff distance. We also compare our results with the recent work [10] that uses GAN for domain adaptation, and our methods obtain better Dice scores for LV, Myo and RV. Finally, we select the solution 1 as our final solution for the hidden testing set.

Table 2. Quantitative results of different solutions on the official hidden validation set. '–' denotes not reported.

Solution	LV			Myo			RV		
	Dice	HD	ASSD	Dice	HD	ASSD	Dice	HD	ASSD
Solution 1	0.9130	**8.089**	1.017	0.8627	**12.19**	**0.710**	**0.8937**	11.88	**0.9078**
Solution 2	0.9131	8.091	**1.015**	0.8627	12.20	**0.710**	0.8935	11.85	0.9101
Solution 3	**0.9166**	8.090	0.958	**0.8660**	12.25	0.734	0.8927	11.35	0.9682
Solution 4	**0.9166**	8.103	0.959	0.8658	12.26	0.736	0.8927	**11.34**	0.9667
Solution 5	**0.9166**	8.099	0.959	0.8659	12.25	0.735	0.8927	11.35	0.9674
GAN [10]	0.903	–	–	0.859	–	–	0.865	–	–

3.4   Testing Set Results

Table 3 presents the average Dice, HD and ASSD for each vendor on testing set. Overall, the performances on vendor C and D are lower than the performances on vendor A and B, because the training set does not have annotated cases from the vendor C and D. It can be observed that the performance on the vendor C is better the performance on the vendor D, especially in HD and ASSD with improvements up to 5 mm, which could[4] demonstrate the effectiveness of our histogram matching data augmentation.

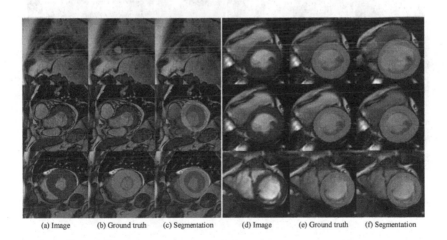

| (a) Image | (b) Ground truth | (c) Segmentation | (d) Image | (e) Ground truth | (f) Segmentation |

Fig. 4. Visual examples of worse cases in vendor A and B.

Worse Cases Analysis. Figure 4 and 5 show some segmentation results with low performance. Basically, it can be observed that most of the segmentation

---

[4] Here we use 'could' because we do not have corresponding testing results of our baseline model where histogram matching data augmentation is not used.

Table 3. Quantitative results on the official hidden testing set

Vendor	LV			Myo			RV		
	Dice	HD	ASSD	Dice	HD	ASSD	Dice	HD	ASSD
A	0.9148	10.78	1.003	0.8435	14.14	0.697	0.8771	12.84	1.142
B	0.9136	7.866	0.971	0.8675	10.14	0.717	0.8792	11.61	1.144
C	0.8943	9.231	1.389	0.8265	11.33	1.066	0.8732	10.82	1.152
D	0.8977	12.11	1.521	0.8243	14.34	0.958	0.8703	15.46	1.513
All	0.9051	9.996	1.221	0.8405	12.49	0.859	0.8749	12.68	1.238

errors occur in the top or bottom of the heart because these regions usually have low contrast and ambiguous boundaries. We find that the worse segmentation results can also have reasonable shapes even for the severe over-segmentation (e.g., Fig. 4 the 2nd row). Some LV segmentation results are significantly smaller or larger than ground truth, which could motivate us to improve our network by imposing size constrain, such as constraining the volume of network outputs to be within a specified range.

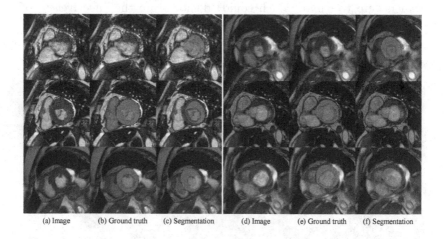

(a) Image   (b) Ground truth   (c) Segmentation   (d) Image   (e) Ground truth   (f) Segmentation

Fig. 5. Visual examples of worse cases in vendor C and D.

Limitation. Although histogram matching is a good pre-processing technique to eliminate intensity distribution differences, it should be noted that histogram matching could cover specific characteristics of other MRI modalities, like for example, LGE MRI. With the histogram changes, relevant informations about scar tissue might be degraded. In the future, we need to verify the effects of histogram matching data augmentation on multi-sequence cardiac MR datasets, such as MyoPS [14].

4   Conclusion

One of the challenging problems of current segmentation CNNs is that the performance would degrade when applying the trained model to a new dataset. In this paper, we introduce histogram matching to augment the training cases that have the similar intensity distributions to the new (unlabelled) dataset set, which is very simple and can be a plug-and-play method to any segmentation CNNs. Based on the quantitative results on the validation set, we believe that our method can be a strong baseline.

Acknowledgement. The authors of this paper declare that the segmentation method they implemented for participation in the M&Ms challenge has not used any pre-trained models nor additional MRI datasets other than those provided by the organizers. We also thank the organizers for hosting the great challenge.

References

1. Bernard, O., et al.: Deep learning techniques for automatic MRI cardiac multi-structures segmentation and diagnosis: is the problem solved? IEEE Trans. Med. Imaging **37**(11), 2514–2525 (2018)
2. Campello, V.M., Palomares, J.F.R., Guala, A., Marakas, M., Friedrich, M., Lekadir, K.: Multi-Centre, Multi-Vendor & Multi-Disease Cardiac Image Segmentation Challenge (2020). https://doi.org/10.5281/zenodo.3715890
3. Chen, C., et al.: Deep learning for cardiac image segmentation: a review. Front. Cardiovasc. Med. **7**, 25 (2020)
4. Chen, C., Dou, Q., Jin, Y., Chen, H., Qin, J., Heng, P.-A.: Robust multimodal brain tumor segmentation via feature disentanglement and gated fusion. In: Shen, D., et al. (eds.) MICCAI 2019. LNCS, vol. 11766, pp. 447–456. Springer, Cham (2019). https://doi.org/10.1007/978-3-030-32248-9_50
5. Dou, Q., de Castro, D.C., Kamnitsas, K., Glocker, B.: Domain generalization via model-agnostic learning of semantic features. In: Advances in Neural Information Processing Systems, pp. 6450–6461 (2019)
6. Ganin, Y., et al.: Domain-adversarial training of neural networks. J. Mach. Learn. Res. **17**(1), 2030–2096 (2016)
7. Geirhos, R., Rubisch, P., Michaelis, C., Bethge, M., Wichmann, F.A., Brendel, W.: Imagenet-trained CNNs are biased towards texture; increasing shape bias improves accuracy and robustness. In: International Conference on Learning Representations (2018)
8. Gonzalez, R.C., Woods, R.E., Eddins, S.L.: Digital Image Processing Using MATLAB. Dorling Kindersley Pvt Ltd (2004)
9. Isensee, F., Jäger, P.F., Kohl, S.A., Petersen, J., Maier-Hein, K.H.: Automated design of deep learning methods for biomedical image segmentation. arXiv preprint arXiv:1904.08128 (2020)
10. Li, H., Zhang, J., Menze, B.: Generalisable cardiac structure segmentation via attentional and stacked image adaptation. arXiv preprint arXiv:2008.01216 (2020)
11. Mach, F., et al.: 2019 ESC/EAS guidelines for the management of dyslipidaemias: lipid modification to reduce cardiovascular risk: the task force for the management of dyslipidaemias of the European Society of Cardiology (ESC) and European Atherosclerosis Society (EAS). Eur. Heart J. **41**(1), 111–188 (2020)

12. Tao, Q., et al.: Deep learning-based method for fully automatic quantification of left ventricle function from cine MR images: a multivendor, multicenter study. Radiology **290**(1), 81–88 (2019)
13. Van der Walt, S., et al.: scikit-image: image processing in Python. PeerJ **2**, e453 (2014)
14. Zhuang, X., Li, L.: Multi-sequence CMR based myocardial pathology segmentation challenge, March 2020. https://doi.org/10.5281/zenodo.3715932

Disentangled Representations for Domain-Generalized Cardiac Segmentation

Xiao Liu[1(✉)], Spyridon Thermos[1], Agisilaos Chartsias[1], Alison O'Neil[1,3], and Sotirios A. Tsaftaris[1,2]

[1] School of Engineering, University of Edinburgh, Edinburgh EH9 3FB, UK
{Xiao.Liu,SThermos,Agis.Chartsias,S.Tsaftaris}@ed.ac.uk
[2] The Alan Turing Institute, London, UK
[3] Canon Medical Research Europe Ltd., Edinburgh, UK
Alison.ONeil@eu.medical.canon

Abstract. Robust cardiac image segmentation is still an open challenge due to the inability of the existing methods to achieve satisfactory performance on unseen data of different domains. Since the acquisition and annotation of medical data are costly and time-consuming, recent work focuses on domain adaptation and generalization to bridge the gap between data from different populations and scanners. In this paper, we propose two data augmentation methods that focus on improving the domain adaptation and generalization abilities of state-to-the-art cardiac segmentation models. In particular, our "Resolution Augmentation" method generates more diverse data by rescaling images to different resolutions within a range spanning different scanner protocols. Subsequently, our "Factor-based Augmentation" method generates more diverse data by projecting the original samples onto disentangled latent spaces, and combining the learned anatomy and modality factors from different domains. Our extensive experiments demonstrate the importance of efficient adaptation between seen and unseen domains, as well as model generalization ability, to robust cardiac image segmentation (The code will be made publicly available).

Keywords: Cardiac image segmentation · Data augmentation · Disentangled factors mixing · Domain adaptation · Domain generalization

1 Introduction

Recent advances in machine and deep learning, as well as in computational hardware, have significantly impacted the field of medical image analysis [14], and more specifically the task of cardiac image segmentation [5]. Since cardiovascular disease continues to be a major public health concern, accurate automatic segmentation of cardiac structures is an important step towards improved analysis

© Springer Nature Switzerland AG 2021
E. Puyol Anton et al. (Eds.): STACOM 2020, LNCS 12592, pp. 187–195, 2021.
https://doi.org/10.1007/978-3-030-68107-4_19

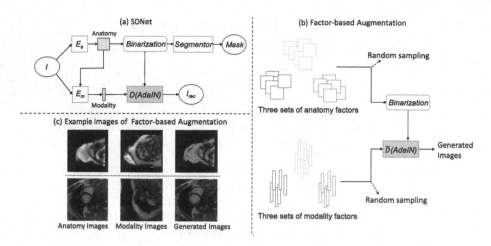

Fig. 1. (a) SDNet: E_a: anatomy encoder, E_m: modality encoder, $D(AdaIN)$: AdaIN decoder. I is the input image to the model and I_{rec} is the output image of the AdaIN decoder *i.e.* the reconstructed image. *Mask* is the predicted segmentation mask for the input image. (b) Illustration of Factor-based Augmentation: $\tilde{D}(AdaIN)$ is a pre-trained AdaIN decoder. (c) Example images produced by Factor-based Augmentation. Anatomy Images provide anatomy factors, Modality Images provide modality factors, and Generated Images are the combination of the anatomy and modality factors.

for prevention and diagnosis. Most existing deep learning methods rely on fully or semi-supervised setups, using pixel-level annotations provided by clinical experts for model training. These methods achieve satisfactory performance in predicting left ventricle (LV), right ventricle (RV), and left ventricular myocardium (MYO) binary masks from unseen data within the same domain. However, such models struggle to achieve similar performance on unseen data from other domains, *e.g.* captured by another scanner or from a different population [19].

Better generalization of models is an open problem in computer vision [6, 7]. Learning a model with good generalization to the target domain is termed *domain adaptation* when labeled samples from the source domain and unlabeled samples from the target domain are given, or *domain generalization* when only labeled samples from the source domain are given.

Annotating more medical images is difficult due to time, cost, and privacy constraints. Therefore, researchers focus on bridging the gap between different domains by learning more general representations from the available data in a semi-supervised manner, *i.e.* domain adaptation [11,20]. Existing methods in cardiac image segmentation achieve domain generalization either by disentangling spatial (anatomy) from imaging (modality) factors [3], or by augmenting the available data [2,4,21]. However, there is limited work [8] focusing on mixing disentangled representations, *i.e.* factors, to augment the original data.

In this paper, we propose two data augmentation methods, termed Resolution Augmentation (RA) and Factor-based Augmentation (FA), which are combined to improve domain adaption and generalization, thus boosting the performance

Fig. 2. Resolution histograms of the M&Ms challenge training data, broken down by vendor (from left to right: Vendors A, B and C).

of state-of-the-art models in Cardiac Magnetic Resonance (CMR) image segmentation. In particular, we use RA to remove the resolution bias by randomly rescaling the training images within a predetermined resolution range, while FA is used to increase diversity in the labeled data through mixing spatial and imaging factors, which we denote as the *anatomy* and *modality* factors, respectively. To extract these factors for FA, we pre-train the SDNet model introduced in [3] using the original (prior to augmentation) data. Experiments on the diverse dataset from the STACOM 2020 Multi-Centre, Multi-Vendor & Multi-Disease Cardiac Image Segmentation Challenge (M&Ms challenge) show the superiority of the proposed augmentation methods when combined with the state-of-the-art U-Net [17] and SDNet models.

2   Method

We train the SDNet model and employ RA and FA to generate a more diverse dataset. The model and FA setup is illustrated in Fig. 1.

2.1   Proposed Augmentation Methods

Resolution Augmentation (RA): It is common in MRI data that the imaging resolution is different for each study due to variation of the scanner parameters. Variation in the imaging resolution can cause the cardiac anatomy to vary significantly in size (*i.e.* area in pixels), beyond normal anatomical variability.

The training dataset contains subjects scanned by scanners from three vendors *i.e.* Vendors A, B and C. In Fig. 2, we show histograms of the training dataset image resolutions (Sect. 3 has more details on the dataset). We observed that the histograms of subjects imaged by scanners from different vendors are distinct from one other *i.e.* this is a bias with respect to the dataset. To reduce this bias, we propose to augment the training dataset such that the resolutions of subjects are equally distributed from $0.954\,\mathrm{mm}^2$ to $2.692\,\mathrm{mm}^2$ per pixel (the minimum and maximum values observed in the data from the 3 vendors), by rescaling the original image to a random resolution in this range. Finally we center-crop the rescaled image to uniform dimensions of 224×224 pixels.

Factor-Based Augmentation (FA): As shown in Fig. 1(a), SDNet decomposes the input image into two latent representations, *i.e.* anatomy and modality factors, respectively. In particular, the anatomy factor contains spatial information about the input, and the modality factor contains non-spatial information only, namely imaging specific characteristics. Ideally, the anatomy factors would not encode the variation caused by different scanners, rather this information would be encoded in the modality factors. Motivated by this, we propose to augment the training dataset by combining different anatomy and modality factors to generate new data.

Considering the three sets of data from Vendors A, B and C, we first pre-train a SDNet model in a semi-supervised manner using the original data. Using this model at inference, we decompose the three sets of data into three sets of anatomy and modality factors. As shown in Fig. 1(b), we randomly sample an anatomy factor from the three anatomy factor sets and a modality factor from the three modality factor sets. A new image can be generated by processing the two factors with the decoder of the pre-trained SDNet model. By repeating this augmentation process, we generate a larger and more diverse dataset that contains segmentation annotations. The segmentation mask of the generated data is the mask of the image providing the anatomy factor, *if* the image is labeled, otherwise the generated data is unlabeled. Some indicative examples of FA are visualized in Fig. 1(c).

2.2 Model Architecture

As depicted in Fig. 1(a), the SDNet model consists of 4 basic modules, namely the anatomy encoder E_a, the modality encoder E_m, the segmentor, and the AdaIN-based decoder $D(AdaIN)$.

The **anatomy encoder** is realized as a U-Net network that consists of 4 downsampling and upsampling convolutional layers coupled with batch normalization [10] layers and ReLU [15] non-linearities. The output feature of the anatomy encoder has 8 channels, while the feature values are thresholded to 0 and 1 by a differentiable rounding operator. By adopting the thresholding constraint and supervision provided by the segmentation masks, the encoded anatomy factor is forced to contain more spatial information about the image.

The **modality encoder** consists of 2 downsampling convolutional layers (4×4 kernel size) that are followed by a global averaging pooling layer, which is used to eliminate the spatial (anatomical) information. The output of the pooling layer is then projected to an 8-dimensional vector (modality factor) using a Multi-Layer Perceptron (MLP) network.

The **segmentor** has 2 convolutional layers using 3×3 kernels, coupled with batch normalization and ReLU activation layers, as well as a 1×1 convolution followed by a channel-wise softmax function. The input to the segmentor is the thresholded anatomy factor. The output target of this segmentor is the ground-truth segmentation masks when the masks are available. In the learning process, this segmentor encourages the anatomy encoder to encode more spatial information about the image such that the segmentor can learn to predict the corresponding masks more efficiently.

Finally, for the **AdaIN decoder**, we use the AdaIN module as in [9], in order to combine the anatomy and modality representations to reconstruct the image. In particular, the decoder consists of 3 convolutional layers (3×3 kernel size) coupled with adaptive instance normalization and ReLU activation layers. A final convolutional layer with 7×7 kernels is used for the reconstruction, followed by a hyperbolic tangent activation function that normalizes the values of the generated image into the [0,1] range. As discussed in [9], the AdaIN decoder normalizes the anatomy factor by firstly applying instance normalization to significantly remove the non-spatial information, then allowing the modality factor to define the new mean and standard derivation of the normalized anatomy factor. In this way, the decoder encourages the anatomy factor to contain spatial information and also force the modality factor to contain non-spatial information. On the other hand, by using AdaIN, the decoder does not simply ignore the modality factor that has a much smaller dimensionality than that of the anatomy factor.

2.3 Objective Function and Model Training

Apart from the original SDNet objective functions, we additionally use focal loss as presented in [13]. Focal loss is widely used in segmentation tasks and helps the model to achieve better accuracy by addressing the class imbalance problem. The augmented overall objective function is defined as:

$$\mathcal{L}_{total} = \lambda_1 \mathcal{L}_{rec} + \lambda_2 \mathcal{L}_{z_{rec}} + \lambda_3 \mathcal{L}_{dice} + \lambda_4 \mathcal{L}_{focal}, \tag{1}$$

where $\mathcal{L}_{rec}$ is the $L1$ distance between the input and the reconstructed image, $\mathcal{L}_{z_{rec}}$ denotes the $L1$ distance between the sampled and the re-encoded modality vector (latent regression), $\mathcal{L}_{dice}$ is the segmentation dice loss [18], and $\mathcal{L}_{focal}$ is the segmentation focal loss. Since we train SDNet in both fully supervised and semi-supervised setups, we set the hyperparameters $\lambda_1 = \lambda_2 = \lambda_3 = \lambda_4 = 1$ when training using labeled data, while when training using unlabeled data we set $\lambda_1 = \lambda_2 = 1$ and $\lambda_3 = \lambda_4 = 0$.

3 Experiment

3.1 Dataset Description and Preprocessing

We train and validate our proposed method on the M&Ms challenge dataset of 350 subjects. Some subjects have hypertrophic and dilated cardiomyopathies (and some are healthy) but disease labels are not provided. Subjects were scanned in clinical centres in 3 different countries using 4 different magnetic resonance scanner vendors *i.e.* Vendor A, B, C and D in this paper. The M&Ms

Challenge training dataset contains 75 **labeled** subjects scanned using technology from Vendor A, 75 **labeled** subjects scanned by Vendor B, and 25 **unlabeled** subjects scanned by Vendor C.[1] The M&Ms challenge test dataset contains 200 subjects (50 from each vendor, including the seen 25 unlabeled subjects from Vendor C). From these, 80 subjects (20 from each vendor) are used to validate the model (results reported in Table 1) and the rest will be used for final challenge rankings. Subjects scanned by Vendor D were unseen during model training.

For each subject, we have 2D cardiac image slice acquisitions captured at multiple phases in the cardiac cycle (including end-systole and end-diastole). We train on the 2D images independently because we adopt 2D convolution neural networks in our model. Following [1], we normalize the training data intensity distribution to a Gaussian distribution with a mean of 0 and a standard deviation of 1. Overall, there are 1,738 pairs of images and masks from A and 1,546 pairs of images and masks from B. Apart from these labeled images, there are 47,346 unlabeled images from A, B and C.

3.2   Model Training

Models are trained using the Adam [12] optimizer with an initial learning rate of 0.001. To stabilize the training, we set the new learning rate to 10% of the previous rate when the validation Dice similarity coefficient between the predicted and the ground-truth masks does not improve for 2 consecutive epochs. Following the original training setting of SDNet, we set the batch size to 4 and train the model for 50 epochs. All models are implemented in PyTorch [16] and trained using an NVidia 1080 Ti GPU.

3.3   Results and Discussions

To verify the effectiveness of the proposed augmentation methods, we train 5 models for the purpose of ablation: **a)** the U-Net model using samples augmented with RA (U-Net+RA), **b)** the original SDNet model trained in a fully supervised fashion (FS SDNet), **c)** the fully supervised SDNet model using samples augmented with RA (FS SDNet+RA), **d)** the SDNet model trained in a semi-supervised fashion, using samples augmented with RA (SS SDNet+RA), and **e)** the semi-supervised SDNet using samples augmented with both FA and RA (SS SDNet+RA+FA). We train the fully supervised models with labeled samples from vendors A and B, while in the semi-supervised scenario we use all available data (labeled and unlabeled) for training. Table 1 reports the per vendor average Dice scores.

Since we are only allowed to validate our model a limited number of times, the results are not comprehensive, therefore we will do pairwise analysis below.

---

[1] We denote subjects scanned by scanners of Vendor A and B as labeled. However, it is notable that only the end-diastole and end-systole phases are labeled.

Table 1. Evaluation of the 5 models. Average Dice similarity coefficients are reported. Bold numbers denote the best performances across the 5 models. LV: left ventricle, MYO: left ventricular myocardium and RV: right ventricle.

Model	Vendor A			Vendor B			Vendor C			Vendor D		
	LV	MYO	RV	LV	MYO	RV	LV	MYO	RV	LV	MYO	RV
U-Net+RA	0.900	0.829	0.811	0.937	0.877	0.907	0.856	0.837	0.852	0.762	0.651	0.503
FS SDNet	0.901	0.837	0.822	0.942	0.877	0.920	0.851	0.826	**0.853**	0.734	0.618	0.474
FS SDNet+RA	0.905	0.846	0.828	**0.945**	0.886	**0.921**	0.855	0.843	0.843	**0.819**	0.749	**0.624**
SS SDNet+RA	**0.909**	**0.854**	**0.841**	**0.945**	**0.887**	0.916	0.843	0.837	0.813	0.811	**0.752**	0.553
SS SDNet+RA+FA	**0.909**	0.846	0.778	0.939	0.882	0.909	**0.863**	**0.847**	0.843	0.812	0.712	0.498

Does RA Help? By inspection of the **FS SDNet** and **FS SDNet+RA** results, we observe that for Vendor A and B, RA helps to achieve better overall performance compared to the respective baseline and RA substantially boosts performance on the unseen Vendor D. Subsequently, for Vendor C, the models have similar performance in the LV class, while FS SDNet+RA achieves the best performance in the MYO class (0.843 Dice score). However, the models using RA do not perform well in the RV class. In the case of Vendor C (no labelled examples), we know that the sample resolution of 24 out of 25 subjects made available at training time from Vendor C is $2.203\,\mathrm{mm}^2$, and the resolution of 32 out of 75 Vendor A training subjects is around $2.000\,\mathrm{mm}^2$, thus we argue that the resolution bias for Vendor A and Vendor C are already similar in the original data, and becoming invariant to this bias by rescaling the Vendor A data does not further help the performance on Vendor C.

Does FA Help? Inspection of **SS SDNet+RA** and **SS SDNet+RA+FA** results shows that FA has mixed performance. It performs well on Vendor C for the scenario of domain adaptation (target samples available but unlabeled), where Vendor C is one of the vendors providing modality factors. However, FA performs poorly on Vendor D for domain generalization (unseen target), where Vendor D is not involved in the augmentation process.

Best Model: Out of our two augmentation methods, we observe that RA has the most reliable performance, and we choose to submit the **FS SDNet+RA** model for final evaluation in the M&Ms challenge. We can see by comparing **U-Net+RA** with **FS SDNet+RA** that our chosen architecture of SDNet has good generalization performance compared to a standard U-Net, and this is particularly evident for the unseen vendor D, supporting our choice of the disentangled SDNet architecture even when FA is not employed.

Improvements to FA: Satisfyingly, our FA method yields a benefit for Vendor C (whilst RA did not give a significant benefit to Vendor C). However, FA does not give considerable improvements on domain generalization. It may be that improvements to the model training would further enhance the benefit of FA. On the other hand, we should also distinguish the effect of simply training on reconstructed images from the effect of FA. In future, we believe that this method can be extended to achieve better domain generalization once we can manipulate the factors realistically to generate out-of-distribution samples.

4   Conclusion

In this paper, two data augmentation methods were proposed to address the domain adaptation and generalization problems in the field of CMR image segmentation. In particular, a geometry-related augmentation method was introduced, which aims to remove the scale and resolution bias of the original data. Further, the second proposed augmentation method aims bridge the gap between populations and data captured by different scanners. To achieve this, the original data is projected onto a disentangled latent space and generates new samples by combining disentangled factors from different domains. The presented experimental results showcase the contribution of the geometry-based method to CMR image segmentation through boosting the domain generalization, while also demonstrating the contribution of the disentangled factors mixing method to the domain adaptation.

Acknowledgement. The authors of this paper declare that the segmentation method they implemented for participation in the M&Ms challenge has not used any pre-trained models nor additional MRI datasets other than those provided by the organizers.

This work was supported by the University of Edinburgh, the Royal Academy of Engineering and Canon Medical Research Europe by a PhD studentship to Xiao Liu. This work was partially supported by the Alan Turing Institute under the EPSRC grant EP/N510129/1. We thank Nvidia for donating a Titan-X GPU. S.A. Tsaftaris acknowledges the support of the Royal Academy of Engineering and the Research Chairs and Senior Research Fellowships scheme.

References

1. Baumgartner, C.F., Koch, L.M., Pollefeys, M., Konukoglu, E.: An exploration of 2D and 3D deep learning techniques for cardiac MR image segmentation. In: Pop, M., et al. (eds.) STACOM 2017. LNCS, vol. 10663, pp. 111–119. Springer, Cham (2018). https://doi.org/10.1007/978-3-319-75541-0_12
2. Chaitanya, K., Karani, N., Baumgartner, C.F., Becker, A., Donati, O., Konukoglu, E.: Semi-supervised and task-driven data augmentation. In: Chung, A.C.S., Gee, J.C., Yushkevich, P.A., Bao, S. (eds.) IPMI 2019. LNCS, vol. 11492, pp. 29–41. Springer, Cham (2019). https://doi.org/10.1007/978-3-030-20351-1_3
3. Chartsias, A., et al.: Disentangled representation learning in cardiac image analysis. Med. Image Anal. **58**, 101535 (2019)
4. Chen, C., et al.: Realistic adversarial data augmentation for MR image segmentation. In: Martel, A.L., et al. (eds.) MICCAI 2020. LNCS, vol. 12261, pp. 667–677. Springer, Cham (2020). https://doi.org/10.1007/978-3-030-59710-8_65

5. Chen, C., et al.: Deep learning for cardiac image segmentation: a review. Front. Cardiovasc. Med. **7**(25), 1–33 (2020)
6. Ganin, Y., Lempitsky, V.: Unsupervised domain adaptation by backpropagation. In: Proceedings of the ICML, pp. 1180–1189 (2015)
7. Ghifary, M., Bastiaan Kleijn, W., et al.: Domain generalization for object recognition with multi-task autoencoders. In: Proceedings of the ICCV, pp. 2551–2559 (2015)
8. Hsu, W., et al.: Disentangling correlated speaker and noise for speech synthesis via data augmentation and adversarial factorization. In: Proceedings of the ICASSP, pp. 5901–5905 (2019)
9. Huang, X., Belongie, S.: Arbitrary style transfer in real-time with adaptive instance normalization. In: Proceedings of the ICCV, pp. 1501–1510 (2017)
10. Ioffe, S., Szegedy, C.: Batch normalization: accelerating deep network training by reducing internal covariate shift. In: Proceedings of the ICML, pp. 448–456 (2015)
11. Jiang, J., et al.: Tumor-aware, adversarial domain adaptation from CT to MRI for lung cancer segmentation. In: Frangi, A.F., Schnabel, J.A., Davatzikos, C., Alberola-López, C., Fichtinger, G. (eds.) MICCAI 2018. LNCS, vol. 11071, pp. 777–785. Springer, Cham (2018). https://doi.org/10.1007/978-3-030-00934-2_86
12. Kingma, D.P., Ba, J.: Adam: a method for stochastic optimization. In: ICLR (2015)
13. Lin, T.Y., Goyal, P., Girshick, R., He, K., Dollár, P.: Focal loss for dense object detection. In: Proceedings of the ICCV, pp. 2980–2988 (2017)
14. Litjens, G., Kooi, T., Bejnordi, B.E., Setio, A.A.A., et al.: A survey on deep learning in medical image analysis. Med. Image Anal. **42**, 60–88 (2017)
15. Nair, V., Hinton, G.E.: Rectified linear units improve restricted Boltzmann machines. In: Proceedings of the ICML, pp. 807–814 (2010)
16. Paszke, A., Gross, S., Massa, F., Lerer, A., et al.: PyTorch: an imperative style, high-performance deep learning library. In: Proceedings of the NeurIPS, pp. 8026–8037 (2019)
17. Ronneberger, O., Fischer, P., Brox, T.: U-Net: convolutional networks for biomedical image segmentation. In: Navab, N., Hornegger, J., Wells, W.M., Frangi, A.F. (eds.) MICCAI 2015. LNCS, vol. 9351, pp. 234–241. Springer, Cham (2015). https://doi.org/10.1007/978-3-319-24574-4_28
18. Sudre, C.H., Li, W., et al.: Generalised dice overlap as a deep learning loss function for highly unbalanced segmentations. In: Proceedings of the ML-CDS, pp. 240–248 (2017)
19. Tao, Q., Yan, W., Wang, Y., et al.: Deep learning-based method for fully automatic quantification of left ventricle function from cine MR images: a multivendor, multicenter study. Radiology **290**(1), 81–88 (2019)
20. Yang, J., Dvornek, N.C., Zhang, F., Zhuang, J., Chapiro, J., et al.: Domain-agnostic learning with anatomy-consistent embedding for cross-modality liver segmentation. In: Proceedings of the ICCV Workshops (2019)
21. Zhao, A., et al.: Data augmentation using learned transformations for one-shot medical image segmentation. In: Proceedings of the CVPR, pp. 8543–8553 (2019)

A 2-Step Deep Learning Method with Domain Adaptation for Multi-Centre, Multi-Vendor and Multi-Disease Cardiac Magnetic Resonance Segmentation

Jorge Corral Acero[1]([✉]), Vaanathi Sundaresan[2], Nicola Dinsdale[2], Vicente Grau[1], and Mark Jenkinson[2,3,4]

[1] Institute of Biomedical Engineering, Department of Engineering Science,
University of Oxford, Oxford OX3 7DQ, UK
jor.corral@eng.ox.ac.uk
[2] Wellcome Centre for Integrative Neuroimaging, Oxford Centre for Functional MRI
of the Brain, Nuffield Department of Clinical Neurosciences,
University of Oxford, Oxford OX3 9DU, UK
[3] Australian Institute of Machine Learning, Department of Computer Science,
University of Adelaide, Adelaide, SA 5005, Australia
[4] South Australian Health and Medical Research Institute (SAHMRI),
North Terrace, Adelaide, SA 5000, Australia

Abstract. Segmentation of anatomical structures from Cardiac Magnetic Resonance (CMR) is central to the non-invasive quantitative assessment of cardiac function and structure. Anatomical variability, imaging heterogeneity and cardiac dynamics challenge the automation of this task. Deep learning (DL) approaches have taken over the field of automatic segmentation in recent years, however they are limited by data availability and the additional variability introduced by differences in scanners and protocols. In this work, we propose a 2-step fully automated pipeline to segment CMR images, based on DL encoder-decoder frameworks, and we explore two domain adaptation techniques, domain adversarial training and iterative domain unlearning, to overcome the imaging heterogeneity limitations. We evaluate our methods on the MICCAI 2020 Multi-Centre, Multi-Vendor & Multi-Disease Cardiac Image Segmentation Challenge training and validation datasets. The results show the improvement in performance produced by domain adaptation models, especially among the seen vendors. Finally, we build an ensemble of baseline and domain adapted networks, that reported state-of-art mean Dice scores of 0.912, 0.857 and 0.861 for left ventricle (LV) cavity, LV myocardium and right ventricle cavity, respectively, on the externally validated Challenge dataset, including several unseen vendors, centers and cardiac pathologies.

Keywords: Segmentation · Cardiac magnetic resonance · Deep learning · Domain adaptation · Data harmonization

J. Corral Acero and V. Sundaresan–Equally contributed to the work.
V. Grau and M. Jenkinson–Equally co-supervised the work.

E. Puyol Anton et al. (Eds.): STACOM 2020, LNCS 12592, pp. 196–207, 2021.
https://doi.org/10.1007/978-3-030-68107-4_20

1 Introduction

Cardiac magnetic resonance (CMR) imaging allows for an accurate non-invasive quantification of cardiac function and structure [1–3]. The valuable information that it provides in cardiovascular disease management has been repeatedly shown [4–9]. Nevertheless, the anatomical variability and the intrinsic complexity of cardiac dynamics and geometry represent a challenge, and CMR analysis remains manual in clinical practice [3, 10–12]. Deep learning (DL) has revolutionized medical image analysis in recent years, progressing towards automatic segmentation. However, these approaches are challenged by (1) the limited data availability due to technical, ethical and financial constraints along with confidentiality issues, especially for specific pathologies and (2) the imaging heterogeneity introduced by anatomical variability and the use of different scanners and protocols [13–16].

Various techniques have been proposed for improved and robust performance of DL models under limited training data. The most common technique is data augmentation, including affine and non-affine transformations to populate the space of shape variability [17–20]. Other techniques incorporate modifications in the network architecture such as reducing the number of network parameters to avoid overfitting or using a 2-step segmentation pipeline that improves class balance by zooming into the anatomical region of interest (ROI) [21, 22]. Scanner-induced variation in datasets echoes 'domain shift' which leads to the performance of models trained on data from one scanner (the source domain) degrading when applied to another (the target domain). Techniques such as transfer learning [23] have proven successful. However, fine-tuning for every unseen domain is still required. Solutions based on domain adaptation (DA) techniques, which aim to create a single feature representation for all domains which is invariant to domain but discriminative for the task of interest [24], have been proposed to overcome this limitation. One of the successful DA approaches is *domain adversarial training of neural networks* (DANN) [25]. DANN assumes that predictions must be based on domain-invariant features, and jointly optimizes the underlying features to simultaneously minimize the loss of the label predictor and maximize the loss of a domain predictor. An alternative approach was proposed using an iterative framework [26] of *domain unlearning* (DU) for adversarial adaptation, creating a classifier which is more uniformly uninformative across domains [27]. This framework has been successfully applied to harmonization of MRI for brain tissue segmentation [28].

The MICCAI 2020 Multi-Centre, Multi-Vendor & Multi-Disease Cardiac Image Segmentation Challenge (M&Ms 2020), including a dataset with a wide variety in centers and scanner vendors and granting external validation, provides a benchmark to assess the generalizability of the segmentation algorithms [29]. We propose a fully automated pipeline to segment CMR short-axis (SAx) stacks, based on the 2-step DL framework that was awarded 1st prize in the LVQuan19 Challenge [22, 30]. We evaluate the segmentation performance of the proposed method, and explore two DA techniques, DANN and DU, for multi-vendor and multi-center applications on the M&Ms 2020 Challenge. The results show that the DA techniques contribute to performance improvement and suggest more experimentation for improvement on unseen vendors applications. A final model ensemble was built, achieving state-of-art performance for both seen and unseen vendors.

2 Materials and Methods

In essence, the proposed pipeline first locates where the heart is (1st neural network - NN) and then focuses on that ROI to produce a fine segmentation (2nd NN). Three alternative implementations are proposed for this latter step. The pipeline is coupled with pre-processing and post-processing stages, as illustrated on Fig. 1.

2.1 Data

We deployed and evaluated our methods on the publicly available M&Ms 2020 Challenge dataset [29], which involves 4 acquisition centers and consists of 150 annotated SAx images from two different MRI vendors, A and B (75 each), and 25 unannotated images from a third vendor, C. The annotated images are segmented only at end-diastole (ED) and end-systole (ES), including left ventricular (LV) cavity, LV endocardium and right ventricle (RV) cavity masks. The image resolution varies from 192×240 to 384×384 pixels, and the pixel spacing, from 0.977 to 1.625 mm. An additional testing set of 200 cases from 6 acquisition centers, including vendors A, B, C and a fourth unseen vendor, D, in equal proportions, is held by the organizers for external validation (20% of the set – 40 cases) and final challenge results (80% of the set). Details on the acquisition and annotation protocols can be checked in [29].

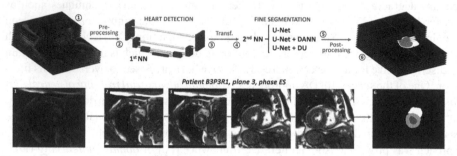

Fig. 1. Scheme of the proposed segmentation pipeline (top) along with a step-by step explanatory illustration of its application on patient B3P3R1 (bottom).

2.2 1st NN: Pre-processing, Heart Detection and Transformation

The 3 most central ES and ED slices of each patient are normalized in intensity and resolution (pre-processing – Fig. 1, step 1–2) and fed to the 1stNN that segments the LV epicardium and RV endocardium (heart detection – Fig. 1, step 2–3), so that a ROI can be defined (transformation – Fig. 1, step 3–4). The pre-processing consists of a linear interpolation to a 2D pre-defined template (256×256 pixels with symmetrical 1.12 mm pixel spacing) centered in the image, followed by an intensity clipping (10th and 96th percentiles, selected empirically to avoid clipping intensities in important structures) and a minmax normalization to 0–255 (uint8). The 2D mass center coordinates of the

LV and RV are provided by the 6 prediction outputs of the 1st NN, and a transformation is calculated from these to align the images to a smaller pre-defined template (144×144 pixels, 1.32 mm pixel spacing – balance between small size and high resolution while image large enough to cover the heart) centered in the heart. This same transformation is applied to each slice of the ES and ED SAx stacks. Details of the architecture and implementation of the 1st NN, based on a standard U-Net, can be checked in [22].

2.3 2nd NN: Fine Segmentation and Postprocessing

The transformed images, linear interpolated to the smaller template and normalized in intensities (5th and 93th percentiles clipping – cutoffs empirically adjusted to the intensities of the structures in the ROI space), are fed to the 2nd NN for a fine segmentation of the LV endo- and epicardium and the RV endocardium (Fig. 1, step 4–5). The LV myocardium is computed as the region between LV endo and epicardium, and the predictions are back interpolated to the original resolution and rearranged into the original 3D setting in a post-processing step (Fig. 1, step 5–6). This final step also accounts for segmentation quality enhancement (binarizing predictions, filling holes and removing stray clusters). Three architectures are proposed for the 2nd NN as described below:

U-Net. Baseline Method
A U-Net, fine-tuned for cardiac segmentations is applied for baseline comparisons. Details are available in [19].

Domain Adversarial Training of Neural Networks (DANN)
The DANN model, proposed by [25], consists of a feature extractor network with a domain predictor and label predictor. The gradient-reversal layer, placed between the feature extractor and the domain predictor, reverses the gradient direction during backpropagation and maximizes the domain prediction loss, thus minimizing the shift between the domains, while simultaneously making the model discriminative towards the main task of label prediction (Fig. 2).

Domain Unlearning (DU)
The DU model[1] [28] is based on the iterative unlearning framework [26] for adversarial adaptation, which rather than using a gradient reversal layer, optimizes two opposing loss functions in a sequential manner: one to maximize the performance of a label predictor given the fixed feature representation, and another to update the feature representation in order to maximally confuse the domain classifier (Fig. 2).

2.4   Implementation Details for DANN and DU

For the label predictor of the DANN and DU models, we used the Adam optimizer, with the same parameters as used in the baseline U-Net method. For the domain predictor in DANN and DU, we trained with the Momentum optimizer (with a momentum value of 0.9) and the Adam optimizer respectively. We used a batch size of 16, with a learning

---

[1] DU code available at: https://github.com/nkdinsdale/Unlearning_for_MRI_harmonisation.

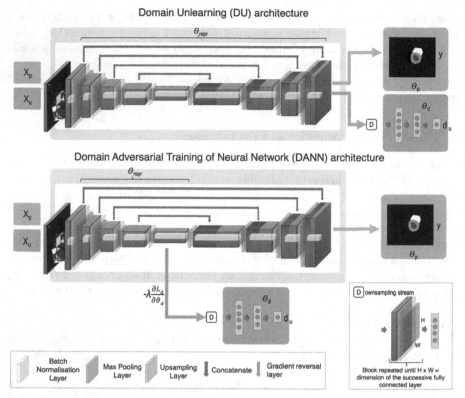

Fig. 2. Architectures of DU (top) and DANN (bottom) models. The models consist of the feature extractor, label predictor and domain predictor with corresponding training parameters θ_{repr}, θ_p and θ_d. The models take input features X_p and input domain information X_u and predicts output labels y, while unlearning output domains d_u. The DU model updates the label predictor, feature extractor and domain predictor in a sequential manner, while in DANN, label prediction and domain unlearning occur simultaneously.

rate of 10^{-3} and 10^{-5} for DANN and DU respectively. In addition, in DU we used a beta value of 1000 (a factor used for weighting the domain confusion a). These training hyper parameters were chosen empirically, and a criterion based on a patience (number of epochs to wait for progress on validation set) value of 25 epochs was used to determine model convergence (early stopping). Both models were run on an NVIDIA Tesla V100 GPU. The DANN and DU models took around 50 s and 45 s per epoch respectively for training.

2.5 Data Augmentation

Standard data augmentation based on random rotations (0 to 360°), translation (1st NN: ±20 mm. 2nd NN: ±8 mm) and flipping was applied during the training of any of the architectures. For each epoch, all the original training images are randomly transformed as described prior to be fed to the NN. Thus, the originals are never used directly, and each instance fed to the NN is only seen once.

2.6 Performance Evaluation

Volumetric Dice scores between predicted and original segmentations in the native image space are calculated to assess the fine segmentation performance of the three proposed approaches.

3 Experiments and Results

The 1st NN was trained on the 150 labeled images (A and B vendors) following a 5-fold cross-validation strategy with a training-validation-testing split ratio of 107-13-30 images. The 5 resulting models were used in combination (majority voting) for the prediction of unseen test subjects.

To assess the quantitative performance of the 3 proposed methods on the 2nd NN fine segmentation, for seen and unseen vendors, we carried out 2 experiments:

- Training on the 150 labeled images from vendors A and B (A+B) and cross-validation evaluation on the same set, A+B (seen vendors assessment).
- Training on vendors A and C (A+C), and evaluation on vendor B (unseen vendor assessment). The baseline method was only trained on A, since unlabeled data (vendor C) cannot be incorporated into training, as done in the other 2 methods.

Table 1. Volumetric Dice results from the two proposed experiments. Data presented as median (interquartile range). The best performing models are highlighted in bold. Significant differences (p < 0.05) of mean values with respect to the baseline are marked with an asterisk (*).

Train	Test	Method	LV cavity	LV myocardium	RV cavity
A+B	5-fold validation A+B	BASELINE	0.927 (0.886−0.954)	0.845 (0.819−0.879)	0.889 (0.833−0.927)
		DANN	0.948 (0.918−0.964)*	0.873 (0.849−0.896)*	0.926 (0.897−0.943)*
		DU	**0.953 (0.927−0.965)***	**0.878 (0.855−0.897)***	**0.928 (0.898−0.945)***
A+C	B	BASELINE	0.913 (0.869−0.956)	0.829 (0.800−0.853)	**0.895 (0.841−0.930)**
		DANN	**0.927 (0.875−0.957)***	**0.838 (0.813−0.866)***	0.891 (0.845−0.936)*
		DU	0.914 (0.864−0.956)	0.835 (0.811−0.862)*	0.892 (0.842−0.934)

In both cases, and for all the 3 methods, we followed a 5-fold cross-validation approach with the split ratio described above. The results are shown in Table 1. While both DA techniques significantly improved the segmentation performance for seen vendors, the improvement in performance decreased for unseen vendors and only DANN reports a significantly better performance on average (*t*-test) than the baseline in segmenting any structure. Indeed, baseline is superior in RV median Dice under unseen vendors.

Finally, the 3 models were trained on A+B+C (only A+B for baseline), following the 5-fold training and split ratio described, and submitted to the M&Ms 2020 Challenge for external validation on unseen samples from vendors A, B, C and D. In addition, an ensemble with the combination in majority voting of the 15-resulting baseline, DANN and DU models (5folds x 3methods) was submitted. The external results, shown in Table 2, are consistent with our experiments and confirm the improvement in performance created by DA, especially for unseen vendors. The differences were not significant due to the

Table 2. Volumetric Dice results from the external validation set, stratified by vendor (median and interquartile range) and aggregated (average). The best performing models are highlighted in bold.

Method	Vendor	LV cavity	LV myocardium	RV cavity
Baseline	A	0.916 (0.857−0.957)	0.835 (0.808−0.879)	0.862 (0.713−0.896)
	B	0.951 (0.922−0.964)	0.864 (0.852−0.902)	0.942 (0.860−0.951)
	C	0.909 (0.828−0.952)	0.860 (0.827−0.888)	**0.870 (0.837−0.921)**
	D	0.911 (0.880−0.935)	0.828 (0.793−0.860)	0.871 (0.803−0.897)
	Average	0.904	0.844	0.852
DANN	A	**0.936 (0.876−0.961)**	**0.863 (0.827−0.889)**	0.878 (0.823−0.909)
	B	0.951 (0.923−0.968)	**0.885 (0.856−0.906)**	0.942 (0.895−0.957)
	C	0.903 (0.810−0.956)	0.860 (0.828−0.886)	0.844 (0.803−0.894)
	D	0.914 (0.874−0.944)	0.843 (0.798−0.869)	0.873 (0.816−0.908)
	Average	0.908	0.856	0.853
DU	A	0.918 (0.876−0.957)	0.857 (0.826−0.875)	0.878 (0.822−0.909)
	B	0.954 (0.922−0.970)	0.879 (0.853−0.902)	0.939 (0.890−0.951)
	C	0.898 (0.828−0.955)	0.862 (0.821−0.891)	0.856 (0.832−0.908)
	D	**0.926 (0.863−0.947)**	0.840 (0.807−0.870)	**0.882 (0.811−0.909)**
	Average	0.906	0.854	0.855
Ensemble	A	0.923 (0.866−0.961)	0.856 (0.825−0.886)	**0.886 (0.807−0.920)**
	B	**0.954 (0.923−0.967)**	0.882 (0.855−0.907)	**0.946 (0.884−0.956)**
	C	**0.911 (0.834−0.956)**	**0.865 (0.839−0.896)**	0.863 (0.833−0.920)
	D	0.922 (0.894−0.944)	**0.844 (0.801−0.867)**	0.878 (0.820−0.904)
	Average	**0.912**	**0.857**	**0.861**

size of the sets (only 10 patients per vendor). The best performance on average was obtained with the ensemble model in any of the structures.

Based on the previous results, the ensemble was submitted for final testing and M&Ms 2020 Challenge participation (remaining 160 unseen cases). Results along with a best and a worst case, provided by the Challenge organizers, in comparison with the rest of the participants are depicted below in Table 3 and Figs. 3 and 4. A drop in performance in comparison to the external validation set is observed, especially in vendor B.

Table 3. Volumetric Dice results from the final testing set, stratified by vendor (median and inter-quartile range) and aggregated (average). Significant differences ($p < 0.05$) of mean values with respect to the ensemble external validation results are marked with an asterisk (*).

Method	Vendor	LV cavity	LV myocardium	RV cavity
Ensemble	A	0.929 (0.887−0.960)	0.853 (0.810−0.875)	0.892 (0.832−0.925)
	B	0.933 (0.887−0.957)	0.866 (0.838−0.893)	0.903 (0.854−0.936)*
	C	0.924 (0.882−0.946)	0.845 (0.806−0.868)	0.887 (0.824−0.929)
	D	0.914 (0.875−0.929)	0.831 (0.784−0.867)	0.892 (0.847−0.922)
	Average	**0.902**	**0.833***	**0.863**

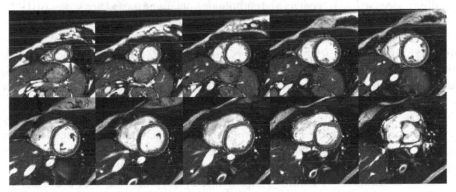

Fig. 3. Best case sample from the test set. Patient G1K1V3, vendor B, ED phase. Ground truth in green, ensemble prediction in red. Dice results of 0.979, 0.891 and 0.954 for LV cavity, LV myocardium and RV cavity, respectively. (Color figure online)

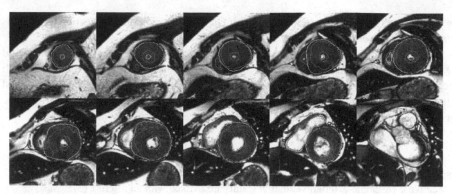

Fig. 4. Worst case sample from the test set. Patient L8N7Z0, vendor B, ES phase. Ground truth in green, ensemble prediction in red. Dice results of 0.653, 0.795 and 0.771 for LV cavity, LV myocardium and RV cavity, respectively. (Color figure online)

4   Discussion and Conclusions

Our contributions in this work are as follows: (1) we have developed a 2-step DL-based completely automatic pipeline for segmenting LV cavity, LV myocardium and RV cavity on CMR images; (2) we have implemented two DA techniques, domain adversarial training and domain unlearning, to overcome image heterogeneity and scanner-related variations; and (3) we have externally validated our methods on a challenging dataset, including unseen scanners, centers and pathologies, obtaining mean Dice scores of 0.902, 0.833 and 0.863 for LV cavity, LV myocardium and RV cavity, respectively.

The results show that the implemented DA methods significantly improve segmentation performance on seen labelled vendors. However, the 2 proposed DA methods only achieved comparable or slightly superior performance to baseline on the unseen vendors, and so the potential improvement that these 2 methods offer remains inconclusive for multi-vendor applications.

The DA model improvements for seen vendors, especially with DU, illustrated by the A+B experiment in Table 1 can be explained by the fact that both DA models learn a generic feature representation (by unlearning domain-specific information) enabling information from both domains to be incorporated. This essentially increases the size of the training dataset and its variability, improving the label prediction. We believe that the DU model provided better results compared to DANN since the minmax optimization (with gradient reversal) could occasionally become unstable and get trapped in local maxima. On the other hand, feature representations are updated to unlearn domain-specific information with each iteration in DU models. Thus, the label predictor is consistently improved using these feature representations, making the DU model comparatively more stable.

However, despite training DANN and DU models with data from vendor C, there was a lack of improvement on that vendor when compared to baseline results, according to the external validation experiments. A plausible explanation is the additional variability introduced by unseen pathological cases and centers within the same vendor domain. Further investigation is required, after the test data is released, to understand this behavior and to propose changes accordingly via domain shift analysis and hyperparameter fine-tuning.

Comparing the A+B and A+C experiment results, the relatively small drop in performance in the baseline models on the prediction of vendor B suggests the robustness of the baseline method for unseen domain prediction. This is further confirmed by the external validation experiments (see predictions on vendor C and D, Table 2). Credit for this robustness should be given to the 2-step implementation, which normalizes for orientation, resolution and appearance and crops the original images to the ROI, levering the label imbalance (background vs structure) towards segmentation enhancement.

In an ensemble model, different models are trained with different initializations and each of them learn different aspects of the data. For instance, while the baseline model learns the salient features for label prediction, DANN and DU learn domain invariant features, still retaining task specific features. Hence, we achieved better overall results in the external validation dataset by combining the predictions using an ensemble model compared to any of the individual models. The dice scores obtained by this model ensemble in the validation dataset (see Table 2) are comparable to those reported in [21], trained on 4875 theoretically healthy subjects, while we only used 150 labelled subjects for training and evaluated on data from multiple centers and vendors, including pathological cases.

While according to the M&Ms 2020 Challenge description the external validation (40 cases) and the final testing set (160 cases) come from the same test set, including samples from different vendors, centers and pathologies in the same proportion, significant differences in performance are obtained for the myocardium structure and for vendor B. A plausible explanation is that the size of the validation set might not be large enough to cover the variability of the test set. We believe that the discrepancies between the testing and validation set across participants is worth exploring to check this hypothesis and to improve the representativeness of the sets in future editions of the Challenge.

To sum up, we have evaluated our method and explored various DA techniques on heterogeneous data from multiple vendors and centers, achieving the best performance with the ensemble of DA and baseline models. We have made our methods publicly available as Singularity containers that may serve as an independent testing tool for the community[2]. Future directions include further exploration of DA models for improving model generalizability.

Acknowledgements. The authors of this paper declare that the segmentation method they implemented for participation in the M&Ms challenge has not used any pre-trained models nor additional MRI datasets other than those provided by the organizers. This work was supported by the EU's Horizon 2020 research and innovation program under the Marie Sklodowska-Curie (g.a. 764738 to JCA). VS is supported by Wellcome Centre for Integrative Neuroimaging, which has core funding from the Wellcome Trust (203139/Z/16/Z). ND is supported by the Engineering and Physical Sciences Research Council and Medical Research Council (EP/L016052/1). MJ is

---

[2] The CMR SAx Segmentation Models presented in this paper are available at: https://figshare.com/articles/software/A_2-step_Deep_Learning_method_with_Domain_Adaptation_for_Multi-Centre_Multi-Vendor_and_Multi-Disease_Cardiac_Magnetic_Resonance_Segmentation/13140554.

supported by the National Institute for Health Research (NIHR), Oxford Biomedical Research Centre (BRC) and Wellcome Trust (215573/Z/19/Z). The computational aspects of this research were supported by the Wellcome Trust Core Award (203141/Z/16/Z) and the NIHR Oxford BRC. The views expressed are those of the authors and not necessarily those of the NHS, the NIHR or the Department of Health.

References

1. Stokes, M.B., Roberts-Thomson, R.: The role of cardiac imaging in clinical practice. Aust. Prescr. **40**(4), 151–155 (2017)
2. Peterzan, M.A., Rider, O.J., Anderson, L.J.: The role of cardiovascular magnetic resonance imaging in heart failure. Card. Fail. Rev. **2**(2), 115–122 (2016)
3. Peng, P., Lekadir, K., Gooya, A., Shao, L., Petersen, S.E., Frangi, A.F.: A review of heart chamber segmentation for structural and functional analysis using cardiac magnetic resonance imaging. Magn. Reson. Mater. Phys., Biol. Med. **29**(2), 155–195 (2016). https://doi.org/10.1007/s10334-015-0521-4
4. Dall'Armellina, E.: From recognized to novel quantitative CMR biomarkers of LV recovery. JACC Cardiovasc. Imaging **10**(9), 1000–1002 (2017)
5. Desch, S., et al.: Cardiac magnetic resonance imaging parameters as surrogate endpoints in clinical trials of acute myocardial infarction. Trials **12**, 204 (2011)
6. Reindl, M., Eitel, I., Reinstadler, S.J.: Role of cardiac magnetic resonance to improve risk prediction following acute ST-elevation myocardial infarction. J. Clin. Med. **9**(4), 1041 (2020)
7. Bulluck, H., Dharmakumar, R., Arai, A.E., Berry, C., Hausenloy, D.J.: Cardiovascular magnetic resonance in acute ST-segment–elevation myocardial infarction. Circulation **137**(18), 1949–1964 (2018)
8. Anderson, J.L., Morrow, D.A.: Acute myocardial infarction. N. Engl. J. Med. **376**(21), 2053–2064 (2017)
9. Zhang, S., Li, L., Guo, W., Wang, L., Yan, X., Zhang, C.: Observation on left ventricular remodeling in acute myocardial infarction. Zhonghua nei ke za zhi **38**(2), 107–109 (1999)
10. Petersen, S.E., et al.: Reference ranges for cardiac structure and function using cardiovascular magnetic resonance (CMR) in Caucasians from the UK Biobank population cohort. J. Cardiovasc. Magn. Reson. **19**(1), 18 (2017)
11. Suinesiaputra, A., et al.: Quantification of LV function and mass by cardiovascular magnetic resonance: multi-center variability and consensus contours. J. Cardiovasc. Magn. Reson. **17**(1), 63 (2015)
12. Ibanez, B., et al.: 2017 ESC Guidelines for the management of acute myocardial infarction in patients presenting with ST-segment elevation: the task force for the management of acute myocardial infarction in patients presenting with ST-segment elevation of the European Socie. Eur. Heart J. **39**(2), 119–177 (2017)
13. Rumsfeld, J.S., Joynt, K.E., Maddox, T.M.: Big data analytics to improve cardiovascular care: promise and challenges. Nat. Rev. Cardiol. **13**(6), 350–359 (2016)
14. Shameer, K., Johnson, K.W., Glicksberg, B.S., Dudley, J.T., Sengupta, P.P.: Machine learning in cardiovascular medicine: are we there yet? Heart **104**(14), 1156–1164 (2018)
15. Litjens, G., et al.: A survey on deep learning in medical image analysis. Med. Image Anal. **42**, 60–88 (2017)
16. Hosny, A., Parmar, C., Quackenbush, J., Schwartz, L.H., Aerts, H.J.W.L.: Artificial intelligence in radiology. Nat. Rev. Cancer **18**(8), 500–510 (2018)

17. Hussain, Z., Gimenez, F., Yi, D., Rubin, D.: Differential data augmentation techniques for medical imaging classification tasks. In: AMIA Symposium, vol. 2017, pp. 979–984 (2017). Accessed 20 Jan 2019
18. Perez, L., Wang, J.: The effectiveness of data augmentation in image classification using deep learning. CoRR, vol. abs/1712.0 (2017)
19. Corral Acero, J., Zacur, E., Xu, H., Ariga, R., Bueno-Orovio, A., Lamata, P., Grau, V.: SMOD - data augmentation based on statistical models of deformation to enhance segmentation in 2D cine cardiac MRI. In: Coudière, Y., Ozenne, V., Vigmond, E., Zemzemi, N. (eds.) FIMH 2019. LNCS, vol. 11504, pp. 361–369. Springer, Cham (2019). https://doi.org/10.1007/978-3-030-21949-9_39
20. Lemley, J., Bazrafkan, S., Corcoran, P.: Smart augmentation learning an optimal data augmentation strategy. IEEE Access **5**, 5858–5869 (2017)
21. Bai, W., et al.: Automated cardiovascular magnetic resonance image analysis with fully convolutional networks. J. Cardiovasc. Magn. Reson. **20**(1), 65 (2018)
22. Corral Acero, J., et al.: Left ventricle quantification with cardiac MRI: deep learning meets statistical models of deformation. In: Pop, M., et al. (eds.) STACOM 2019. LNCS, vol. 12009, pp. 384–394. Springer, Cham (2020). https://doi.org/10.1007/978-3-030-39074-7_40
23. Pan, S.J., Yang, Q.: A survey on transfer learning. IEEE Trans. Knowl. Data Eng. **22**(10), 1345–1359 (2010)
24. Ben-David, S., Blitzer, J., Crammer, K., Kulesza, A., Pereira, F., Vaughan, J.W.: A theory of learning from different domains. Mach. Learn. **79**(1–2), 151–175 (2010)
25. Ganin, Y., Ustinova, E., Ajakan, H., Germain, P., Larochelle, H., et al.: Domain-adversarial training of neural networks. J. Mach. Learn. Res. (2015)
26. Hoffman, J., Tzeng, E., Darrell, T., Saenko, K.: Simultaneous deep transfer across domains and tasks. In: Csurka, G. (ed.) Domain Adaptation in Computer Vision Applications. ACVPR, pp. 173–187. Springer, Cham (2017). https://doi.org/10.1007/978-3-319-58347-1_9
27. Alvi, M., Zisserman, A., Nellåker, C.: Turning a blind eye: explicit removal of biases and variation from deep neural network embeddings. In: Leal-Taixé, L., Roth, S. (eds.) ECCV 2018. LNCS, vol. 11129, pp. 556–572. Springer, Cham (2019). https://doi.org/10.1007/978-3-030-11009-3_34
28. Dinsdale, N.K., Jenkinson, M., Namburete, A.I.L.: Unlearning Scanner Bias for MRI Harmonisation in Medical Image Segmentation, pp. 15–25. Springer, Cham (2020). https://doi.org/10.1007/978-3-030-59713-9_36
29. Campello, V.M.: Multi-centre, multi-vendor & multi-disease cardiac image segmentation challenge (M&Ms) MICCAI 2020. https://www.ub.edu/mnms/. Accessed 1 July 2020
30. Yang, G., Hua, T., Xue, W., Shuo, L.: Left ventricle full quantification challenge MICCAI 2019. https://lvquan19.github.io/. Accessed 1 July 2020

Random Style Transfer Based Domain Generalization Networks Integrating Shape and Spatial Information

Lei Li[1,2,3], Veronika A. Zimmer[3], Wangbin Ding[4], Fuping Wu[2,5], Liqin Huang[4], Julia A. Schnabel[3], and Xiahai Zhuang[2(✉)]

[1] School of Biomedical Engineering, Shanghai Jiao Tong University, Shanghai, China
zxh@fudan.edu.cn
[2] School of Data Science, Fudan University, Shanghai, China
[3] School of Biomedical Engineering and Imaging Sciences, King's College London, London, UK
[4] College of Physics and Information Engineering, Fuzhou University, Fuzhou, China
[5] School of Management, Fudan University, Shanghai, China

Abstract. Deep learning (DL)-based models have demonstrated good performance in medical image segmentation. However, the models trained on a known dataset often fail when performed on an unseen dataset collected from different centers, vendors and disease populations. In this work, we present a random style transfer network to tackle the domain generalization problem for multi-vendor and center cardiac image segmentation. Style transfer is used to generate training data with a wider distribution/heterogeneity, namely domain augmentation. As the target domain could be unknown, we randomly generate a modality vector for the target modality in the style transfer stage, to simulate the domain shift for unknown domains. The model can be trained in a semi-supervised manner by simultaneously optimizing a supervised segmentation and a unsupervised style translation objective. Besides, the framework incorporates the spatial information and shape prior of the target by introducing two regularization terms. We evaluated the proposed framework on 40 subjects from the M&Ms challenge2020, and obtained promising performance in the segmentation for data from unknown vendors and centers.

Keywords: Domain generalization · Random style transfer · Multi-center and multi-vendor

1 Introduction

Quantification of volumetric changes during the cardiac cycle is essential in the diagnosis and therapy of cardiac diseases, such as dilated and hypertrophic cardiomyopathy. Cine magnetic resonance image (MRI) can capture cardiac motions

© Springer Nature Switzerland AG 2021
E. Puyol Anton et al. (Eds.): STACOM 2020, LNCS 12592, pp. 208–218, 2021.
https://doi.org/10.1007/978-3-030-68107-4_21

and presents clear anatomical boundaries. For quantification, accurate segmentation of the ventricular cavities and myocardium from cine MRI is generally required [16].

Recently, deep learning (DL) based methods have obtained promising results in cardiac image segmentation. However, the generalization capability of the DL-based models is limited, i.e., the performance of a trained model will be drastically degraded for unseen data. It is mainly due to the existence of the domain shift or distribution shift, which is common among the data collected from different centers and vendors. To tackle this problem, domain adaptation and domain generalization are two feasible schemes in the literature. Domain adaptation algorithms normally assume that the target domain is accessible in the training stage. The models will be trained using labeled source data and a few labeled or unlabeled target data, and learn to align the source and target data onto a domain-invariant feature space. For example, Yan et al. adopted an unpaired generative adversarial network (GAN) for vendor adaptation without using manual annotation of target data [11]. Zhu et al. proposed a boundary-weight domain adaptive network for prostate MRI segmentation [15]. Domain generalization, on the other hand, is more challenging due to the absence of target information. It aims to learn a model that can be trained using multi-domain source data and then directly generalized to an unseen domain without retraining. In clinical scenarios, it is impractical to retrain a model each time for new domain data collected from new vendors or centers. Therefore, improving the generalization ability of trained models would be of great practical value.

Existing domain generalization methods can be summarized into three categories, 1) domain-invariant feature learning approaches, which focus on extracting task-specific but domain-invariant features [6,7]; 2) model-agnostic meta-learning algorithms, which optimize on the meta-train and meta-test domain split from the available source domain [4,8]; 3) data augmentation methods, which enlarge the scope of distribution learned by models for model generalization [2,14]. For example, Volpi et al. utilized adversarial data augmentation to iteratively augment the dataset with examples from a fictitious target domain [10]. Zhang et al. propose a deep stacked data augmentation (BigAug) approach to simulate the domain shift for domain generalization [14]. Chen et al. employed some data normalization and augmentation for cross-scanner and cross-site cardiac MR image segmentation [2]. However, data augmentation in the source domain is not enough to mitigate the generalization gap, as the domain shift is not accidental but systematic [11].

In this work, we propose novel random style transfer based domain generalization networks (STDGNs) incorporating spatial and shape information. The framework focuses on improving the networks' ability to deal with the modality-level difference instead of structure-level variations, i.e., domain augmentation. Specifically, we first transfer the source domain images into random styles via the style transfer network [12]. The generated new domain and known source domain images are then employed to train the segmentation network. We combine the two steps by connecting a style transfer and style recover network, to

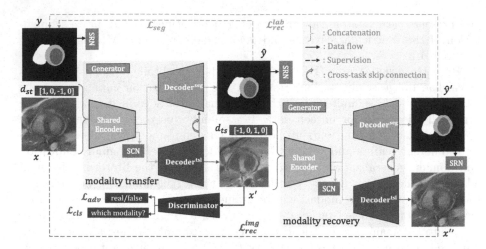

Fig. 1. The proposed STDGNs incorporating shape and spatial information. The generator is consist of one shared encoder and two decoders for segmentation and style translation. Note that the skip connections between the encoder and two decoders are omitted here.

constrain the two domains with the same structure but different styles. With the domain augmentation strategy, the segmentation network is supposed to be domain-invariant and therefore can be adapted to unknown target domains. The framework can be trained in a semi-supervised manner, as it consists of an unsupervised style transfer task and a supervised segmentation task. Besides, we employ a shape reconstruction network (SRN) to include shape priors as a constrain, and utilize a spatial constrain network (SCN) to consider the large variety of 2D slices from various slice positions of 3D cardiac images [13].

2   Method

Figure 1 provides an overview of the proposed framework. The proposed network is a cascaded modified U-Net consisting of two decoders for cardiac segmentation and style transfer, respectively. The generators consist of a shared encoder and two specific decoders, while the discriminator includes two parts, i.e., real/ fake discriminator and modality discriminator. Sect. 2.1 presents the unified segmentation and style transfer framework based on GANs. To use unlabeled data, semi-supervised learning is employed, followed by a cross-task attention module to efficiently connect the segmentation and style transfer tasks (see Sect. 2.2). In Sect. 2.3, two losses based on the shape and spatial information are introduced as regularization terms.

2.1   Deep Domain Generalization Framework

Given N source domains $\mathcal{D}_s = \{\mathcal{D}_1, \mathcal{D}_2, \cdots, \mathcal{D}_N\}$, we aim to construct a model $f(X)$,

$$f(X) \rightarrow Y, X \in \mathcal{D}_s \cup \mathcal{D}_t, \tag{1}$$

where $\mathcal{D}_t$ are unknown target domains, and $X, Y \in \mathbb{R}^{1 \times H \times W}$ denote the image set and corresponding segmentation set. To achieve this, we propose to employ the style transfer to randomly generate new domains $\mathcal{D}'_s$ with the same structure as the input source domain. Specifically, we assign a one-hot encoding modality vector $v \in [0,1]^{N+1}$ for source domain image $x \in \mathbb{R}^{1 \times H \times W}$, and then randomly generate a modality vector $v' \in [0,1]^{N+1}$ for target domain image. The modality difference vector $d_{st} = v' - v, d_{st} \in [-1,0,1]^k$ is broadcasted from $\mathbb{R}^k$ to $\mathbb{R}^{k \times H \times W}$ and then concatenated with the input image.

As Fig. 1 shows, in the modality transfer phase, the generator G translates the source domain image x into a new target domain image x' and predicts the segmentation $\hat{y}$, i.e., $G(x, d_{st}) \rightarrow (x', \hat{y})$. The generated x' is then inputted into a discriminator $D = D_{src}, D_{cls}$, where D_{src} recognizes it as real/ fake and D_{cls} predicts its modality. In the modality recovery phase, $d_{ts} = v - v'$ is concatenated with x' as the input of G, to obtain the reconstructed original source image x'' and its label $\hat{y}'$ for structure preservation. Therefore, the adversarial loss of the GAN is adopted as follows,

$$\mathcal{L}_{adv}^D = \mathbb{E}_x \left[D_{src}(x) \right] - \mathbb{E}_{x'} \left[D_{src}(x') \right] - \lambda_{gp} \mathbb{E}_{\hat{x}} \left[(\|\nabla_{\hat{x}} D_{src}(\hat{x})\|_2 - 1)^2 \right], \tag{2}$$

$$\mathcal{L}_{adv}^G = \mathbb{E}_{x'} \left[D_{src}(x') \right], \tag{3}$$

where $\hat{x}$ is sampled from the real and fake images and λ_{gp} is the balancing parameter of gradient penalty. For style transfer, D_{cls} learns to classify x into its source modality v based on $\mathcal{L}_{cls}^r$, while G tries to transfer source modality into target modality v' via $\mathcal{L}_{cls}^f$ with

$$\mathcal{L}_{cls}^r = \mathbb{E}_{x,v} \left[-\log D_{cls}(v \mid x) \right], \tag{4}$$

$$\mathcal{L}_{cls}^f = \mathbb{E}_{x',v'} \left[-\log D_{cls}(v' \mid x') \right]. \tag{5}$$

The cycle consistency losses include $\mathcal{L}_{rec}^{lab}$ and $\mathcal{L}_{rec}^{img}$. $\mathcal{L}_{rec}^{img}$ is defined as

$$\mathcal{L}_{rec}^{img} = \mathbb{E}_{x,x''} \left[\|x - x''\|_1 \right], \tag{6}$$

where x'' is the predicted source domain image. $\mathcal{L}_{rec}^{lab}$ is formulated in Sect. 2.3. Hence, the final overall loss to optimize the generator and discriminator are defined as,

$$\min_D \mathcal{L}_D = -\mathcal{L}_{adv}^D + \lambda_{cls} \mathcal{L}_{cls}^r, \tag{7}$$

$$\min_G \mathcal{L}_G = -\mathcal{L}_{adv}^G + \lambda_{cls} \mathcal{L}_{cls}^f + \lambda_{rec}^{img} \mathcal{L}_{rec}^{img} + \lambda_{seg} \mathcal{L}_{seg} + \lambda_{rec}^{lab} \mathcal{L}_{rec}^{lab}, \tag{8}$$

where all λ are balancing parameters, and $\mathcal{L}_{seg}$ and $\mathcal{L}_{rec}^{lab}$ are introduced in Sect. 2.3.

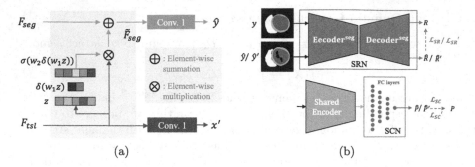

Fig. 2. Two auxiliary modules for the proposed framework: (a) cross-task skip connection module: the auxiliary features from the style transfer task F_{tsl} should be recalibrated before connected with the features of the segmentation task F_{seg}. (b) shape reconstruction and spatial constrain module: SRN reconstructs the label images to encode the shape prior information, and SCN predict the slice positions to incorporate the spatial information.

2.2   Semi-supervised Learning with a Cross-Task Attention

To employ unlabeled data, we utilize a semi-supervised method via multi-task attention. Specifically, the proposed two-task networks include a supervised segmentation task and an unsupervised style transfer task, and the two tasks are connected via an attention mechanism. The style transfer decoder is trained without requirements to manually labeled data, and therefore the model can learn more discriminating features for segmentation from unlabeled images.

The assumption is that combining the features from the style transfer task can be beneficial to encode more representative and robust structural information for the segmentation task. However, due to the diverse feature spaces of the two tasks, a feature recalibration module is required to discard unrelated information. As Fig. 2(a) shows, we employ a cross-task skip connection for the last layers of the two decoders based on the squeeze and excitation (SE) module [5]. The recalibration feature is defined as follows,

$$\tilde{F}_{seg} = F_{seg} + F_{tsl} \otimes \sigma \left(W_2 \delta \left(W_1 z \right) \right), \qquad (9)$$

where F_{seg} and F_{tsl} are the feature maps before the final output layers of the two tasks, W_1 and W_2 are fully connected layers, and $z = \text{AvgPool} \left(F_{tsl} \right) \in \mathbb{R}^{C \times 1 \times 1}$. Specifically, W_1 is employed to squeeze the spatial information into a channel vector, i.e., $W_1 z \in \mathbb{R}^{r \times 1 \times 1}$, while W_2 is used to recover the channel number from r to C for attention.

2.3   Shape Reconstruction and Spatial Constraint for Segmentation

To employ the shape prior, we use the SRN to reconstruct the label image to learn an intermediate representation, see Fig. 2(b). Specifically, the SRN model is pre-trained using available label images and corresponding predicted label

Table 1. The distribution of training and test data. I^{lab}: images with gold standard labels; I^{unl}: images without labels.

Vendor	Training data		Test data	
	Center	Cine MRI	Center	Cine MRI
A	1	75 I^{lab}	1/ 6	5/ 5 I^{unl}
B	2/ 3	50/ 25 I^{lab}	2/ 3	5/ 5 I^{unl}
C	4	25 I^{unl}	4	10 I^{unl}
D	N/A	N/A	5	10 I^{unl}

images from basic U-Net [9]. It can be regarded as a constraint to regularize the predicted segmentation into a desired realistic shape. As Fig. 1 shows, SRN is connected with the output of Decoder^{seg} to refine the shape of the prediction. The loss function of SRN is defined as,

$$\mathcal{L}_{SR} = \mathbb{E}_{y,\hat{y},\hat{y}'} \left[\left\| R - (\hat{R} \text{ or } \hat{R}') \right\|_2 \right], \qquad (10)$$

where R denotes the reconstructed gold standard label from y, while $\hat{R}$ and $\hat{R}'$ indicate the reconstructed predicted label from $\hat{y}$ and $\hat{y}'$, respectively.

Considering the appearance of basal and apical slices can differ significantly, we utilize the SCN to predict the spatial information of each slice, i.e, its normalized distance to the central slice. As Fig. 2(b) shows, the SCN is connected to the bottom of U-Net via fully connected (FC) layers to predict the slice position. The SC loss is formulated as,

$$\mathcal{L}_{SC} = \mathbb{E}_{x,x'} \left[\left\| P - (\hat{P} \text{ or } \hat{P}') \right\|_2 \right], \qquad (11)$$

where P is the ground truth position, while $\hat{P}$ and $\hat{P}'$ are the predicted positions of x and x', respectively. By incorporating the SR and SC loss, the segmentation loss and the label reconstruction loss both can be defined as,

$$\mathcal{L}_{seg} = L_{rec}^{lab} = \mathcal{L}_{CR} + \lambda_{SR}\mathcal{L}_{SR} + \lambda_{SC}\mathcal{L}_{SC}, \qquad (12)$$

where $\mathcal{L}_{CR}$ is the cross-entropy loss, and λ_{SR} and λ_{SC} are the balancing parameters.

3   Experiments

3.1   Materials

Data Acquisition and Pre-processing. The data is from the M&Ms2020 challenge [1], which provides 375 short-axis cardiac MRIs from patients with hypertrophic and dilated cardiomyopathies as well as healthy subjects. All subjects were collected from three different countries (Spain, Germany and Canada)

Table 2. Summary of the quantitative evaluation results of cardiac image segmentation on the test data including four vendors. SR and SC refer to the two auxiliary regularization terms; semi denotes that the model employed both labeled and unlabeled data.

Method	Dice (LV)	Dice (Myo)	Dice (RV)	Mean
U-Net	0.700 ± 0.322	0.675 ± 0.272	0.590 ± 0.369	0.655 ± 0.326
MASSL^{semi}	0.727 ± 0.292	0.691 ± 0.253	0.567 ± 0.371	0.662 ± 0.315
STDGNs^{semi}	0.738 ± 0.277	0.694 ± 0.238	0.601 ± 0.356	0.678 ± 0.299
STDGNs^{semi}-SR	0.754 ± 0.302	$\mathbf{0.718 \pm 0.261}$	0.619 ± 0.370	0.697 ± 0.318
STDGNs^{semi}-SRSC	$\mathbf{0.767 \pm 0.256}$	0.716 ± 0.237	$\mathbf{0.636 \pm 0.355}$	$\mathbf{0.706 \pm 0.292}$

using four different MRI vendors (Siemens, General Electric, Philips and Canon). As Table 1 shows, the training data includes 150 labeled images from vendor A and B and 25 unlabeled images from vendor C. The test data in this study contains 40 images from each vendor, i.e., A, B, C and a novel unseen vendor D. All slices were cropped into a unified size of 144×144 centering at the heart region and were normalized using z-score.

Gold Standard and Evaluation. The images have been manually segmented by experienced clinicians with labels of the left ventricle (LV), right ventricle (RV) blood pools, and left ventricular myocardium (Myo). Note that manual annotations are provided for only two cardiac time frames, i.e., end-diastole (ED) and end-systole (ES). These manual segmentations were considered as the gold standard. For segmentation evaluation, Dice volume overlap and Hausdorff distance (HD) were applied.

Implementation. The framework was implemented in PyTorch, running on a computer with 1.90 GHz Intel(R) Xeon(R) E5-2620 CPU and an NVIDIA TITAN X GPU. We used the SGD optimizer to update the network parameters (weight decay $= 0.0001$, momentum $= 0.9$). The initial learning rate was set to 3e-4 and divided by 10 every 5000 iterations, and batch size was set to 20. Note that the generator and discriminator are alternatively optimized every 50 epochs. The balancing parameters were set as follows, $\lambda_{seg} = 100$, $\lambda_{rec}^{img} = 100$, $\lambda_{SC} = 1$, $\lambda_{SR} = 100$, $\lambda_{gp} = 10$, $\lambda_{cls} = 10$, and λ_{rec}^{lab} was initial set to 0 and then gradually increase to 100.

3.2   Result

Comparisons with Literature and Ablation Study. Table 2 presents the quantitative results of different methods for cardiac segmentation in terms of Dice. Two algorithms were used as baselines for comparisons, i.e., fully supervised U-Net [9] and multi-task attention-based semi-supervised learning (MASSL) network [3]. MASSL is a two-task network consisting of two decoders,

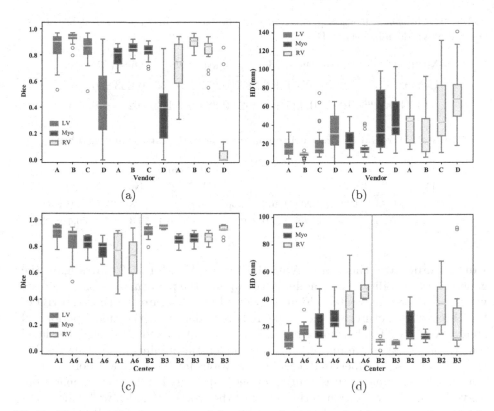

Fig. 3. The quantification results of the three substructures by the proposed method on each vendor and center: (a) Dice of the four vendors; (b) HD of the four vendors; (c) Dice of the two centers on vendor A and B, respectively; (d) HD of the two centers on vendor A and B, respectively. Here, A1 and A6 refer to data of center 1 and 6 from vendor A, while B2 and B3 denote the data of center 2 and 3 from vendor B.

which are used for segmentation prediction and image reconstruction, respectively. Compared to U-Net, MASSL employed the unlabeled data for training, and therefore obtained better mean Dice though with worse Dice on RV. STDGNs is also semi-supervised and it fully utilized the supervised and unsupervised task relationship via a cross-task attention module. It obtained evidently better Dice scores in all sub-regions compared to U-Net and MASSL.

To explore the effectiveness of the proposed shape and spatial constrain, we also adopted an ablation study. One can see that STDGNs-SR outperformed STDGNs in terms of Dice, and incorporating the SC module further improved the performance on LV and RV segmentation. Furthermore, one can see that the performance on the RV segmentation were all worse than that on the LV and Myo for these algorithms. It may due to the fact that the shape of the RV is generally more irregular than that of the LV and Myo.

Table 3. The generalization performance of the baseline and proposed algorithm on the unknown domain vendor D.

Method	Dice (LV)	Dice (Myo)	Dice (RV)
U-Net	0.241 ± 0.284	0.277 ± 0.253	0.063 ± 0.183
STDGNs^{semi}-SRSC	0.425 ± 0.282	0.380 ± 0.103	0.240 ± 0.355

Table 4. Dice and HD for cardiac image segmentation at ED and ES phase.

Phase	Dice			HD (mm)		
	LV	Myo	RV	LV	Myo	RV
ED	0.818 ± 0.236	0.724 ± 0.209	0.671 ± 0.361	19.9 ± 17.7	32.8 ± 23.1	51.0 ± 32.5
ES	0.715 ± 0.268	0.708 ± 0.265	0.601 ± 0.351	20.3 ± 16.4	33.3 ± 27.2	50.4 ± 34.6

Performance on the Data from Different Vendors. To investigate the generalization ability of our model, we compared the performances on data from various vendors and centers. Figure 3(a) and (b) illustrate the Dice and HD of the three substructures on each vendor. It is evident that the model performed generally well on the known domains, i.e., vendor A, B and C. Though the label images of vendor C are unavailable, our model still obtained promising results, thanks to the semi-supervised learning. However, for the unknown domain vendor D, the performance drastically dropped, especially for the RV segmentation. To verify the generalization performance of the proposed model, we further compared the results of our model with that of U-Net on vendor D. As Table 3 shows, the proposed model performed evidently better in generalization compared to the baseline, though there is still a huge scope for improvements.

Performance on the Data from Different Centers. Besides the generalization on different vendors, we also explored the effectiveness of the proposed model on different centers. As vendors C and D include one single center, we only present the results of two centers of vendor A and B, respectively (see Fig. 3(c) and (d)). One can see that the results of center A6 were generally worse than that of center A1, as the data of center A6 is not included in the training dataset. However, compared to the performance decline on the new vendor D, the performance decrease on the new center is evidently less. For the centers already included in the training dataset (centers B2 and B3), the model obtained promising results in terms of both Dice and HD. Note that the performance on center B3 is evidently better than that of center B2, though the training data includes more center B2 data (see Table 1). This may be because the center B2 data is more challenging to segment or center B3 has more similar appearance with other centers in the training data.

Performance on the ED and ES Phase. Table 4 presents the quantitative results of the proposed method for cardiac segmentation at the ED and ES phase.

One can see that the performance was substantially higher in ED than in ES for all substructures in terms of Dice, but was similar in HD. This might be caused by the larger ventricular shape in the ED phase than the ES phase.

4 Conclusion

In this work, we have proposed an end-to-end learning framework for domain generalization via random style transfer. The proposed algorithm has been applied to 40 cine images and obtained promising results on the unknown domain images from new vendors and centers. Especially, the model could generalize better on new centers than new vendors. The results have demonstrated the effectiveness of the proposed spatial and shape constrain schemes, which encourage the network to learn more domain-invariant features. Our technique can be easily extended to other segmentation tasks on data from new vendors and centers. A limitation of this work is that the definition of the modality vector requires the number of domains, which is normally unknown. Besides, the style transfer only can modify modality globally, but sometimes the target modality can be complex with high heterogeneity. In future work, we will replace the modality vector by other more flexible modality representations, and develop a style transfer module that can focus on the local difference among different domains.

Acknowledgement. This work was supported by the National Natural Science Foundation of China (61971142), and L. Li was partially supported by the CSC Scholarship. JA Schnabel and VA Zimmer would like to acknowledge funding from a Wellcome Trust IEH Award (WT 102431), an EPSRC program Grant (EP/P001009/1), and the Wellcome/EPSRC Center for Medical Engineering (WT 203148/Z/16/Z).

References

1. Campello, M.V., Lekadir, K.: Multi-centre, multi-vendor & multi-disease cardiac image segmentation challenge (M&Ms) (2020). https://www.ub.edu/mnms/
2. Chen, C., et al.: Improving the generalizability of convolutional neural network-based segmentation on CMR images. arXiv preprint arXiv:1907.01268 (2019)
3. Chen, S., Bortsova, G., García-Uceda Juárez, A., van Tulder, G., de Bruijne, M.: Multi-task attention-based semi-supervised learning for medical image segmentation. In: Shen, D., et al. (eds.) MICCAI 2019. LNCS, vol. 11766, pp. 457–465. Springer, Cham (2019). https://doi.org/10.1007/978-3-030-32248-9_51
4. Dou, Q., de Castro, D.C., Kamnitsas, K., Glocker, B.: Domain generalization via model-agnostic learning of semantic features. In: Advances in Neural Information Processing Systems, pp. 6450–6461 (2019)
5. Hu, J., Shen, L., Sun, G.: Squeeze-and-excitation networks. In: Proceedings of the IEEE Conference on Computer Vision and Pattern Recognition, pp. 7132–7141 (2018)
6. Li, H., Pan, S.J., Wang, S., Kot, A.C.: Domain generalization with adversarial feature learning. In: Proceedings of the IEEE Conference on Computer Vision and Pattern Recognition, pp. 5400–5409 (2018)

7. Li, Y., et al.: Deep domain generalization via conditional invariant adversarial networks. In: Proceedings of the European Conference on Computer Vision (ECCV), pp. 624–639 (2018)
8. Liu, Q., Dou, Q., Heng, P.A.: Shape-aware meta-learning for generalizing prostate MRI segmentation to unseen domains. arXiv preprint arXiv:2007.02035 (2020)
9. Ronneberger, O., Fischer, P., Brox, T.: U-Net: convolutional networks for biomedical image segmentation. In: Navab, N., Hornegger, J., Wells, W.M., Frangi, A.F. (eds.) MICCAI 2015. LNCS, vol. 9351, pp. 234–241. Springer, Cham (2015). https://doi.org/10.1007/978-3-319-24574-4_28
10. Volpi, R., Namkoong, H., Sener, O., Duchi, J.C., Murino, V., Savarese, S.: Generalizing to unseen domains via adversarial data augmentation. In: Advances in Neural Information Processing Systems, pp. 5334–5344 (2018)
11. Yan, W., et al.: The domain shift problem of medical image segmentation and vendor-adaptation by Unet-GAN. In: Shen, D., et al. (eds.) MICCAI 2019. LNCS, vol. 11765, pp. 623–631. Springer, Cham (2019). https://doi.org/10.1007/978-3-030-32245-8_69
12. Yuan, W., Wei, J., Wang, J., Ma, Q., Tasdizen, T.: Unified generative adversarial networks for multimodal segmentation from unpaired 3D medical images. Med. Image Anal. 101731 (2020)
13. Yue, Q., Luo, X., Ye, Q., Xu, L., Zhuang, X.: Cardiac segmentation from LGE MRI using deep neural network incorporating shape and spatial priors. In: Shen, D., et al. (eds.) MICCAI 2019. LNCS, vol. 11765, pp. 559–567. Springer, Cham (2019). https://doi.org/10.1007/978-3-030-32245-8_62
14. Zhang, L., et al.: Generalizing deep learning for medical image segmentation to unseen domains via deep stacked transformation. IEEE Trans. Med. Imaging (2020)
15. Zhu, Q., Du, B., Yan, P.: Boundary-weighted domain adaptive neural network for prostate MR image segmentation. IEEE Trans. Med. Imaging 39(3), 753–763 (2019)
16. Zhuang, X.: Multivariate mixture model for myocardial segmentation combining multi-source images. IEEE Trans. Pattern Anal. Mach. Intell. 41(12), 2933–2946 (2018)

Semi-supervised Cardiac Image Segmentation via Label Propagation and Style Transfer

Yao Zhang[1,2(✉)], Jiawei Yang[3], Feng Hou[1,2], Yang Liu[1,2], Yixin Wang[1,2], Jiang Tian[4], Cheng Zhong[4], Yang Zhang[5], and Zhiqiang He[6]

[1] Institute of Computing Technology, Chinese Academy of Sciences, Beijing, China
`zhangyao215@mails.ucas.ac.cn`
[2] University of Chinese Academy of Sciences, Beijing, China
[3] Southeast University, Nanjing, China
[4] AI Lab, Lenovo Research, Beijing, China
[5] Lenovo Research, Beijing, China
[6] Lenovo Group, Beijing, China

Abstract. Accurate segmentation of cardiac structures can assist doctors to diagnose diseases, and to improve treatment planning, which is highly demanded in the clinical practice. However, the shortage of annotation and the variance of the data among different vendors and medical centers restrict the performance of advanced deep learning methods. In this work, we present a fully automatic method to segment cardiac structures including the left (LV) and right ventricle (RV) blood pools, as well as for the left ventricular myocardium (MYO) in MRI volumes. Specifically, we design a semi-supervised learning method to leverage unlabelled MRI sequence timeframes by label propagation. Then we exploit style transfer to reduce the variance among different centers and vendors for more robust cardiac image segmentation. We evaluate our method in the M&Ms challenge (https://www.ub.edu/mnms/), ranking 2nd place among 14 competitive teams.

1 Introduction

Cardiac disease is one of the leading threats to human health and causes massive death every year. In the clinical routine, advanced medical imaging techniques (such as MRI, CT, ultrasound) are used for the diagnosis of cardiac disease. Accurate segmentation of cardiac structures from medical images is an essential step to quantatively evaluate cardiac function and improve therapy planning, which is highly demanded in the clinical practice.

In the recent years, deep learning models have been widely used for cardiac image segmentation and achieved promising results [4]. However, these models could be failed on unseen datasets acquired from distinct clinical centers or medical imaging scanners [8]. The M&Ms challenge is motivated to contribute to the effort of building generalizable models that can be applied consistently

ⓒ Springer Nature Switzerland AG 2021
E. Puyol Anton et al. (Eds.): STACOM 2020, LNCS 12592, pp. 219–227, 2021.
https://doi.org/10.1007/978-3-030-68107-4_22

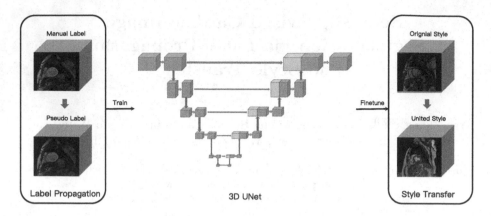

Fig. 1. The pipeline of our method.

across clinical centers [2]. In this challenge, the cohort is composed of 350 patients with hypertrophic and dilated cardiomyopathies as well as healthy subjects. All subjects were scanned in clinical centers in three different countries (Spain, Germany and Canada) using four different magnetic resonance scanner vendors (Siemens, General Electric, Philips and Canon). The variance of data among multiple centers and vendors poses extreme challenge to the generalization of machine/deep learning models.

The success of these deep learning models usually relies on large-scale manually annotated datasets. It requires expensive resources to manually label the images. In contrast to 3D MRI or CT images for other organs, cardiac MRI sequence timeframes has 4 dimensions (i.e., height, depth, width, and time series), which leads to much more workload for annotation. In M&Ms challenge, only the End of Systole (ES) and the End of Diastole (ED) of the cardiac MRI sequence timeframes are annotated. Plenty of images between ES and ED stay unlabelled, which limits the performance of supervised learning.

In this paper, we propose to leverage unlabelled MRI sequence timeframes to improve cardiac segmentation by label propagation. Then we exploit style transfer to reduce the variance among different centers and vendors for more robust cardiac image segmentation.

2   Method

In this section, we will describe our method in detail. We employ a semi-supervised method to achieve effective cardiac image segmentation using unlabelled time-frames from the MRI sequence timeframes. The pipeline of our method is illustrated in Fig. 1. In contrast to [3], we exploit label propagation to leverage unlabelled cardiac sequence. Firstly, a set of pseudo labels of the unlabelled images between ES and ED are generated. Secondly, a 3D UNet [7] is trained on the data with both manual and pseudo labels. At last, as those

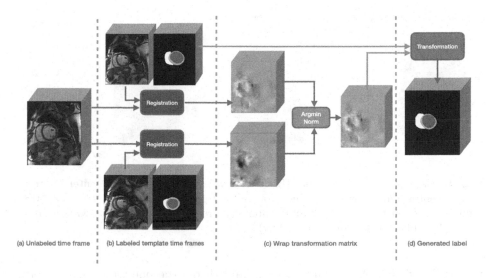

Fig. 2. The procedure of label propagation.

pseudo labels is not as accurate as manual ones, the trained UNet is fine-tuned on the manually labelled data. Furthermore, in order to reduce the gap of the data from different vendors, we augment the training data through style transfer to improve the generalization of semi-supervised learning.

2.1 Label Propagation

Here we augment our training set from unlabeled time frames by leveraging the insights of label propagation. Label propagation is a family of semi-supervised algorithms. It advocates the exploitation of similarity between labeled and unlabeled data, granting us the ability to assign labels to previously unlabeled data.

Fortunately, this task is a well-formulated problem for label propagation for its subject-level time-series prosperity. Firstly, the annotations of End of Systole (ES) and End of Diastole (ED) are good priors as they capture the most extreme scenarios and thus can cover the transition frames between ES and ED. Meanwhile, data from different time frames within a patient share almost identical distribution, which will alleviate propagation errors caused by inter-subject variabilities. Hence, with ES and ED frames as priors, label propagation algorithms can propagate their labels to non-ES and non-ED time frames in a intra-subject manner. By doing so, the propagated labels are better constrained with specificity compared with inter-subject propagation.

Specifically, we use registration algorithm to propagate labels, as shown in Fig. 2. For each patient, the ES and ED frames are the template frames. Given a target time frame, two template frames are first registered to T, resulting two *Warp* matrix, called ES_{wrap} and ED_{wrap}. Naturally, we want the registered label smooth and not elasticizing from the template too much. So we then compare

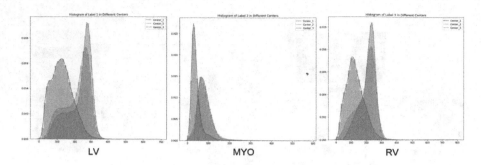

Fig. 3. Histogram distribution of MRI from different clinical centers. Center 1, center 2, and center 3 are marked in blue, orange, and green, respectively. Note that MRI scanners used in both center 2 and center 3 is from the same vendor, while that in center 1 is different. (Color figure online)

the norm of two ES_{wrap} and ED_{wrap} and choose the one with smaller norm as T_{wrap}. Finally, the generated label is obtained by applying the T_{warp} to the corresponding template label.

2.2   Style Transfer

Instead of involving a complicated deep learning network for style transfer [5], we utilize a simple yet effective method, histogram matching, to achieve this. Histogram matching is a process where a time series, image, or higher dimension scalar data is modified such that its histogram matches that of reference dataset. A common application of this is to match the images from two sensors with slightly different responses, or from a sensor whose response changes. We analyze the histogram distribution between different vendors and medical centers, and find that the histogram difference between vendors is much more remarkable while that between centers is insignificant (see Fig. 3). Herein, we apply it to match MRI images from different vendors.

The procedure is as follows. The cumulative histogram is computed for each dataset from a vendor. For any particular value x_i in the data to be adjusted has a cumulative histogram value given by $S(x_i)$. This in turn is the cumulative distribution value in the reference dataset, denoted as $T(x_j)$. The input data value x_i is replaced by x_j.

Specifically, we sample 100 volumes from training set and then randomly select a slice from each volume as the unified reference data. Then other data is matched to the reference one. As large scale dataset benefits the deep learning method, we also augment the training data by transferring the data from one vendor to another (i.e., from vendor A to vendor B or vice versa).

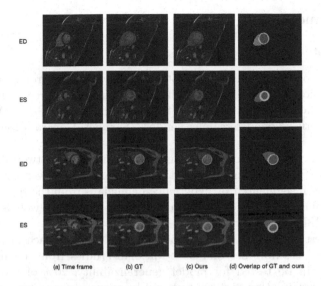

Fig. 4. Visual examples of accurate segmentation. From left to right, each column shows the timeframe, ground truth, prediction of the proposed approach, and the difference between ground truth and prediction. LV, MYO, and RV are marked in red, green, and blue, respectively. (Color figure online)

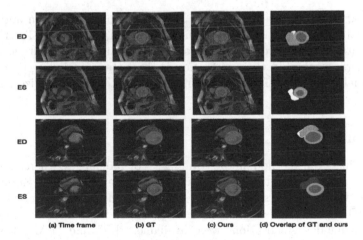

Fig. 5. Visual examples of inaccurate segmentation. From left to right, it shows the timeframe, ground truth, prediction of the proposed approach, and the difference between ground truth and prediction. LV, MYO, and RV are marked in red, green, and blue, respectively. (Color figure online)

3 Experiments and Results

3.1 Dataset

The challenge cohort is composed of 350 patients with hypertrophic and dilated cardiomyopathies as well as healthy subjects. All subjects were scanned in clinical centres in three different countries (Spain, Germany and Canada) using four different magnetic resonance scanner vendors (Siemens, General Electric, Philips and Canon).

The training set will contain 150 annotated images from two different MRI vendors (75 each) and 25 unannotated images from a third vendor. The CMR images have been segmented by experienced clinicians from the respective institutions, including contours for the LV and RV blood pools, as well as for the left ventricular MYO.

The 200 test cases correspond to 50 new studies from each of the vendors provided in the training set and 50 additional studies from a fourth unseen vendor, that will be tested for model generalizability. 20% of these datasets will be used for validation and the rest will be reserved for testing and ranking participants.

3.2 Implementation Details

Our method is built upon nnUNet [6], a powerful implementation and out-of-the-box tool for medical image segmentation. Then, we use ANTs [1] as the implementation of label propagation algorithm and using the default parameters. The transform method in the registration is three-stage, including rigid, affine, and deformable SyN transform. As for histogram matching, we utilize the scikit-image [9].

3.3 Results

We first evaluate our method on the training set of 150 images, where 120 images are used for training and the rest for validation. Dice similarity is employed for evaluation. Table 1 collects the results. Baseline is the fully supervised method (i.e., the 3D UNet) trained solely on end-systole and end-diastole time-frames for which manual segmentations were available. "LP" and "HM" denote the

Table 1. Results on training set

Method	Dice similarity [%]			
	LV	MYO	RV	Average
Baseline	92.88	86.48	87.78	89.05
LP	92.43	86.73	89.59	89.58
LP+HM	92.46	87.00	90.94	90.13

proposed semi-supervised method that exploits label propagation and histogram matching. It is observed that label propagation excels baseline by 0.53% and histogram matching further increases 0.55% in terms of Dice per case.

Table 2. Results on validation set

Method	Dice similarity [%]			Hausdorff distance [mm]		
	LV	MYO	RV	LV	MYO	RV
Baseline	91.15 ± 7.35	86.00 ± 4.28	88.26 ± 6.59	8.83 ± 4.27	13.08 ± 12.19	12.70 ± 6.72
Proposed	91.75 ± 7.09	86.58 ± 4.25	89.32 ± 6.20	8.10 ± 4.27	12.66 ± 12.30	11.39 ± 6.12

We then validate our method on the validation set, which is held out by the organizers. The docker image with our method is submitted to the organizers to get the results. As shown in Table 2, our method consistently outperforms baseline on LV, MYO and RV. Specifically, it improves 0.75% in terms of Dice per case, and 0.82 mm in terms of Hausdorff Distance on average of the three targets. The results on different vendors are collected in Table 3.

Table 3. Results of different vendors on validation set

Method	Vendor	Dice [%]			HD [mm]		
		LV	MYO	RV	LV	MYO	RV
Baseline	A	91.70	86.31	87.46	10.05	11.28	14.21
	B	94.43	88.79	92.90	7.31	11.16	8.55
	C	88.05	86.02	86.26	9.25	12.07	13.16
	D	90.39	82.74	86.34	8.73	18.06	14.99
Proposed	A	91.80	86.53	87.95	9.26	11.47	13.58
	B	94.64	89.07	93.78	6.50	10.25	7.39
	C	89.58	87.20	88.12	8.83	12.35	11.32
	D	91.00	83.54	87.44	7.80	16.57	13.26

We also submit our method as docker image to the organizers for online test[1]. Please note that neither post-processing nor ensemble strategy is employed in our evaluation. Table 4 shows the top 5 teams and our method ranks 2nd place, demonstrating the effectiveness of our method for cardiac segmentation from multiple vendors and centers. Table 5 presents the detailed results of our method on the images from 4 different vendors. It is observed that the proposed method obtains consistently promising results on both seen and unseen vendors. Figure 4 and Fig. 5 show some accurate and inaccurate predictions generated by our method.

---

[1] The test results are presented by organizers at https://www.ub.edu/mnms/.

Table 4. Results on test set

Method	Dice similarity [%]		
	LV	MYO	RV
Peter M. Full	91.0	84.9	88.4
Yao Zhang (Ours)	90.6	84.0	87.8
Jun Ma	90.2	83.5	87.4
Mario Parreño	91.2	83.8	85.3
Fanwei Kong	90.2	82.8	85.7

Table 5. Detailed results of our method on 4 vendors of the test set

Vendor	Dice similarity [%]			Hausdorff distance [mm]		
	LV	MYO	RV	LV	MYO	RV
A	91.87 ± 5.92	84.83 ± 4.37	88.47 ± 5.90	10.72 ± 16.32	11.85 ± 18.76	12.46 ± 9.90
B	91.58 ± 7.36	87.24 ± 4.58	88.65 ± 8.10	7.85 ± 3.72	10.13 ± 3.72	11.41 ± 5.82
C	89.87 ± 7.32	83.38 ± 6.22	87.65 ± 6.36	7.97 ± 4.80	9.97 ± 4.58	10.71 ± 4.76
D	90.29 ± 5.54	82.67 ± 4.30	87.07 ± 10.29	11.16 ± 19.01	13.11 ± 19.97	16.03 ± 21.20
Overall	90.90 ± 6.61	84.53 ± 5.21	87.96 ± 7.84	9.42 ± 12.92	11.85 ± 14.07	12.65 ± 12.40

4 Conclusion

In this paper, we design and develop a semi-supervised method for cardiac image segmentation from multiple vendors and medical centers. We exploit label propagation and iterative refinement to leverage unlabelled data in a semi-supervised manner. We further reduce distribution gap between MRI images from different vendors and centers by histogram matching. The results show that our framework is able to achieve superior performance for robust LV, MYO, and RV segmentation. The proposed method ranks 2nd place among 14 competitive teams in the M&M Challenge.

References

1. Avants, B.B., Tustison, N., Song, G.: Advanced normalization tools (ants). Insight **2**, 1–35 (2009)
2. Campello, V.M., Palomares, J.F.R., Guala, A., Marakas, M., Friedrich, M., Lekadir, K.: Multi-Centre, multi-vendor & multi-disease cardiac image segmentation. Challenge (2020). https://doi.org/10.5281/zenodo.3715890
3. Chen, C., et al.: Unsupervised multi-modal style transfer for cardiac MR segmentation, pp. 209–219. arXiv preprint arXiv:1908.07344 (2019)
4. Chen, C., et al.: Deep learning for cardiac image segmentation: a review. Front. Cardiovasc. Med. **7**, 25 (2020)
5. Chen, L.C., et al.: Leveraging semi-supervised learning in video sequences for urban scene segmentation. arXiv preprint arXiv:2005.10266 (2020)

6. Isensee, F., Petersen, J., Kohl, S.A.A., Jäger, P.F., Maier-Hein, K.H.: nnu-net: breaking the spell on successful medical image segmentation. arXiv preprint arXiv:1904.08128 (2019)

7. Ronneberger, O., Fischer, P., Brox, T.: U-Net: convolutional networks for biomedical image segmentation. In: Navab, N., Hornegger, J., Wells, W.M., Frangi, A.F. (eds.) MICCAI 2015. LNCS, vol. 9351, pp. 234–241. Springer, Cham (2015). https://doi.org/10.1007/978-3-319-24574-4_28

8. Tao, Q., et al.: Deep learning-based method for fully automatic quantification of left ventricle function from cine MR images: a multivendor, multicenter study. Radiology **290**(1), 81–88 (2019)

9. van der Walt, S., et al.: scikit-image: image processing in python. PeerJ **2**(1), e453 (2014)

Domain-Adversarial Learning for Multi-Centre, Multi-Vendor, and Multi-Disease Cardiac MR Image Segmentation

Cian M. Scannell[1]([✉]) [ID], Amedeo Chiribiri[1], and Mitko Veta[2]

[1] School of Biomedical Engineering and Imaging Sciences, King's College London, London, UK
cian.scannell@kcl.ac.uk
[2] Department of Biomedical Engineering, Eindhoven University of Technology, Eindhoven, The Netherlands

Abstract. Cine cardiac magnetic resonance (CMR) has become the gold standard for the non-invasive evaluation of cardiac function. In particular, it allows the accurate quantification of functional parameters including the chamber volumes and ejection fraction. Deep learning has shown the potential to automate the requisite cardiac structure segmentation. However, the lack of robustness of deep learning models has hindered their widespread clinical adoption. Due to differences in the data characteristics, neural networks trained on data from a specific scanner are not guaranteed to generalise well to data acquired at a different centre or with a different scanner. In this work, we propose a principled solution to the problem of this domain shift. Domain-adversarial learning is used to train a domain-invariant 2D U-Net using labelled and unlabelled data. This approach is evaluated on both seen and unseen domains from the M&Ms challenge dataset and the domain-adversarial approach shows improved performance as compared to standard training. Additionally, we show that the domain information cannot be recovered from the learned features.

Keywords: Domain-adversarial learning · Cardiac MRI segmentation

1 Introduction

The characterisation of cardiac structure and function is an important step in the diagnosis and management of patients with suspected cardiovascular disease. Cardiac magnetic resonance (CMR) is the method of choice for the non-invasive assessment of cardiac function and allows the accurate quantification of structural and functional parameters such as the chamber volumes and ejection fraction. A limiting factor for the routine analysis of these parameters in the clinic is that it requires the tedious manual delineation of the anatomical structures.

A. Chiribiri and M. Veta—Joint last authors.

© Springer Nature Switzerland AG 2021
E. Puyol Anton et al. (Eds.): STACOM 2020, LNCS 12592, pp. 228–237, 2021.
https://doi.org/10.1007/978-3-030-68107-4_23

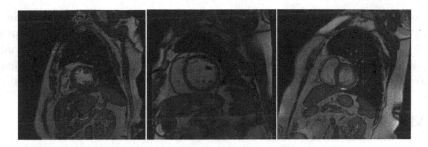

Fig. 1. A comparison of images from different MR scanner vendors showing varying levels of signal, noise and contrast.

Deep learning has become the state-of-the-art approach for image segmentation and convolutional neural network (CNN)-based approaches have shown huge potential for the automated analysis of cardiac magnetic resonance (CMR) images [1–4]. Ruijsink et al. [5] recently showed that it is further possible to automatically compute a wide range of advanced indices of cardiac function, including shape, motion, and strain, from short-axis cine CMR images. This is an important step as it makes it feasible to analyse large numbers of patients and it reduces the dependency of the analysis on the operator [6].

However, the aforementioned studies are all limited by the homogeneity of the data that was used. They were either single/few centre studies or they used publicly available databases such as the UK Biobank which have uniform imaging protocols and few pathological cases. As such, the datasets do not match the variability seen in clinical practice and although the studies report good performance on their data, it is not guaranteed that they will generalise well to real-world settings. Images acquired using different scanners can have widely-varying levels of signal, noise, and contrast. Images acquired at different centres may be planned differently resulting in differing locations of the heart in the images, and distinct cardiovascular diseases can alter the shape of the heart. Some of these variations are shown in Fig. 1. These variations introduce a so-called *domain shift*, a change between the distribution of training data and the distribution of the data that the model is being applied to [7].

Since deep learning models are known to not generalise well in the presence of such a domain shift [7], this leads to the need for techniques for domain adaption. The most simple technique for domain adaption is use of data augmentation to try to account for the variability between domains during training [8]. It is, however, difficult to account for all possible variations in this manner. Other approaches include fine-tuning previously learned weights on a new domain and, more recently, conditional generative models, such as cycleGAN [9], have been used to translate images from a new domain to the domain of the model. However, training data would be required from the new domain for these approaches and thus they cannot be applied to completely unseen domains.

Ganin et al. [10] proposed an approach to the domain-adversarial training of neural networks based on the assumption that for a model to generalise well to new domains the features learned by the model must be unable to discriminate between domains. This approach has been recently applied to domain adaption in histological images and shows promising results [11]. In this work, domain-adversarial training is used for cine CMR segmentation and is shown to learn features that are domain invariant and generalise well to new unseen domains.

2 Materials and Methods

2.1 Dataset

The dataset for this study was provided by the Multi-Centre, Multi-Vendor & Multi-Disease (M&Ms) Cardiac Image Segmentation Challenge[1] [12]. The datasets consisted of 350 patients with a mix of healthy controls and patients with hypertrophic and dilated cardiomyopathies. It was acquired at six different clinical centres using MRI scanners from four different vendors. From this dataset 150 annotated patients from two different MRI vendors were made available for training as well as 25 patients from a third MRI vendor without annotations. There were 40 patients, 10 from each vendor, for validation and 160 patients, 40 from each vendor, for testing. These images were not available but results could be obtained from the challenge organisers. Since the test set for the challenge could only be used once, the comparison of our approach to the baseline is performed on this challenge validation set and 32 patients from the training set were used for internal validation. Expert annotations are provided for the left (LV) and right ventricle (RV) blood pools, as well as for the left ventricular myocardium (MYO).

2.2 Domain-Adversarial Learning

Domain-adversarial learning attempts to improve the generalisation of neural networks by encouraging them to learn features that do not depend on the domain of the input. In this situation, as well as learning a network for the segmentation task, we concurrently learn a classifier which attempts to discriminate between the input domains using the activations of the segmentation network. An adversarial training step is then introduced which is a third training step after the optimisation of the segmentation network and the optimisation of the domain discriminator. This adversarial step uses a gradient reversal to update the weights of the segmentation network to maximise the loss of the domain discriminator and thus prevents the domain information being recovered from the learned representations of the segmentation network. The three optimisation steps, using a standard (stochastic) gradient descent algorithm would be:

$$\boldsymbol{\theta}_S \leftarrow \boldsymbol{\theta}_S - \lambda_S \frac{\partial \mathcal{L}_S}{\partial \boldsymbol{\theta}_S} \tag{1}$$

---

[1] https://www.ub.edu/mnms/.

$$\boldsymbol{\theta}_D \leftarrow \boldsymbol{\theta}_D - \lambda_D \frac{\partial \mathcal{L}_D}{\partial \boldsymbol{\theta}_D} \tag{2}$$

$$\boldsymbol{\vartheta}_S \leftarrow \boldsymbol{\vartheta}_S + \alpha \lambda_S \frac{\partial \mathcal{L}_D}{\partial \boldsymbol{\vartheta}_S} \tag{3}$$

where $\boldsymbol{\theta}_S, \lambda_S, \& \mathcal{L}_S$ and $\boldsymbol{\theta}_D, \lambda_D, \& \mathcal{L}_D$ are the parameters, learning rate, and loss function of the segmentation network and domain discriminator, respectively. $\boldsymbol{\vartheta}_S \subset \boldsymbol{\theta}_S$ are the parameters of the convolutional layers of the segmentation network and $\alpha \in [0,1]$ controls the strength of the adversarial update. The proposed training pipeline is illustrated in Fig. 2.

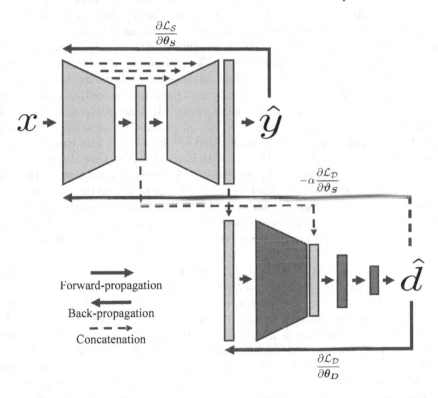

Fig. 2. The flowchart of the domain-adversarial training pipeline. On top, the segmentation network is updated to minimise the segmentation loss. The intermediate representations of this network are then feed to a classification network which is optimised to discriminate between input domains. Finally, the adversarial back-propagation is used to update the parameters of the segmentation network to encourage domain invariance.

2.3   Implementation Details

Our TensorFlow implementation and trained model weights are provided on github[2].

The baseline model used in this work was a 2D U-Net model [13]. The model consists of four down-sampling blocks followed by max-pooling with each down-sampling block being made of up two convolution blocks containing a convolutional layer, batch normalisation, and ReLU activation.

The domain discriminator takes as input the activations from the penultimate and lowest resolution layers of the U-Net. The model consists of 4 convolutional layers with ReLU activations, batch normalisation, and max-pooling and three fully-connected layers also with ReLU activation. For the purpose of this work, different MRI scanner vendors were considered as distinct domains.

Images are resampled to a uniform resolution of 1.25×1.25 mm and the network is trained on images cropped to 192×192 pixels. Pixel intensity values are normalised to the range $[0, 1]$ before data augmentation. In this work, we use extensive spatial augmentation consisting of stochastic scaling, translation, rotation, and B-spline deformations using *gryds* [14]. Intensity augmentation is performed by adding Gaussian noise and random intensity shifts. The intensities are further normalised using contrast limited adaptive histogram equalisation.

The segmentation network is trained with batch size of 16 to optimise the sum of (1 - DSC) and the cross-entropy, where DSC is the Dice similarity coefficient. The domain discriminator is trained with a batch size of 20 and the cross-entropy loss. Optimisation is performed using the ADAM optimiser in both cases. Early-stopping is used on the domain classification accuracy to chose a model that has the least dependence on domain knowledge after the segmentation accuracy has plateaued.

Images from the two annotated domains are used to train the segmentation networks while images from all three available domains are used to train the domain discriminator and compute the adversarial update, including the unlabelled images, as no ground-truth segmentations are required. Since the domain-adversarial training is unstable [11], the models are trained in stages. For the first 150 epochs, the segmentation U-Net is trained only with a learning rate of 1e−4. For the next 150 epochs, the domain discriminator is trained only with a learning rate of 1e−3. This initial training without the segmentation network updating allows it to achieve a baseline level of performance without its inputs (the activations of the segmentation network) changing. After this, both networks are trained together, including the adversarial update. The weighting of the adversarial update increases linearly from 0 to 1 over the next 150 epochs. The learning rate for the segmentation network (1e−3) is higher than the learning rate for the domain discriminator (1e−4) to allow the segmentation network to learn more quickly than the domain discriminator. At test time, predictions are made in a sliding window fashion over a 3×3 grid of patches with input size (192×192).

---

[2] https://github.com/cianmscannell/da_cmr.

This is compared to a baseline model trained in the exact same way, using the same intensity and data augmentation, except for the domain-adversarial update. The comparison was made using the nonparametric Mann-Whitney U test in SciPy [15].

3 Results

The loss and accuracy curves are shown for the domain adversarial training in Fig. 3. The mean (SD) Dice similarity coefficient for the baseline model is 0.8 (0.18), 0.76 (0.13) and 0.77 (0.2) for the LV, MYO, and RV, respectively. The equivalent scores for the domain-adversarial model are 0.9 (0.07), 0.83 (0.05), and 0.87 (0.07). The Hausdorff distances for the three classes, in mm, are 17.6 (12.8), 26.3 (27.7), and 19.0 (12.1) for the baseline model and 12 (15.2), 17.4 (19.8), and 18.4 (33.4) for the domain-adversarial model. Figure 4 further compares the baseline and domain-adversarial models showing the Dice for all three classes and how it varies across domains. The final test set results for our approach give Dice scores of 0.88 (0.1), 0.8 (0.09), and 0.84 (0.14) and Hausdorff distances (mm) of 14.54 (19.12), 17.36 (20.86), and 17.51 (19.01) for the LV, MYO, and RV, respectively. There is no performance drop-off across domains and a summary of all performance metrics is shown in the Table 1.

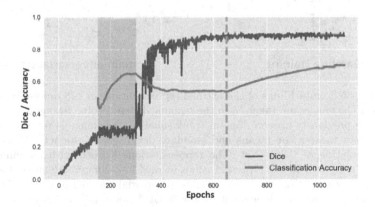

Fig. 3. The loss curve, depicting the process of the domain-adversarial training. For the first 150 epochs, the segmentation network is trained only, then the domain discriminator is trained only (dark blue shaded area). Finally, both networks are trained together including adversarial updates. The orange dotted line indicates the network chosen using early-stopping. (Color figure online)

The t-SNE embeddings of the learned representations of the baseline model are compared to the domain-adversarial learning approach in Fig. 5. With the baseline training there are clear clusters in the learned features indicating that they are domain dependent. For the domain-adversarial approach the learned

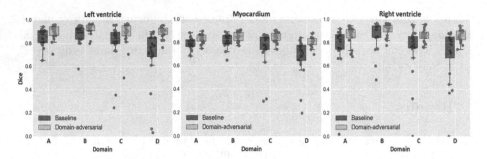

Fig. 4. The two distributions of the Dice scores for the three classes for the different training domains. The domain-adversarial model has higher Dice with no drop-off in performance on the unseen domains (C and D). A drop-off in performance is seen from the training domains (A and B) to the unseen domains (C and D) for the baseline model.

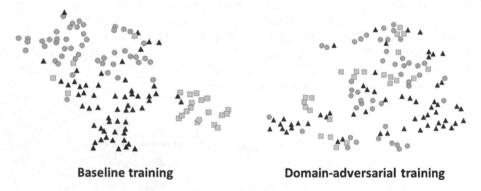

Baseline training   **Domain-adversarial training**

Fig. 5. The t-SNE embeddings of the learned representations of 128 randomly selected image slices. The images are taken from the training domains (blue triangles and orange circles) and a previously unseen domain (pink squares). An image is represented as a vector by concatenating the means and standard deviations of the activations at the minimum resolution of the U-Net. The representations learned with baseline training (left) are compared to the representations learned in a domain-adversarial manner (right). This shows, qualitatively, that domain-adversarial training promotes the learning of features that are less dependent on the input domain. (Color figure online)

features are more widely distributed in the embedding space, including for the domain not used for the training of the segmentation network (pink squares). Figure 6 shows the output segmentations for example images from one of the unseen domains, comparing the baseline model to the domain-adversarial approach.

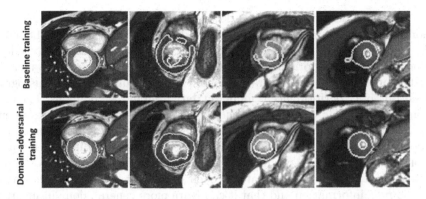

Fig. 6. Example segmentations of images from a previously unseen domain, comparing standard training (top row) to domain-adversarial training (bottom row). Ground-truth segmentations are not available in this case but qualitatively superior performance is demonstrated with domain-adversarial training.

Table 1. A summary, mean (standard deviation), of performance metrics across validation and test sets. For the validation set, a statistical comparison is further made (and p-value reported) for the Dice and Hausdorff (mm) metrics, respectively.

Domain		Test		Validation				
		Domain-adversarial		Baseline		Domain-adversarial		
		Dice	Hausdorff	Dice	Hausdorff	Dice	Hausdorff	p-value
A	LV	0.89 (0.08)	16.78 (21.03)	0.85 (0.07)	15.23 (5.89)	0.9 (0.06)	14.5 (11.5)	0.02	0.05
	MYO	0.81 (0.06)	19.51 (22.18)	0.79 (0.05)	23.22 (21.19)	0.83 (0.04)	17.37 (8.3)	0.01	0.2
	RV	0.86 (0.08)	16.86 (14.9)	0.77 (0.19)	21.55 (8.53)	0.86 (0.08)	16.55 (6.05)	0.03	0.06
B	LV	0.89 (0.01)	11.65 (11.97)	0.88 (0.08)	10.7 (3.9)	0.95 (0.05)	6.35 (2.45)	0.01	0.02
	MYO	0.82 (0.08)	13.94 (12.16)	0.79 (0.04)	17.88 (25.77)	0.82 (0.06)	9.53 (3.1)	0.07	0.01
	RV	0.84 (0.16)	17.5 (19.06)	0.86 (0.13)	11.36 (6.08)	0.91 (0.05)	9.98 (4.92)	0.2	0.47
C	LV	0.85 (0.13)	17.39 (22.4)	0.8 (0.18)	18.07 (9.5)	0.88 (0.11)	18.18 (26.37)	0.02	<0.01
	MYO	0.75 (0.13)	21.28 (26.07)	0.75 (0.16)	29.55 (28.01)	0.84 (0.04)	24.96 (29.29)	<0.01	0.06
	RV	0.8 (0.2)	17.29 (18.58)	0.74 (0.22)	17.29 (9.21)	0.86 (0.06)	12.16 (4.56)	0.01	0.03
D	LV	0.88 (0.05)	12.16 (18.56)	0.68 (0.24)	26.5 (19.49)	0.89 (0.05)	9.12 (3.45)	<0.01	<0.01
	MYO	0.8 (0.05)	14.69 (19.53)	0.67 (0.16)	34.35 (32)	0.8 (0.04)	17.79 (22.72)	<0.01	0.01
	RV	0.87 (0.11)	18.31 (22.7)	0.7 (0.23)	25.88 (16.52)	0.86 (0.05)	34.89 (63.21)	<0.01	0.02
All	LV	0.88 (0.1)	14.54 (19.12)	0.8 (0.18)	17.6 (12.8)	0.9 (0.07)	12 (15.2)	<0.01	<0.01
	MYO	0.8 (0.09)	17.36 (20.86)	0.76 (0.13)	26.3 (27.7)	0.83 (0.05)	17.4 (19.8)	<0.01	<0.01
	RV	0.84 (0.14)	17.51 (19.01)	0.77 (0.2)	19 (12.1)	0.87 (0.07)	18.4 (33.4)	<0.01	<0.01

4 Discussion

This paper deals with the challenge of the generalisation of deep learning models across different data domains. In particular, it deals with the segmentation of short-axis cine CMR images which vary significantly across MRI scanner vendors.

We have demonstrated that domain-adversarial learning can improve the generalisation of a U-Net model trained to segment short-axis cine CMR images.

We trained the model using annotated data from two vendors and incorporated the domain information using unannotated data from a third vendor. The model can be applied on data from unseen domains (i.e. from different MRI vendors) with no significant performance decrease. Although baseline training includes intensity normalisation and contrast normalisation through the use of histogram equilisation, lower performance is found on data from unseen domains. The comparison over the whole validation set shows statistically significant improvements using domain-adversarial learning, for the segmentation of all structures. This indicates that there are further variations across vendors, in addition to intensity and contrast, that the network relies on for the segmentation. The improved performance of the domain-adversarial approach shows that it can avoid over-fitting to this domain information and that it can learn more general, domain-invariant features. The performance of the baseline model is also worse on the training domains. This is likely because even data from the same vendor can be from different clinical centres which in itself may constitute a domain shift and because the domain-adversarial update has a regularising effect on training.

Domain-adversarial learning has the benefit, over alternative domain-adaption strategies, of not requiring any data from the domains that the model is being applied on and not requiring any re-training. It also allows the incorporation of data without annotations in the domain discriminator and adversarial update so that unlabelled data can still be used to encourage domain invariance.

As a limitation, the domain-adversarial training of neural networks is an unstable process. It relies on the ad-hoc hyper-parameter tuning more than convention training and work to address this would be needed for the technique to be used more commonly. Though the dataset used does include some pathological cases, these are not with a wide range of conditions and it remains to be determined if the domain-adversarial approach can still adapt to patients with different conditions that alter the geometry of the heart, such as ischaemic heart disease.

In conclusion, domain-adversarial learning improves the ability of deep learning approaches for CMR image segmentation to generalise to unseen data from new domains. It does not require data from the new domains or to fine-tune/adapt the network.

Acknowledgements. The authors of this paper declare that the segmentation method they implemented for participation in the M&Ms challenge has not used any pre-trained models nor additional MRI datasets other than those provided by the organizers.

The authors acknowledge financial support from the King's College London & Imperial College London EPSRC Centre for Doctoral Training in Medical Imaging (EP/L015226/1); Philips Healthcare; and The Centre of Excellence in Medical Engineering funded by the Wellcome Trust and EPSRC under grant number WT 203148/Z/16/Z.

References

1. Bernard, O., et al.: Deep learning techniques for automatic MRI cardiac multi-structures segmentation and diagnosis: is the problem solved? IEEE Trans. Med. Imaging **37**(11), 2514–2525 (2018)
2. Bai, W., et al.: Automated cardiovascular magnetic resonance image analysis with fully convolutional networks. J. Cardiovasc. Magn. Reson. **20**(65), 1–12 (2018)
3. Tao, Q., et al.: Deep learning-based method for fully automatic quantification of left ventricle function from cine MR images: a multivendor. Multicenter Study. Radiology **290**(1), 81–88 (2019)
4. Scannell, C.M., et al.: Deep-learning-based preprocessing for quantitative myocardial perfusion MRI. J. Magn. Reson. Imaging **51**(6), 1689–1696 (2020)
5. Ruijsink, B., et al.: Fully automated, quality-controlled cardiac analysis from CMR: validation and large-scale application to characterize cardiac function. JACC Cardiovasc. Imaging **13**(3), 684–695 (2020)
6. Villa, A.D.M., et al.: Importance of operator training and rest perfusion on the diagnostic accuracy of stress perfusion cardiovascular magnetic resonance. J. Cardiovasc. Magn. Reson. **20**(1), 74 (2018)
7. Dou, Q., et al.: Unsupervised cross-modality domain adaptation of ConvNets for biomedical image segmentations with adversarial loss. In: Proceedings of the Twenty-Seventh International Joint Conference on Artificial Intelligence, IJCAI-18, pp. 691–697 (2018)
8. Chen, C., et al.: Improving the generalizability of convolutional neural network-based segmentation on CMR images. Front. Cardiovasc. Med. **7**, 105 (2020)
9. Yan, W., et al.: MRI manufacturer shift and adaptation: increasing the generalizability of deep learning segmentation for MR images acquired with different scanners. Radiol. Artif. Intell. **2**, e190195 (2020)
10. Ganin, Y., et al.: Domain-adversarial training of neural networks. J. Mach. Learn. Res. **17**(59), 1–35 (2016)
11. Lafarge, M.W., Pluim, J.P.W., Eppenhof, K.A.J., Veta, M.: Learning domain-invariant representations of histological images. Front. Med. **6**, 162 (2019)
12. Campello, V.M., et al.: Multi-Centre, Multi-Vendor & Multi-Disease Cardiac Image Segmentation (In preparation)
13. Ronneberger, O., Fischer, P., Brox, T.: U-Net: convolutional networks for biomedical image segmentation. In: Navab, N., Hornegger, J., Wells, W.M., Frangi, A.F. (eds.) MICCAI 2015. LNCS, vol. 9351, pp. 234–241. Springer, Cham (2015). https://doi.org/10.1007/978-3-319-24574-4_28
14. Eppenhof, K.A.J., Pluim, J.P.W.: Pulmonary CT registration through supervised learning with convolutional neural networks. IEEE Trans. Med. Imaging **38**(5), 1097–1105 (2019)
15. Virtanen, P., et al.: SciPy 1.0 fundamental algorithms for scientific computing in Python. Nat. Methods **17**, 261–272 (2020)

Studying Robustness of Semantic Segmentation Under Domain Shift in Cardiac MRI

Peter M. Full[1,2(✉)] (iD), Fabian Isensee[1], Paul F. Jäger[1], and Klaus Maier-Hein[1]

[1] Division of Medical Image Computing, German Cancer Research Center (DKFZ),
Heidelberg, Germany
p.full@dkfz-heidelberg.de
[2] Medical Faculty Heidelberg, Heidelberg University, Heidelberg, Germany

Abstract. Cardiac magnetic resonance imaging (cMRI) is an integral part of diagnosis in many heart related diseases. Recently, deep neural networks have demonstrated successful automatic segmentation, thus alleviating the burden of time-consuming manual contouring of cardiac structures. Moreover, frameworks such as nnU-Net provide entirely automatic model configuration to unseen datasets enabling out-of-the-box application even by non-experts. However, current studies commonly neglect the clinically realistic scenario, in which a trained network is applied to data from a different domain such as deviating scanners or imaging protocols. This potentially leads to unexpected performance drops of deep learning models in real life applications. In this work, we systematically study challenges and opportunities of domain transfer across images from multiple clinical centres and scanner vendors. In order to maintain out-of-the-box usability, we build upon a fixed U-Net architecture configured by the nnU-net framework to investigate various data augmentation techniques and batch normalization layers as an easy-to-customize pipeline component and provide general guidelines on how to improve domain generalizability abilities in existing deep learning methods. Our proposed method ranked first at the *Multi-Centre, Multi-Vendor & Multi-Disease Cardiac Image Segmentation Challenge (M&Ms)*.

Keywords: Heart diseases · Cine MRI · Deep learning · Robustness · Semantic segmentation · Annotation · nnU-Net

1 Introduction

Cardiac MRI (cMRI) is frequently used in clinical workflow to allow for radiation-free, reliable and reproducible diagnostics regarding heart diseases. Intermediate steps, in which contours of the heart's ventricles are manually delineated for at least two time steps of the cardiac cycle, are necessary to diagnose potential heart diseases. These steps are time-consuming and prone to interrater-variability [1].

© Springer Nature Switzerland AG 2021
E. Puyol Anton et al. (Eds.): STACOM 2020, LNCS 12592, pp. 238–249, 2021.
https://doi.org/10.1007/978-3-030-68107-4_24

To mitigate this problem fully automatic segmentation approaches have been proposed [2]. However, so far approaches using deep neural networks (DNN) are commonly validated in a setting where the test data is from the same domain as the training data. A drop in performance is often observed when a trained network is applied to data from another domain (i.e. different scanner vendor or acquisition protocol from other hospitals) [9]. The implied requirement for annotated data from all vendors and centres limits the application of DNNs in real-life clinical scenarios. Recent studies propose to alleviate this burden with domain adaptation [11] or transfer learning [7], where knowledge from a source domain is used to estimate a target domain. Limitations of these approaches are the concomitant technical complexity and the fact that unlabeled or partially labeled data from the target domain is required during training. One approach that circumvents these constraints is domain generalization, which aims to learn domain agnostic features given multiple source domains [6] at training time. In this work, we hypothesize that domain shifts in cardiac segmentation such as between different vendors or centres are less distinct than in typical domain adaptation or domain generalization tasks (e.g. match clip art with photography) and could potentially be bridged with less specialized techniques such as extensive data augmentation.

nnU-Net is a fully automated segmentation toolkit that allows experts as well as non-experts in the field of medical image analysis to achieve state-of-the-art results in a variety of segmentation tasks without manual intervention [3]. Here, we aim to enhance nnU-Net's generalization abilities under target domain shifts without degrading its out-of-the-box usability. To this end, we apply extensive data augmentation (DA) and study the interaction with different normalization layers within the nnU-Net framework. The effect of DA on cross-domain performance on medical images [10,11] and cMRI [8] has been explored before, but these approaches have not been evaluated on publicly accessible datasets, did not consider interdependencies with regularization techniques such as normalization layers, and did not provide results for 2D as well as 3D models.

Our contributions are the following: 1) We investigate the effect of common DA techniques and batch normalization (BN) on model robustness under domain shifts in cMRI. 2) We investigate whether pooling data from multiple domains during training improves test time performance on individual domains. 3) Our proposed method builds upon the publicly available nnU-Net ensuring out-of-the-box usability to allow for a kick start into domain-invariant fully automatic cardiac segmentation. Our final models will be integrated into the nnU-Net framework and will be freely accessible[1], as well as our final submission singularity container.[2]

---

[1] https://doi.org/10.5281/zenodo.4134721.
[2] https://doi.org/10.5281/zenodo.4134879.

2   Methods and Experiments

2.1   Data and Evaluation Metrics

The Multi-Centre, Multi-Vendor & Multi-Disease Cardiac Image Segmentation Challenge (*M&Ms*) dataset [4] is used for our experiments. The dataset was acquired with four different vendors, namely Siemens, Philips, General Electric, and Canon (in the following as A, B, C, and D, respectively) and at six different centres. The training dataset comprises short axis cMRI images from 175 patients, of which 150 are annotated. For each annotated patient, pixel-wise manual segmentations of the left ventricular myocardium (LVM), the left ventricular blood pool (LV), and the right ventricular blood pool (RV) are given for the end-diastolic (ED) and end-systolic (ES) frames. The 150 annotated samples were acquired with two different vendors (A and B) at three different centres (1, 2, and 3). The test set comprises 200 patients from four different vendors A, B, C, and D and six different centres. Note that vendor C and D are not represented in the training data. Due to the different vendors and centres, domain gaps within the data are expected. While different annotation guidelines at different centres may inject a bias in ground truth segmentation, we assume that this effect is negligible in a highly standardized setting of a challenge. We expect the main difference in data being due to different MR scanner vendors or cardiac diseases being unevenly represented in the different datasets. In the following we will refer to different data domains, when data was acquired with different vendors (e.g. *domain A* when the data was acquired with a scanner from vendor A). For a brief summary of the training and test data see Table 1 for a detailed description we refere to [4].

All results will be reported on hold-out validation sets using the dice score as target metric.

Table 1. Training and test dataset description.

	Vendor	Centre	# studies	Annotations	Used for training
Training	A	1	75	Yes	Yes
	B	2	50	Yes	Yes
		3	25	Yes	Yes
	C	4	25	No	No
Testing	A	1	21	No	No
		6	29	No	No
	B	2	24	No	No
		3	26	No	No
	C	4	50	No	No
	D	5	50	No	No

2.2 Data Augmentation

A default nnU-Net version (*default nnU-Net*), with no code changes at all, is compared to a, what we refer to as, *cMRI baseline nnU-Net (BL)*, which adopts all DA settings, that have shown promising results for the field of cMRI, as proposed by [8]. *BL* comprises image scaling within a range of [0.7; 1.4], random rotation within a range of [± 30 degrees], and random flipping horizontally and vertically. Random cropping is automatically applied by nnU-Net when the image size is larger than the patch size. On top of *BL* different DA transformations are added and their effect to the models' performance is measured. For this we consider transformations in image appearance (e.g. brightness, contrast), image shape (e.g. elastic deformations), noise, and orientation (e.g. rotation, random flipping). The specifications of different DA settings can be seen in Table 2.

All of the utilized transformations for our experiments are included in the batchgenerators [12] library and do not require manual implementation. DA is performed on-the-fly during training, no additional hard drive storage is needed.

Table 2. Data augmentation configurations for initial cross domain experiments. The table is thought so complement Fig. 1 and give quantitative information about DA methods. Settings for [8] are presented as baseline *BL*. We extend DA by *inv. gamma*: inverse gamma , *gaussian*: Gaussian blur and noise, *multi./additive br.*: multiplicative/additive brightness, *all*: all of the aforementioned, *BL enhanced*: ranges of *BL* enhanced, *BL enhanced + br.*: ranges of *BL* enhanced and brightness enhanced, *heavy data augmentation*: combination of different DA settings.

Experiment	Rotation	Gamma	Inv. gamma	Gaussian	Add. brightness	Multi. brightness	Contrast
Default nnU-Net	±30	(0.7, 1.5)	–	Yes		$\mu = 0$, $\sigma = 0.1$	No
BL	±30	–	–	No		–	No
BL enhanced	±60	–	–	No	–	–	No
BL enhanced + br	±60	–	–	No	(0.6, 1.5)	$\mu = 0$, $\sigma = 0.2$	No
BL + all	±30	(0.7, 1.5)	(0.7, 1.5)	Yes	(0.7, 1.3)	$\mu = 0$, $\sigma = 0.1$	Yes
Heavy DA	±180	(0.6, 1.6)	(0.6, 1.6)	Yes	(0.7, 1.3)	$\mu = 0$, $\sigma = 0.3$	Yes

2.3 Model Architecture

Our experimental setup was built with nnU-Net, a dynamic fully automatic segmentation framework for medical images which is based on the widely used U-Net architecture. nnU-Net has shown impressive results in segmentation tasks for different organs and comes with great out-of-the-box usability.[2]

2.4 Experiments and Results

In the following, we refer to *intra-domain* performance when a network was trained and evaluated on the same domain (e.g. $A(B) \rightarrow A(B)$) and we refer

---

[2] Code available at: https://github.com/MIC-DKFZ/nnUNet

to *cross-domain* performance when a network trained on data from one domain and evaluated on another domain (e.g. $A(B) \rightarrow B(A)$). Further a *mixed domain* scenario is studied, in which data from both vendor A and B is used during training time. We investigate the following research questions:

Effect of Data Augmentation on Cross-domain Performance for both *cross-domain* scenarios $A(B) \rightarrow B(A)$. All 75 patients from vendor A(B) were used for training(evaluation). All nnU-Net configurations are fixed through-out these experiments except for DA settings, as described above. This experiment serves as a baseline for the nnU-Net in a typical clinical setting, in which inference is done on a data domain that differs from the training data domain.

Cross-domain results are presented in Fig. 1. While the dice scores for $A \rightarrow B$ are constantly >0.8 (best dice score 0.8548 for *BL enhanced + br*), $B \rightarrow A$ performs significantly worse for all DA settings (best dice score 0.6622 for *BL enhanced*). This result indicates that DA alone is not able to ensure robust cross-domain performance, especially for $B \rightarrow A$.

Results from the ACDC challenge suggest that higher dice scores should be achievable [2]. However, the ACDC challenge reports *intra domain* results. It remains to be investigated whether lower dice scores are to be expected on the M&Ms data in general or whether the performance gap between ACDC and M&Ms is due to the differing scenario (*intra* vs *cross domain*). An evaluation on the same validation set is desirable to allow for a fair comparison between models.

Gap Between Intra- and Cross-domain Performance: 25 patients from each vendor are held-out as a validation set. All results will be reported on this fixed evaluation sets. The remaining 50 patients from each vendor are utilized to generate four different training datasets: 1) 50 patients from vendor A, 2) 50 patients from vendor B, 3+4) 25 patients from vendor A and 25 patients from vendor B (*mixed I+II*). With this experimental setup, we study a potential performance gap of $A \rightarrow B$ vs $B \rightarrow A$ on the same hold-out validation set for different DA settings. Furthermore, we investigate whether pooling data from multiple domains during training improves test time performance on individual domains.

The results are shown in Table 3. Our main observations are that 1) no significant deviations of performance are observed between models trained under the *intra-domain* setting ($A(B) \rightarrow A(B)$) and the *mixed domain* setting ($AB \rightarrow A(B)$). 2) although DA increases the performance for both intra- and cross-domain tasks, the domain gap for $B \rightarrow A$ cannot be closed with our training pipeline.

Our findings render two possibilities how to improve segmentation results on the target domain: 1) if available add images from the target domain to the training data or 2) if no data from the target domain is available during training, more configurations than just changing DA will have to be optimized. Both will be investigated in the following experiments.

Effect of target domain data in the training set is studied by steadily increasing the number of images from the target domain in the training data. The previous experiment showed that a good segmentation performance was achieved on domain A when a *mixed domain* approach ($AB \rightarrow A$) instead of the *cross domain* approach was used. It remains to be investigated which proportion of images from the initial training set and the target domain are needed to achieve desired segmentation results. Note, in this study we aim to keep the utilized architecture and training procedure fixed throughout all experiments and thus leave more sophisticated approaches such as transfer learning to future research.

Results are shown in Fig. 2. When images from the target domain are added to the training set, cross-domain performance increases until it saturates at $\sim$30% (15) target domain patients. Beneficial effects of target domain training data are most apparent for $B + \% * A \rightarrow A$, where certain proportions of images from domain A are added to a fixed stack of training images from domain B to predict on images from domain A. Furthermore we observe that the performance for $A + \% * B \rightarrow A$ and $B + \% * A \rightarrow B$ is not compromised even for large proportions of images from a domain that is not the same as the target domain.

Interdependent effects of data augmentation and batch normalization were studied in a strict *cross domain* scenario, in which no data from the target was used during training. Instead adding data from the target domain during training, further changes to the nnU-Net configuration were applied. Other studies suggest that introducing BN is beneficial to DNNs generalizability [5].

To this end, we train a models with little DA (*default nnU-Net*) and extensive DA (*mnms nnU-Net*). Both settings will be trained with 1) the default IN layers of nnU-Net and 2) BN layers. Otherwise the architecture of the network will stay untouched. All results will be reported on a fixed evaluation set of 15 patients per vendor. Our experiments comprise three data scenarios: 1+2) *cross domain* scenarios: Training on 60 patients from vendor A (B) and 3) *mixed domain*: Training on 60 patients from vendor A **plus** 60 patients from vendor B. All results are provided for 2D networks as well as 3D networks. Configuration details are given in Table 4

The results are presented in Table 5. Our main observations are: 1) the use of BN layer is especially beneficial, when applied with extensive DA in *cross domain* scenarios (in particular in the previous problematic case $B \rightarrow A$) and 2) *mixed domain* scenarios are on a par with *intra domain* scenarios independent of the chosen normalization technique and DA settings, once DA is extensive enough (compare with Table 3).

2.5   Final Model Selection

The following observations were considered for the final model submission at the M&Ms challenge: 1) Table 5 suggests that doubling training data from 60 to 120 patients (*mixed D*) by combining data from vendor A and B, the segmentation performance is on a par with the *intra D* setting. Therefore we will mix the training data set to achieve good results on the known domains A and B.

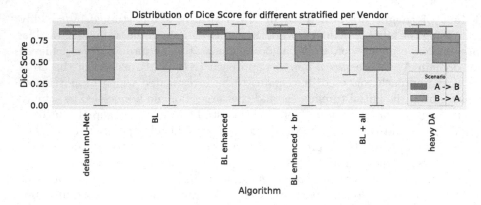

Fig. 1. *Cross-domain* dice scores for different data augmentation (DA) settings in 2D networks are presented. The blue boxes (left) represent $A \to B$, with constantly higher dice scores and lower dice score spread for all DA settings compared to $B \to A$. Note, that for all DA settings some patients achieved a dice score of 0 for $B \to A$. A detailed description of DA configuration can be found in Table 2

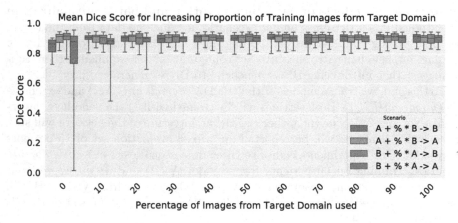

Fig. 2. Boxplot of mean dice scores for all three classes (LV, LVM, RV). For percentage = 0, only images from the cross-domain setting are included into the training set (100 cross-domain images, 0 images from the target domain). For percentage = 100 all images from the cross-domain setting as well as all images from the target domain are used during training. $B \to A = B + \% * A \to A$ for percentage = 0, has been the most problematic scenario in the experiments above. We observe that the mean dice score increases with an increasing number of images from the target domain. For 30% of images from the target domain, the domain gap between $A \to B$ and $B \to A$ is not observable anymore.

Table 3. Results for networks trained on one domain only (A or B) vs networks trained on images from both domains (mixed I or II), where mixed I/II use different, mutually exclusive subsets from both A and B. The columns hold the mean dice score achieved when evaluated for a specific target domain (A/B) or both domains (A & B). The four tables represent different data augmentation (DA) setting: upper left: *no DA*; upper right: *BL* DA; lower left and right: two DA settings that go beyond *BL*

Train on	Test on			Train on	Test on		
	A	B	A & B		A	B	A & B
A	**0.6417**	0.5212	0.5815	A	**0.8902**	0.8414	0.8658
B	0.1637	**0.6429**	0.4033	B	0.6126	0.9022	0.7574
Mixed I	0.6115	0.6052	0.6083	Mixed I	0.8845	0.8996	0.8918
Mixed II	0.6353	0.6202	**0.6277**	Mixed II	0.8857	**0.9049**	**0.8953**

Train on	Test on			Train on	Test on		
	A	B	A & B		A	B	A & B
A	**0.8869**	0.8412	0.8641	A	**0.8921**	0.8374	0.8647
B	0.5879	**0.9020**	0.7449	B	0.5669	0.9044	0.7356
Mixed I	0.8825	0.8940	0.8883	Mixed I	0.8807	0.899	0.8899
Mixed II	0.8800	0.8984	**0.8892**	Mixed II	**0.8921**	**0.9054**	**0.8987**

Assigning images randomly from both vendors, results in a ratio of $\sim 1{:}1$, which is well above the threshold of saturation (1:0.3), seen in Fig 2. 2) Table5 suggests that extensive DA combined BN yields large performance improvements especially for the cross domain setting (see *cross D*). Therefore we train our models with BN layers and extensive DA in order to optimize our models for the unseen domains C and D.

For the final submission we train an ensemble of five 2D and five 3D nnU-Net models using the BN layers together with extensive DA. The trained data is split into five folds comprising 80% of images from vendor A and B, while the left-out 20% are mutually exclusive for each fold. For 3D networks we will further adjust the resampling configuration. We will set the z-spacing to be the minimal z-spacing from all training cases instead of the nnU-Net default, which (for anisotropic dataset) uses the 10th percentile of the z-spacings found across all training cases.

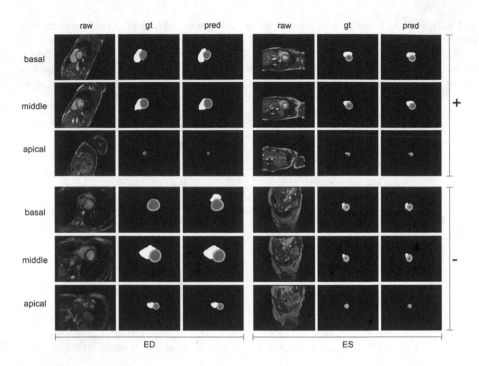

Fig. 3. Examples of good (+) and poor (−) segmentation results from our final submission for end-diastolic (ED) and end-systolic (ES) time points for basal, mid and apical slices for four different patients from the test set. The definition of *good* and *poor* was given by the challenge organizers. The following vendors are presented: ED(+): vendor D, ES(+): vendor B, ED(−): vendor A, ES(−): vendor C. While the ground truth (gt) and our predictions (pred) are very well aligned for the good cases, some deviations are observed for the poor cases. Comparing the segmentation for ED in row four, we observed that our model predicted a label for the RV, while the human raters did not agree. Note, that automatic and human raters in general disagree the most in the most basal slide, as ambiguities are considered to be most distinct in this part of the heart. Overall good and poor predictions are visually plausible.

2.6   Final Results on Test Set

The scores for vendor A and vendor B (domains known from training) are slightly above the scores from vendor C and vendor D for LV, LVM, and RV. However we observe that our model achieves a similar score on all four vendors indicating that our submitted model is robust to cMRI data from different vendors and centres in the $M\&Ms$ challenge. Quantitiaive results are shown in Table 6. Qualitative segmentation examples from our test set submission can be seen in Fig. 3. Note, test set ground truth information was not accessible during the challenge and only made available to us for the purpose of this figure.

Table 4. Main differences in data augmentation and normalization layer settings for the baseline *nnU-Net default* and the final submission model *mnms nnU-Net*. Values in brackets represent ranges for the given data augmentation technique. Probabilities that the data augmentation technique will be applied to an image is given as $p = X$. All other architecture features of the nnU-Net were kept unmodified.

Setting	Default nnU-Net	mnms nnU-Net
Rotation	$p = 0.2$	$p = 0.7$
Elastic deformations	–	$p = 0.1$
Independent scale factor per axis	–	$p = 0.3$
Elastic deformation alpha	(0, 200)	(0, 300)
Elastic deformation sigma	(9, 13)	(9, 15)
Scale	$p = 0.2$	$p = 0.3$
Gamma range	(0.7, 1.5)	(0.5, 1.6)
Additive brightness mu	–	0
Additive brightness sigma	–	0.2
Normalization layers	IN	BN

Table 5. The effect of batch normalization (BN) compared to instance normalization (IN) for two different DA settings. The results for 2D and 3D are presented. The nnU-net's default DA scheme is presented as *default nnU-Net*. Our model final submission model (*mnms nnU-Net*) showed the best overall performance for 2D and 3D. The three lowest rows in each tabular summarize the six above scenarios into three categories: *intra D* ($A \rightarrow A$ and $B \rightarrow B$), *cross D* ($A \rightarrow B$ and $B \rightarrow A$) and *mixed D* ($AB \rightarrow A$ and $AB \rightarrow B$). For *mixed D* 120 instead of only 60 patients were used for training. Note, that all networks were evaluated on the same hold-out dataset of 15 patients per vendor.

2D	Default nnU-Net (IN)	Default nnU-Net (BN)	mnms nnU-Net (IN)	mnms nnU-Net (BN)
$A \rightarrow A$	0.8718	**0.8832**	0.8804	0.8800
$A \rightarrow B$	0.8502	0.8594	0.8512	**0.8735**
$B \rightarrow A$	0.6287	0.8046	0.5809	**0.8457**
$B \rightarrow B$	0.9121	**0.9125**	0.9063	0.905
$AD \rightarrow A$	**0.8948**	0.8911	0.8877	0.8880
$AB \rightarrow B$	0.9066	**0.9074**	0.9069	0.9063
Intra D	0.8949	**0.8983**	0.8947	0.8941
Cross D	0.7394	0.8320	0.7161	**0.8596**
Mixed D	**0.9007**	0.8993	0.8973	0.8972
3D	Default nnU-Net (IN)	Default nnU-Net (BN)	mnms nnU-Net (IN)	mnms nnU-Net (BN)
$A \rightarrow A$	0.8900	**0.8914**	0.8738	0.8838
$A \rightarrow B$	0.8513	0.8640	0.8568	**0.8761**
$B \rightarrow A$	0.5668	0.6203	0.5525	**0.7955**
$B \rightarrow B$	**0.9112**	0.9110	0.90859	0.9098
$AB \rightarrow A$	0.8860	0.8917	0.8832	**0.8931**
$AB \rightarrow B$	0.9090	**0.9104**	0.90677	0.9072
Intra D	0.8995	**0.9012**	0.8925	0.8979
Cross D	0.7090	0.7421	0.7046	**0.8358**
Mixed D	0.8975	**0.9011**	0.8950	0.9001

Table 6. Mean Dice Score from the final test set stratified for vendors.

	Vendor A	Vendor B	Vendor C	Vendor D
LV	0.9232	0.9146	0.9032	0.9091
LVM	0.8571	0.8761	0.8418	0.8384
RV	0.8870	0.8876	0.8838	0.8822

3   Conclusion

We presented an approach to increase robustness against domain shifts in cMRI data from different vendors and centres while at the same time maintaining out-of-the-box usability. This was achieved by 1) building on nnU-Net, a publicly available segmentation framework that achieves state-of-the-art performance on intra-domain segmentation tasks and 2) systematically investigating improvements by means of different data augmentation schemes as an easy-to-customize pipeline component and 3) combining BN with extensive DA to achieve significant cross-domain performance gains.

A limitation of this work is, that we cannot predict whether a data domain will be robustly predicted by a model or not (i.e. why $A \rightarrow B$ works better than $B \rightarrow A$). We will leave this to future research.

We expect these insights to be helpful for typical clinical scenarios in which expert-annotations covering all data domains are not available, but deep learning approaches are to be deployed to minimize cumbersome manual annotation effort.

Acknowledgement. The authors of this paper declare that the segmentation method they implem- ented for participation in the $M\&Ms$ challenge has not used any pre-trained models nor additional MRI datasets other than those provided by the organizers. Peter M. Full holds a Kaltenbach scholarship from the German Heart Foundation (Deutsche Herzstiftung). Fabian Isensee is funded by the Helmholtz Imaging Platform (HIP).

References

1. Suinesiaputra, A., et al.: A collaborative resource to build consensus for automated left ventricular segmentation of cardiac MR images. Med. Image Anal. **18**, 50–62 (2014). https://doi.org/10.1016/j.media.2013.09.001
2. Bernard, O., et al.: Deep learning techniques for automatic MRI Cardiac multi-structures segmentation and diagnosis: is the problem solved? IEEE Trans. Med. Imaging **37**, 2514–2525 (2018). https://doi.org/10.1109/TMI.2018.2837502
3. Isensee, F., Jäger, P.F., Kohl, S.A.A., Petersen, J., Maier-Hein, K.H.: Automated design of deep learning methods for biomedical image segmentation. arXiv preprint arXiv:1904.08128 [cs]. (2020)
4. Campello, Víctor M. et al.: Multi-Centre, Multi-Vendor & Multi-Disease Cardiac Image Segmentation. (in preparation)

5. Bjorck, N., Gomes, C.P., Selman, B., Weinberger, K.Q.: Understanding batch normalization. In: Bengio, S., Wallach, H., Larochelle, H., Grauman, K., Cesa-Bianchi, N., Garnett, R. (eds.) Advances in Neural Information Processing Systems 31, pp. 7694–7705. Curran Associates, Inc. (2018)
6. Volpi, R., Namkoong, H., Sener, O., Duchi, J.C., Murino, V., Savarese, S.: Generalizing to unseen domains via adversarial data augmentation. In: Bengio, S., Wallach, H., Larochelle, H., Grauman, K., Cesa-Bianchi, N., Garnett, R. (eds.) Advances in Neural Information Processing Systems 31, pp. 5334–5344. Curran Associates, Inc. (2018)
7. Karani, N., Chaitanya, K., Baumgartner, C., Konukoglu, E.: A lifelong learning approach to brain MR segmentation across scanners and protocols. In: Frangi, A.F., Schnabel, J.A., Davatzikos, C., Alberola-López, C., Fichtinger, G. (eds.) MICCAI 2018. LNCS, vol. 11070, pp. 476–484. Springer, Cham (2018). https://doi.org/10.1007/978-3-030-00928-1_54
8. Chen, C., et al.: Improving the Generalizability of Convolutional Neural Network-Based Segmentation on CMR Images. (2020). https://doi.org/10.3389/fcvm.2020.00105
9. Zech, J.R., Badgeley, M.A., Liu, M., Costa, A.B., Titano, J.J., Oermann, E.K.: Variable generalization performance of a deep learning model to detect pneumonia in chest radiographs: a cross-sectional study. PLOS Med. 15, e1002683 (2018). https://doi.org/10.1371/journal.pmed.1002683
10. Zhang, L., et al.: Generalizing deep learning for medical image segmentation to unseen domains via deep stacked transformation. IEEE Trans. Med. Imaging 39, 2531–2540 (2020). https://doi.org/10.1109/TMI.2020.2973595
11. Sandfort, V., Yan, K., Pickhardt, P.J., Summers, R.M.: Data augmentation using generative adversarial networks (CycleGAN) to improve generalizability in CT segmentation tasks. Sci. Rep. 9, 16884 (2019). https://doi.org/10.1038/s41598-019-52737-x
12. Isensee, F., et al.: batchgenerators - a python framework for data augmentation (2020). https://github.com/MIC-DKFZ/batchgenerators, https://doi.org/10.5281/zenodo.3632567 (2020)

A Deep Convolutional Neural Network Approach for the Segmentation of Cardiac Structures from MRI Sequences

Adam Carscadden[1,2], Michelle Noga[1,2], and Kumaradevan Punithakumar[1,2(✉)]

[1] Deparment of Radiology and Diagnostic Imaging, University of Alberta, Edmonton, Canada
punithak@ualberta.ca
[2] Servier Virtual Cardiac Centre, Mazankowski Alberta Heart Institute, Edmonton, Canada

Abstract. Cardiac magnetic resonance imaging typically generates hundreds of images in each scan, and manual delineation of structures from these scans is tedious and time-consuming. Automating the process is of great interest as it will allow for generating clinical measurements such as ejection fraction and stroke volume. However, automated segmentation poses several challenges due to the difference in scanning parameters and patient characteristics, which lead to high variability in image statistics and quality. In this study, we propose a neural network approach based on U-Net++ and ResNet101 architectures to segment the endocardial and epicardial borders of the left ventricle as well as the right ventricle. The method relied on the Dice metric as the loss function for optimizing the trainable network parameters. We also applied a preprocessing step to retain only the largest connected component of the segmented labels in the predicted binary image by the neural network. The proposed algorithm was trained using the datasets provided by Multi-Centre, Multi-Vendor & Multi-Disease Cardiac Image Segmentation Challenge hosted by MICCAI 2020 conference. The evaluation of the method was performed by the challenge organizers using a test set consisting of 200 datasets, where 160 were used for testing and ranking.

Keywords: Cardiac magnetic resonance imaging · Deep convolutional neural networks · Left ventricle · Right ventricle · Image segmentation

1 Introduction

Cardiac magnetic resonance imaging (MRI) is a crucial diagnostic tool in the management of cardiac patients as it allows for the assessment of the heart function [16]. One of the challenges in utilizing the full potential of cardiac MRI is the requirement for intense manual post-processing for the generation of clinical parameters. Cardiac MRI typically produces 200 to 300 cine images per scan and manual contouring to delineate the cardiac structures such as left

© Springer Nature Switzerland AG 2021
E. Puyol Anton et al. (Eds.): STACOM 2020, LNCS 12592, pp. 250–258, 2021.
https://doi.org/10.1007/978-3-030-68107-4_25

ventricle endocardium (LV), myocardium (MYO), and right ventricle (RV) is tedious and time-consuming. Regardless, clinical parameters such as ejection fractions of left and right ventricles as well as myocardial mass require accurate delineation of the LV, MYO and RV. Automating the segmentation process to delineate cardiac structures is of great importance, and it has been an active area of research [4].

Over the past few decades, many traditional computer vision algorithms were proposed for the cardiac MRI segmentation problem [11]. Majority of the earlier methods focused on the segmentation of the left ventricular structures such as endocardium and epicardium. A wide range of solutions were proposed, including thresholding [6], knowledge-based models [10], deformable models [15], atlas-based models [7], active contours [1,14], graph cuts [2,3] and continuous max-flow min-cut [9] to delineate the LV. Some researchers also proposed methods to delineate the RV [12,13], which is considered more challenging than the delineation of the LV due to its thin borders and complex morphology.

Recently, deep convolutional neural network-based approaches have been proposed for the automated segmentation of cardiac structures from MRI sequences [4,8,17]. The convolutional neural network approaches remove the need for hand-crafted features and are capable of learning features using the information available on the training data.

In this study, we propose a neural network approach based on U-Net++ architecture [19] to delineate the LV, MYO and RV from cardiac MRI. The proposed method relies on three separate neural network algorithm that segments LV, LV+MYO, LV+MYO+RV regions to avoid performing semantic segmentation of the ring-shaped structure MYO. The ResNet101 was used as the backbone for all three neural network architectures. The proposed method was trained using 150 MRI scans provided by the Multi-Centre, Multi-Vendor & Multi-Disease Cardiac Image Segmentation Challenge [5]. The method was evaluated by the challenge organizers on a test set consisting of 160 scans, which contain images acquired from scanners that are not included in the training set.

2   Methodology

2.1   Training Data

The training set consists of cardiac MRI images from two different vendors, denoted by A and B, with the corresponding manual labels at the end-diastolic (ED) and end-systolic (ES) phases. The annotated training sets consist of scans acquired from 75 patients for vendor A and 75 patients for vendor B. Each ground truth label image consists of pixel-wise labeling of LV, MYO, RV and background where the corresponding label values were 1, 2, 3 0. A total of 6636 2D slices and the corresponding labels were used in the training process, where 80% and 20% of the total set were designated as the training and validation sets, respectively. Although 25 additional image datasets from another vendor, C, was provided by the challenge organizers, the additional datasets were not included in the training process since the corresponding ground truth labels were not provided.

2.2 Neural Network

The proposed method utilized the U-Net++ architecture with ResNet101 as the backbone for the segmentation task [18,19]. Three separate neural networks were constructed and trained using ground truth labels corresponds to LV, LV+MYO and LV+MYO+RV, respectively. We used $(1 - \text{Dice metric})$ as the loss function for all three networks. The Adam optimizer was used for optimizing the trainable neural network parameters. Each neural network consists of 56.6 million trainable parameters. The following neural network options were used during the training process: Initial learning rate $= 1 \times 10^{-5}$, batch size $= 8$, number of epochs $= 300$. Dice metric is used for evaluating the accuracy of neural networks on the validation set during the training process.

2.3 Preprocessing

All images were normalized using 5th and 95th percentile values, I_{05} and I_{95}, of the intensity distribution of the 4D data to obtain relatively uniform training sets. The normalized intensity value, I_n, is computed using $I_n = \dfrac{I - I_{05}}{I_{95} - I_{05}}$ where I denotes the original pixel intensity. We also perform cropping to obtain images of size 224×224, where the center of a cardiac MRI image was used as the center point for cropping. For any image with height or width smaller than 224 pixels, zero-padding is applied on the respective dimension to obtain a final image size of 224×224.

2.4 Implementation and Training

The proposed approach was implemented using Python programming language with Keras and Tensorflow neural network modules. Image manipulations and array computations were performed using Scikit-Image and Numpy modules. All three networks took around 10 h and 30 min to complete the training process consisting of 300 epochs on NVIDIA Tesla–P100 graphical processing units with 12 GB memory. The trained neural network model with the highest validation accuracy was saved to the disk.

2.5 Post-processing

We applied post-processing to retain only the largest connected component for the predictions by each network. The operation was performed in 3D space instead of performing the operation on a slice-by-slice basis. In addition, we applied a post-processing operation to remove holes that appear inside the foreground masks, which was also applied in 3D space.

3 Results

The proposed method was evaluated over images acquired from a total of 160 patients. The test set consisted of 40 MRI scans from vendors A, B, C, and D each [5]. The evaluation of the proposed method on test sets was performed by the challenge organizers. The images acquired from vendor D was not shared with the participants.

The agreement between the segmentation using the proposed method and manual ground truth delineations was quantitatively evaluated using the Dice metric, Jaccard index, Hausdorff distance and average symmetric surface distance (ASSD). The overall performance of the proposed algorithm over the entire test set consisting of 160 scans is reported in Table 1. The highest performance was observed for the LV with a mean Dice metric of 88.2%. The Dice metrics corresponding to the segmentation of the MYO and RV were 79.3% and 76.2%, respectively. Figure 1 shows the overall performance of the proposed method using box plots.

Table 1. Overall performance of the proposed method evaluated over MRI test datasets acquired from 160 patients using scanners produced by vendors A, B, C and D.

Region of interest	Dice metric (%)	Jaccard index (%)	Hausdorff distance (mm)	ASSD (mm)
LV	88.2 ± 10.0	80.0 ± 12.2	13.7 ± 13.9	1.65 ± 3.38
MYO	79.3 + 9.0	66.5 ± 11.4	16.2 ± 15.2	1.23 ± 1.44
RV	76.2 ± 19.6	64.7 ⊥ 20.8	31.9 ± 19.6	3.36 ± 6.46

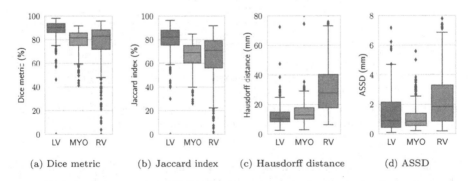

(a) Dice metric (b) Jaccard index (c) Hausdorff distance (d) ASSD

Fig. 1. The boxplots showing the performance of the proposed approach over test sets acquired from 160 patients. The evaluations were performed using Dice metric, Jaccard index, Hausdorff distance and ASSD.

Table 2 reports the performance of the proposed method measure in terms of Dice metric over the test images from vendors A, B, C and D. The proposed method yielded best overall results for the images from vendor B. Despite not included in the training set, the method performs well on the test images from vendors C and D, especially for the segmentation of the LV. The proposed method yielded an average Dice metric of 87.5% and 86.9% for the LV segmentation for the images from vendors C and D, respectively. Figure 2 shows the reliability assessment [2] of the proposed method for the segmentation of the LV, MYO and RV for images from different vendors. Figure 2(a) indicate that the proposed method yielded similar reliability for the segmentation of the LV regardless of the vendor type. The highest variability in the reliability is observed for the RV segmentation as shown in Fig. 2(c). Yea

Table 2. The performance of the proposed method evaluated in terms of Dice metric (%) on 4 separate test sets each consisting of scans acquired from 40 patients on scanners produced by vendors A, B, C and D.

Region of interest	Vendor A	Vendor B	Vendor C	Vendor D
LV	88.5 ± 8.4	89.9 ± 7.3	87.5 ± 12.2	86.9 ± 11.1
MYO	78.1 ± 10.2	84.6 ± 5.2	78.7 ± 9.1	75.8 ± 8.3
RV	77.8 ± 14.7	84.6 ± 11.5	77.3 ± 16.7	65.0 ± 26.9

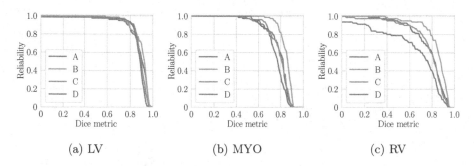

(a) LV   (b) MYO   (c) RV

Fig. 2. The reliability curves for the results by the proposed approach for the images acquired from MRI scanners produced by vendors A, B, C and D.

Table 3 reports the segmentation accuracy of the proposed approach in terms of Hausdorff distance in comparison to the ground truth delineations. The reported values indicate that the proposed method yielded the best overall performance for scans from vendor B. Figure 3 depicts the segmentation accuracy for scans from each vendor in terms of Hausdorff distance.

Overall, the proposed method yielded higher agreement with manual delineation for all three regions of interest considered in this study. Figure 4 shows example segmentation results where the proposed method achieved the highest agreement with the ground truth delineations. The contours for the LV, MYO,

Table 3. The performance of the proposed method evaluated in terms of Hausdorff distance (mm) on 4 separate test sets each consisting of scans acquired from 40 patients on scanners produced by vendors A, B, C and D.

Region of interest	Vendor A	Vendor B	Vendor C	Vendor D
LV	14.9 ± 16.6	9.8 ± 4.7	12.5 ± 9.0	17.7 ± 19.2
MYO	18.4 ± 18.9	12.4 ± 5.0	16.1 ± 11.6	17.7 ± 20.0
RV	36.8 ± 18.7	26.7 ± 15.1	29.3 ± 14.2	34.7 ± 26.5

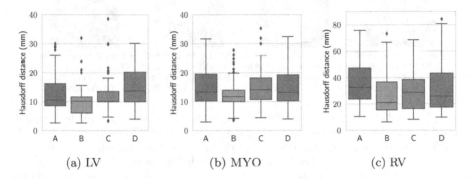

(a) LV   (b) MYO   (c) RV

Fig. 3. The performance of the proposed approach with images acquired from MRI scanners produced by vendors A, B, C and D.

and RV regions are respectively given in yellow, orange, and purple colors, where the darker and lighter colored contours correspond to the manual and automated delineations, respectively.

The cases where the proposed method deviated the most from the manual delineations are given in Fig. 5. Significant deviations are observed for the automated delineation of the ventricles from images acquired from vendor D.

4   Conclusion

We proposed a U-Net++ based neural network architecture to segment the left and right ventricular regions from cardiac MRI images. The proposed approach relied on three separate neural networks trained to predict the LV, LV+MYO, and LV+MYO+RV regions. The ResNet101 was used as the backbone for all three neural network, and the algorithms were trained using images acquired from 150 patients using scanners by two different vendors. An image intensity normalization and cropping operations were utilized as the preprocessing steps. Evaluation of the proposed method was performed by the M&M challenge orgnizers over 160 scans. The proposed method yielded higher conformance with manual delineations of the regions despite the test set consisted of images from vendors that were not included as a part of the training set. The proposed

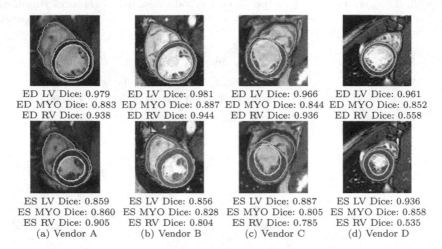

ED LV Dice: 0.979 ED LV Dice: 0.981 ED LV Dice: 0.966 ED LV Dice: 0.961
ED MYO Dice: 0.883 ED MYO Dice: 0.887 ED MYO Dice: 0.844 ED MYO Dice: 0.852
ED RV Dice: 0.938 ED RV Dice: 0.944 ED RV Dice: 0.936 ED RV Dice: 0.558

ES LV Dice: 0.859 ES LV Dice: 0.856 ES LV Dice: 0.887 ES LV Dice: 0.936
ES MYO Dice: 0.860 ES MYO Dice: 0.828 ES MYO Dice: 0.805 ES MYO Dice: 0.858
ES RV Dice: 0.905 ES RV Dice: 0.804 ES RV Dice: 0.785 ES RV Dice: 0.535
   (a) Vendor A   (b) Vendor B   (c) Vendor C   (d) Vendor D

Fig. 4. Examples showing ground truth and predicted contours where the proposed method had a higher agreement with the manual delineations. Manual contours for the LV, MYO and RV are given in darker colors of yellow, orange, and purple, respectively. The corresponding automated contours are given in lighter colors of yellow, orange, and purple. The first and second rows correspond to end-diastolic and end-systolic phases. (Color figure online)

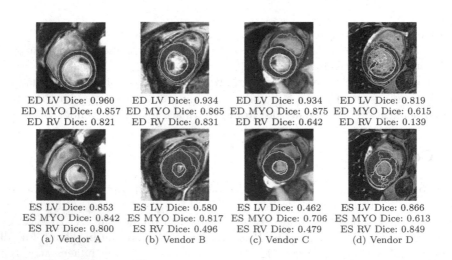

ED LV Dice: 0.960 ED LV Dice: 0.934 ED LV Dice: 0.934 ED LV Dice: 0.819
ED MYO Dice: 0.857 ED MYO Dice: 0.865 ED MYO Dice: 0.875 ED MYO Dice: 0.615
ED RV Dice: 0.821 ED RV Dice: 0.831 ED RV Dice: 0.642 ED RV Dice: 0.139

ES LV Dice: 0.853 ES LV Dice: 0.580 ES LV Dice: 0.462 ES LV Dice: 0.866
ES MYO Dice: 0.842 ES MYO Dice: 0.817 ES MYO Dice: 0.706 ES MYO Dice: 0.613
ES RV Dice: 0.800 ES RV Dice: 0.496 ES RV Dice: 0.479 ES RV Dice: 0.849
   (a) Vendor A   (b) Vendor B   (c) Vendor C   (d) Vendor D

Fig. 5. Examples showing ground truth and predicted contours where the proposed method had larger deviations from the manual delineations. Manual contours for the LV, MYO, and RV are given in darker colors of yellow, orange, and purple, respectively. The corresponding automated contours are given in lighter colors of yellow, orange, and purple. The first and second rows correspond to end-diastolic and end-systolic phases. (Color figure online)

method yielded overall mean Dice metrics of 88.2%, 79.3% and, 76.2% for the LV, MYO and RV delineations.

Acknowledgment. The authors wish to thank the challenge organizers for providing train and test datasets as well as performing the algorithm evaluation. The authors of this paper declare that the segmentation method they implemented for participation in the M&Ms challenge has not used any pre-trained models nor additional MRI datasets other than those provided by the organizers.

References

1. Ben Ayed, I., Li, S., Ross, I.: Embedding overlap priors in variational left ventricle tracking. IEEE Trans. Med. Imaging **28**(12), 1902–1913 (2009)
2. Ben Ayed, I., Mei Chen, H., Punithakumar, K., Ross, I., Li, S.: Max-flow segmentation of the left ventricle by recovering subject-specific distributions via a bound of the Bhattacharyya measure. Med. Image Anal. **16**(1), 87–100 (2012)
3. Ben Ayed, I., Punithakumar, K., Li, S., Islam, A., Chong, J.: Left ventricle segmentation via graph cut distribution matching. In: Yang, G.-Z., Hawkes, D., Rueckert, D., Noble, A., Taylor, C. (eds.) MICCAI 2009. LNCS, vol. 5762, pp. 901–909. Springer, Heidelberg (2009). https://doi.org/10.1007/978-3-642-04271-3_109
4. Bernard, O., et al.: Deep learning techniques for automatic MRI cardiac multi-structures segmentation and diagnosis: is the problem solved? IEEE Trans. Med. Imaging **37**(11), 2514–2525 (2018)
5. Campello, V.M., et al.: Multi-centre, multi-vendor and multi-disease cardiac image segmentation (In preparation)
6. Liu, H., Hu, H., Xu, X., Song, E.: Automatic left ventricle segmentation in cardiac MRI using topological stable-state thresholding and region restricted dynamic programming. Acad. Radiol. **19**(6), 723–731 (2012)
7. Lorenzo-Valdés, M., et al.: Segmentation of 4D cardiac MR images using a probabilistic atlas and the EM algorithm. Med. Image Anal. **8**(3), 255–265 (2004)
8. Martin-Isla, C., et al.: Image-based cardiac diagnosis with machine learning: a review. Front. Cardiovasc. Med. **7**, 1–19 (2020)
9. Nambakhsh, C.M., et al.: Left ventricle segmentation in MRI via convex relaxed distribution matching. Med. Image Anal. **17**(8), 1010–1024 (2013)
10. Paragios, N.: A level set approach for shape-driven segmentation and tracking of the left ventricle. IEEE Trans. Med. Imaging **22**(6), 773–776 (2003)
11. Petitjean, C., Dacher, J.N.: A review of segmentation methods in short axis cardiac MR images. Med. Image Anal. **15**(2), 169–184 (2011)
12. Petitjean, C., et al.: Right ventricle segmentation from cardiac MRI: a collation study. Med. Image Anal. **19**(1), 187–202 (2015)
13. Punithakumar, K., Boulanger, P., Noga, M.: A GPU-accelerated deformable image registration algorithm with applications to right ventricular segmentation. IEEE Access **5**, 20374–20382 (2017)
14. Punithakumar, K., et al.: Detection of left ventricular motion abnormality via information measures and Bayesian filtering. IEEE Trans. Inf. Technol. Biomed. **14**(4), 1106–1113 (2010)
15. Punithakumar, K., et al.: Regional heart motion abnormality detection: an information theoretic approach. Med. Image Anal. **17**(3), 311–324 (2013)

16. Raisi-Estabragh, Z., et al.: Cardiac magnetic resonance radiomics: basic principles and clinical perspectives. Eur. Heart J. - Cardiovasc. Imaging **21**(4), 349–356 (2020)
17. Tran, P.V.: A fully convolutional neural network for cardiac segmentation in short-axis MRI. arXiv preprint arXiv:1604.00494 (2016)
18. Zhou, Z., Rahman Siddiquee, M.M., Tajbakhsh, N., Liang, J.: UNet++: a nested U-net architecture for medical image segmentation. In: Stoyanov, D., et al. (eds.) DLMIA/ML-CDS -2018. LNCS, vol. 11045, pp. 3–11. Springer, Cham (2018). https://doi.org/10.1007/978-3-030-00889-5_1
19. Zhou, Z., Siddiquee, M.M.R., Tajbakhsh, N., Liang, J.: Unet++: redesigning skip connections to exploit multiscale features in image segmentation. IEEE Trans. Med. Imaging **39**, 1856–1867 (2019)

Multi-center, Multi-vendor, and Multi-disease Cardiac Image Segmentation Using Scale-Independent Multi-gate UNET

Mina Saber[1(✉)], Dina Abdelrauof[1(✉)], and Mustafa Elattar[1,2(✉)]

[1] Research and Development Division, Intixel Co. S.A.E., Cairo, Egypt
{mina.saber,dina.abdelrauof}@intixel.com, melattar@nu.edu.eg
[2] Medical Imaging and Image Processing Group, Information Technology and
Computer Science School, Nile University, Giza, Egypt

Abstract. Heart segmentation in Cardiac MRI images is a fundamental step to quantify myocardium global function. In this paper, we introduce a pipeline for heart localization and segmentation that is fast and robust even in the apical slices that have small myocardium. Also, we propose an enhancement to the popular U-Net architecture for segmentation. The proposed method utilizes the aggregation of different feature scales from the image by using the inception block along with the multi-gate block that propagates the multi-scale context of the supplied data where the heart is subject to changes in scale, contrast and resolution due to the different imaging parameters from different vendor images. The model is trained using the data provided in Multi-center, Multi-vendor, and Multi-Disease Cardiac Image Segmentation Challenge (M&Ms challenge) with no prior training or transfer learning. The model achieved an average dice of 90%,84.5%, and 82.5% for the left ventricle blood pool, right ventricle, and Myocardium respectively. Moreover, the proposed pipeline took an average inference time of 0.0022 s per image when running on NVIDIA GTX 1080 TI GPU.

Keywords: Cardiac segmentation · Magnetic resonance imaging · Segmentation

1 Introduction

Heart segmentation is an initial step to get measures like Left Ventricle (LV) mass and global function which is an essential step for the diagnosis and therapy of patients with CAD [4]. Using magnetic resonance cardiac imaging is considered the gold standard to quantify the global LV function since it is non-invasive procedure producing high resolution images [4,10]. However, manually annotating the images is a daunting task and highly time-consuming since it requires the iteration over large number of slices per patient then go through a

© Springer Nature Switzerland AG 2021
E. Puyol Anton et al. (Eds.): STACOM 2020, LNCS 12592, pp. 259–268, 2021.
https://doi.org/10.1007/978-3-030-68107-4_26

large number of images per slice to select the end-systolic and end-diastolic time frames then start delineation. Moreover, manual annotation suffers from inter and intra-observer variability between annotators. So, a fast, robust and result reproducing method is required to achieve this task. A significant number of approaches have been proposed to this specific task. Some classical approaches used deformable models and level sets [2], also, some used graph cuts and active shape models [5], and multi-level Otsu thresholding used [6]. However, the main issue with those models is the lack of robustness when tested with new cases [12]. Recently, fully convolutional (FCN) encoder-decoder neural networks are gaining popularity in the medical image segmentation problems. U-Net; proposed by Ronneberger et al.; is an extension for encoder-decoder famous architecture [13]. The FCN architecture enhanced the segmentation accuracy across different datasets, however; in several studies; it was not capable of properly extracting the low-level features in the short-axis apical slices [16]. Here, we present a two-tire approach where we first do heart localization and extract a tile around this location then segment the resultant crop. For the segmentation network we propose an enhancement to the encoding path in the native U-Net to better encode the context and heart variability between images through the use of the dilated inception block and the multi-gate block.

2 Related Work

A significant number of approaches have been proposed in the lines of automatic LV segmentation. Tran et al. have proposed fully convolutional neural network architecture for semantic segmentation in cardiac MRI. However, this approach was unable to perform well with segmenting the apical slices, as the dice index drops by 20% when moving from basal to apical slices [16]. Khened et al. proposed fully convolutional multi-scale residual dense nets; a variant of U-Net with dense and residual connections; to train the region of interest (ROIs) [8]. Those ROIs were extracted using Fourier analysis alongside Hough transform making use of the fact that a heart in the short-axis view is almost circular. Also, Li Kuo et al. detected the center of the heart then calculated discrete Fourier Transform of the temporal information along the cardiac cycle to guide segmentation through different time frames [15]. This approach was limited to cine images where all time frames are available and will not perform properly on single or few time-frames such as end-systolic (ES) and end-diastolic (ED) only. Isensee et al. have utilized the 3D U-Net by ensembling it with conventional 2D U-Net, however, such a large number of parameters may lead to poor generalization [7]. Most of the mentioned techniques have common limitations such as being trained and validated only on the short-axis view, sub-optimal performance in segmenting apical slices and segmenting the papillary muscles as a part of the left ventricle. Kernel size has a great impact on the network performance [11]. The Heart and its chambers appear in the images with different scales due to different field of view scan settings as well as intra-population differences. This can be tackled by using both small and big size kernels feature extractors. Increasing the kernel size enables the network to extract more context and global features and reducing kernel size would help in extracting local features [11]. Moreover, increasing the field of view using large kernel size would dramatically increase the model parameters, quadratic relationship, which would lead to the risk of model over-fitting. Using sparse kernels, i.e dilated convolutions, introduced in [17] can lead to enlarged receptive field of view with significantly smaller number of parameters, also, the authors proposed the multi-scale context aggregation module which consists of consecutive layers of expanding dilated kernels. However, the increase in dilation factor would lead to poor context aggregation for small objects in the image [3]. The authors here proposed an enhancement to the exploding dilation context issue using dilation rate reduction (local feature extraction). However, image resolution is small in our case unlike the high resolution satellite images used in that paper, also the heart size compared to the total image area and anatomical structure can not go very low, otherwise the image would be challenging for humans to see. Inception networks [14] introduced the idea of combining multi-scale features in parallel rather than sequentially by utilizing parallel kernels with different scales followed by a 1×1 convolution layer to reduce the network width (number of features). Here, we utilize the parallel multi-scale aggregation with the use of dilated kernels as well. Unlike inception networks, we use the 1×1 convolution only after scale aggregation not before since we do not have large number of features to reduce. We also introduce another scaling mechanism in which we use the average pooling to produce multiple scales of the input image. That allows better use of object context and also decreases the total network size. In this study, we introduce a light-weight fully automated left ventricle segmentation technique that handles long-axis and short-axis views. The proposed architecture can be considered as an extension for the U-Net architecture that utilizes the power of combining Multi-Gate Input block and dilated convolution in the encoding path to aggregate the most representative features from the input image without assuming any prior information.

3   Methods

3.1   Data

The data-set is a composition of cine MRI images collected from four vendors and scanned in six different centers. Training data consisted of 175 patients, however, there were only 150 annotated patients and 25 were not annotated. The annotation provided was only for the end-systolic and end-diastolic time frames for each slice. Table 1 shows the distribution of the cases per vendor and center. We split the annotated cases 90:10 train and validation respectively, specifically, we split each center with this ratio to get a good distribution between vendors and centers. Additionally, the data-set contained 200 cases for testing and we report based on these cases' results. Image dimensions, height and width, range from 200 to 340 with a median of 240×255. As a preprocessing step for both of the networks we normalize the input image using min-max normalization to scale the value range between 0 and 1. For the segmentation network, training data was generated by extracting a crop of size 128×128 around the geometric center of the segmentation contour. Moreover, multiple crops were extracted around the heart center with random small shifts in the horizontal and vertical directions in order to increase the model's ability to generalize.

Table 1. Distribution of the training set among vendors and centers. Vendors A,B had annotation but C did not. Centers are identified by numbers from 1 to 4.

Vendor	Center	Number of cases
A	1	75
B	2	50
B	3	25
C	4	25

3.2   Heart Localization

As a pre-processing step, we trained a very light weight network to detect the heart location within the image [1]. The network is a simple encoder-decoder like network enabling very fast detection of the heart. Here, we try to get a rough segmentation of the heart then extract the center of the heart. Also, here we treat the output as a binary mask making no differentiation between left ventricle blood pool, right ventricle or myocardium , i.e the mask is two classes foreground (the heart) and background. The identified location is then made the center of a crop which is to be fed to the segmentation network. Ideally, the output of this stage should be a square of 128×128. However, if the detected location was close to the borders of the image we attempt to pad the this tile with zeros to obtain a square of 128×128.

3.3   Heart Segmentation Network

The proposed architecture for segmentation builds on the architecture of U-Net. The vanilla U-Net consisted of repeated blocks of two consecutive 3×3 convolutions in the encoding path. Here we refer to this block as the standard block. In the decoding path we use up-sampling followed by a 1×1 convolution and we refer to this as the up-sampling block. Furthermore, two mechanisms of scale aggregation are incorporated in this architecture. First, it utilizes the dilated inception block which consists of four parallel kernels with different of different sizes and dilation rates as depicted in Fig. 2. The second mechanism, which is called multi-gate block, we apply average pooling with multiple window sizes on the input image. Hence, we get the input image with multiple scales and different context information since the bigger the pooling window the smaller the output and the blurrier it would be. The purpose of this mechanism is to pass the image in different scales which would give the network the chance to scan the input with different resolutions and different level of details. We combine the multi-gate block with the inception block to be used with the input of the network. As depicted in Fig. 1, the multi-gate block is used three times to produce three image scales of 64×64, 32×32 and 16×16 before the use of dilated inception block. This setup would increase the likelihood of the network to detect the object's features since these features are present with multiple scales allowing the learned kernels for some scales to work.

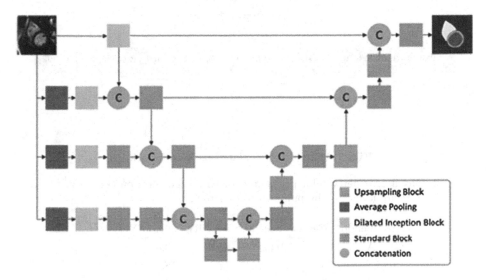

Fig. 1. Schematic diagram of the proposed architecture.

3.4   Training

For both networks we used Adam optimizer [9] for loss optimization, also we used the cross entropy as the objective function to our network. Training started with initial learning rate of 0.001 and a decay of 0.5 whenever the loss function did not improve for 3 consecutive epochs. For localization network, binary cross entropy was used a loss function for optimization. Moreover, we used image augmentation for both networks. We used transformations such as random rotation within 15 degrees, flipping and brightness adjustment. We trained each model separately then we combined them both for the testing phase and result reporting.

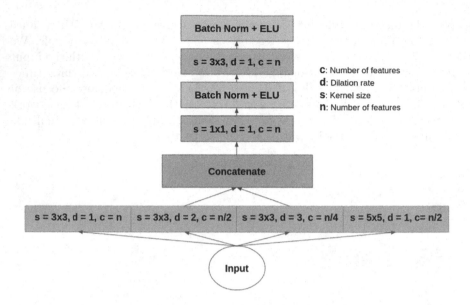

Fig. 2. Details of the dilated inception block.

Table 2. Dice scores per vendor per center for each segmentation class; LV, RV, and MYO. Centers are identified by numbers from 1 to 6.

Vendor	Center	Dice LV	Dice RV	Dice MYO
A	1	93.3	88.5	83.8
	6	86.2	77.7	78.1
B	2	88.6	86.4	84.1
	3	92	87.6	87.6
C	4	90.1	84.8	82.4
D	5	90.8	86	81.7

4   Results

Here we report the results for the test-set of the challenge, also it is worth noting that we did not post-process the output of the network; resulted segmentation. The final output of the whole pipeline, which is a cropped image, was mapped back to the original location in the input image. For the localization model, we report the results from the validation set since the groundtruth of the test-set was not supplied. Our localization network achieved 86% dice score when segmenting the whole heart as a foreground in the ground-truth masks. For the segmentation network, Fig. 3 shows the final distribution of the Dice scores for each segmented class per both end-systolic and end-diastolic time frames, the average dice scores are 90%, 84.5%, and 82.5% for LV, RV, and MYO respectively. Also, It shows a decrease in the end-systolic time frames for both RV and LV scores but an

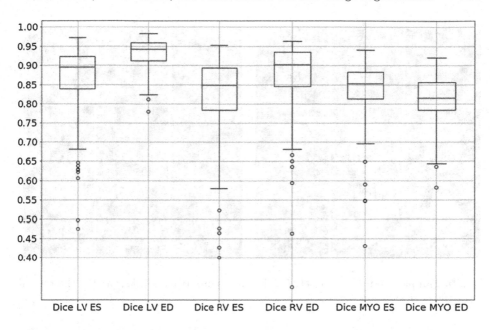

Fig. 3. Dice score per category for each time frame; end-systolic and end-diastolic.

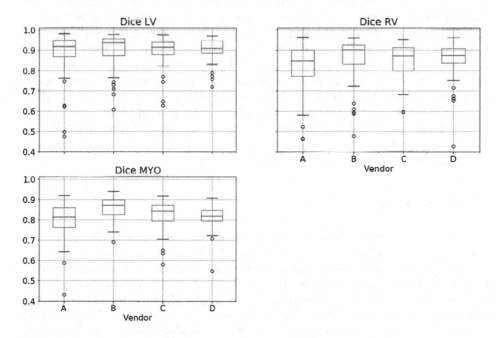

Fig. 4. Average dice score per vendor for each category; LV, RV and MYO.

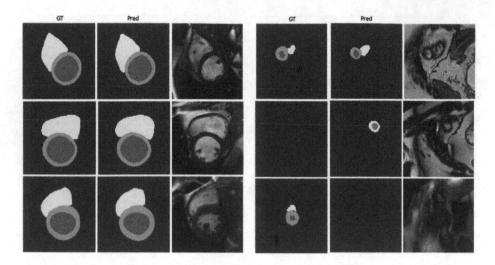

Fig. 5. Samples from the test set. GT is the ground truth masks, pred is the output from the segmentation model.

increase in MYO dice. Moreover, the model performance is consistent across the four vendors as depicted in Fig. 4 despite the fact that the model only had been trained on the vendors A and B, proving the model's ability to generalize well. Table 2 shows stability in the segmentation scores per center; however; center 6 has lower dice scores compared to the other five centers. Additionally, both the localization and segmentation models are very light weight. For instance, the localization model average run time per image is 0.0006 s and the segmentation model achieves 0.0016 s per image when tested on NVIDIA GTX 1080 TI GPU. Using the localization network and cropping the image lead to an enhancement of the final result over the baseline (using the full image) with 4,4,2.5% for the LV, RV, and MYO dice scores respectively proving the benefit of using the localization in the segmentation pipeline.

5   Conclusion

In this work we presented the multi-gate and dilated inception blocks to increase context awareness of the model and enhance the segmentation results by utilizing the multi-scaling of an image. We also used heart localization as a pre-processing before the final segmentation to enhance the segmentation results and speed up the segmentation time. We obtained a light weight pipeline running in total of 0.0022 s per image. Also, the model is robust and generalizable when tested on two unseen images from two vendors. Finally, the model achieves high dice scores which indicates the potential usefulness in real-world medical applications.

Acknowledgement. The authors declare that the segmentation method implemented for participation in the M&Ms challenge has not used any pre-trained models nor additional MRI datasets other than those provided by the organizers.

References

1. Abdelrauof, D., Essam, M., Elattar, M.: LVLNET: lightweight left ventricle localizer using encoder-decoder neural network. In: 2019 Novel Intelligent and Leading Emerging Sciences Conference (NILES), vol. 1, pp. 235–238. IEEE (2019)
2. Alattar, M.A., Osman, N.F., Fahmy, A.S.: Myocardial segmentation using constrained multi-seeded region growing. In: Campilho, A., Kamel, M. (eds.) ICIAR 2010. LNCS, vol. 6112, pp. 89–98. Springer, Heidelberg (2010). https://doi.org/10.1007/978-3-642-13775-4_10
3. Hamaguchi, R., Fujita, A., Nemoto, K., Imaizumi, T., Hikosaka, S.: Effective use of dilated convolutions for segmenting small object instances in remote sensing imagery. In: 2018 IEEE Winter Conference on Applications of Computer Vision (WACV), pp. 1442–1450. IEEE (2018)
4. Hazirolan, T., et al.: Comparison of short and long axis methods in cardiac MR imaging and echocardiography for left ventricular function. Diagn. Interv. Radiol. (Ankara, Turkey) **13**(1), 33–8 (2007)
5. Honghai Zhang, Wahle, A., Johnson, R., Scholz, T., Sonka, M.: 4-D cardiac MR image analysis: left and right ventricular morphology and function. IEEE Trans. Med. Imaging **29**(2), 350–364 (2010). https://doi.org/10.1109/TMI.2009.2030799
6. Huang, D.Y., Wang, C.H.: Optimal multi-level thresholding using a two-stage Otsu optimization approach. Patt. Recogn. Lett. **30**(3), 275–284 (2009). https://doi.org/10.1016/j.patrec.2008.10.003
7. Isensee, F., Jaeger, P.F., Full, P.M., Wolf, I., Engelhardt, S., Maier-Hein, K.H.: Automatic cardiac disease assessment on cine-MRI via time-series segmentation and domain specific features. In: Pop, M., et al. (eds.) STACOM 2017. LNCS, vol. 10663, pp. 120–129. Springer, Cham (2018). https://doi.org/10.1007/978-3-319-75541-0_13
8. Khened, M., Alex, V., Krishnamurthi, G.: Densely connected fully convolutional network for short-axis cardiac Cine MR image segmentation and heart diagnosis Using random forest. In: Pop, M., et al. (eds.) STACOM 2017. LNCS, vol. 10663, pp. 140–151. Springer, Cham (2018). https://doi.org/10.1007/978-3-319-75541-0_15
9. Kingma, D.P., Ba, J.: Adam: a method for stochastic optimization. arXiv preprint arXiv:1412.6980 (2014)
10. Nabel, E.G., Braunwald, E.: A tale of coronary artery disease and myocardial infarction. New England J. Med. **366**(1), 54–63 (2012). https://doi.org/10.1056/NEJMra1112570
11. Peng, C., Zhang, X., Yu, G., Luo, G., Sun, J.: Large kernel matters-improve semantic segmentation by global convolutional network. In: Proceedings of the IEEE Conference on Computer Vision and Pattern Recognition, pp. 4353–4361 (2017)
12. Petitjean, C., Dacher, J.N.: A review of segmentation methods in short axis cardiac MR images. Med. Image Anal. **15**(2), 169–184 (2011)
13. Ronneberger, O., Fischer, P., Brox, T.: U-Net: convolutional networks for biomedical image segmentation. In: Navab, N., Hornegger, J., Wells, W.M., Frangi, A.F. (eds.) MICCAI 2015. LNCS, vol. 9351, pp. 234–241. Springer, Cham (2015). https://doi.org/10.1007/978-3-319-24574-4_28

14. Szegedy, C., et al.: Going deeper with convolutions. In: Proceedings of the IEEE Conference on Computer Vision and Pattern Recognition, pp. 1–9 (2015)
15. Tan, L.K., Liew, Y.M., Lim, E., McLaughlin, R.A.: Convolutional neural network regression for short-axis left ventricle segmentation in cardiac cine MR sequences. Med. Image Anal. **39**, 78–86 (2017). https://doi.org/10.1016/j.media.2017.04.002
16. Tran, P.V.: A fully convolutional neural network for cardiac segmentation in short-axis MRI. arXiv preprint arXiv:1604.00494 (2016)
17. Yu, F., Koltun, V.: Multi-scale context aggregation by dilated convolutions. arXiv preprint arXiv:1511.07122 (2015)

Adaptive Preprocessing for Generalization in Cardiac MR Image Segmentation

Firas Khader[1(✉)], Justus Schock[2], Daniel Truhn[1,3], Fabian Morsbach[1], and Christoph Haarburger[1]

[1] ARISTRA GmbH, Dusseldorf, Germany
{firas.khader,christoph.haarburger}@aristra.com
[2] Department of Diagnostic and Interventional Radiology,
University Hospital Dusseldorf, Dusseldorf, Germany
[3] Department of Diagnostic and Interventional Radiology,
University Hospital Aachen, Aachen, Germany

Abstract. Recent advances in deep learning have shown the capability to accurately segment cardiac structures in magnetic resonance images. However, while these models provide a good segmentation performance for the specified datasets, their generalization with respect to unseen data across different MRI scanners, vendors or clinics is still under investigation. Previous work that aims to increase the generalization performance provides proof that emphasizing the model design on a uniform preprocessing step may be more beneficial than searching for a better neural architecture. In this paper we build upon this idea and show that a carefully designed preprocessing pipeline plays an important role in enabling the neural network to generalize to the large variety in MRI images. We evaluate our model in the context of the Multi-Centre, Multi-Vendor & Multi-Disease Cardiac Image (M&Ms) Segmentation Challenge.

Keywords: Cardiac MRI · Cardiac segmentation

1 Introduction

In recent years, cardiac magnetic resonance imaging (MRI) has increasingly been used to assess cardiac function. To quantify parameters such as the ejection fraction and stroke volume, an accurate segmentation of the left ventricle, right ventricle and myocardium in both diastolic and systolic phase is necessary. When manually performed by a medical expert, the segmentation is time-consuming and subject to intra- and inter-rater variability. Therefore, automated segmentation approaches would help to improve reproducibility of segmentations and derived parameters, as well as save valuable time. A number automated segmentation approaches have been proposed [2], many of which have shown impressive performance in terms of typical segmentation performance measures [1,3]. It has even been discussed whether the problem is solved technically [2].

© Springer Nature Switzerland AG 2021
E. Puyol Anton et al. (Eds.): STACOM 2020, LNCS 12592, pp. 269–276, 2021.
https://doi.org/10.1007/978-3-030-68107-4_27

Clinically, the diversity in image acquisition parameters, sequences and reconstruction parameters is much higher than captured by most (public) research datasets. Therefore, the goal of the Multi-Centre, Multi-Vendor and Multi-Disease Cardiac Image Segmentation Challenge (M&M) was to develop generalizable models that show consistent performance across institutions and scanner vendors. As the U-Net [5] still represents the current state-of-the-art method for medical image segmentation [4] our work is based on that approach. Previous work has shown that applying such preprocessing techniques can largely contribute to an improvement of the models' capability to generalize. The nn-UNet proposed by Isensee et al. [4] makes use of such a preprocessing pipeline by introducing a resampling step together with a normalization routine in order to mitigate the large variability in medical datasets. Additionally, by performing a quantitative analysis of the training data, important architectural parameters such as the depth and kernel sizes as well as the input patch size can be extracted automatically. Along with the benefit of providing a way to automatically adapt to the variability of the datasets, this approach also reduces effort put into manually finding an appropriate set of hyperparameters for each dataset. In our challenge approach, we particularly focused on appropriate preprocessing to tackle differences between scanner vendors, institutions and acquisition parameters.

2   Materials

The provided dataset used to train our models consists of 150 annotated images that originate from two MRI vendors and were scanned in clinical centers across three different countries (Canada, Germany, Spain). Additionally, the challenge organizers provided 25 unannotated images from a third vendor. The images are four dimensional short-axis cardiac MRI images, given in the form (x, y, z, t), where t denotes the time. The resolution of the in-plane axis across all annotated training samples range from 0.97×0.97 mm to 1.625×1.625 mm, while the resolution for the through-plane axis ranges from 5 mm to 10 mm. Besides many healthy subjects, the dataset includes patients with hypertrophic and dilated cardiomyopathies. Creation of the respective ground-truth has been performed by experienced clinicians that were instructed to label the left- (LV) and right ventricle (RV), as well as the left ventricular myocardium (MYO) for the end-diastolic (ED) and end-systolic (ES) cardiac phases. The time points between these two phases were left unlabeled. Along with the three known vendors provided in the training set, one unseen vendor is included in the test set, resulting in 50 new studies from each of the vendors. To allow for appropriate model selection, 20% of this unseen dataset was used for validation purposes, while the rest is used to rank the challenge participants.

3   Methods

Building up on the premise that data-preprocessing and hyperparameter search for the training routine and architecture can be automated to a great extent by a quantitative analysis of the underlying datasets, we were inspired by the steps taken by Isensee et al. [4] to build an appropriate preprocessing pipeline that ensures better generalization performance. In the following, we will briefly outline the most important steps of the implemented pipeline.

3.1 Preprocessing

Preprocessing is used in our approach to facilitate the learning task for our neural network. This is achieved by performing basic image transformations that aim at harmonizing the variability of voxel geometries and intensity distributions in the training set. The steps listed below are thus incorporated into our preprocessing pipeline and were carried out in the same order as presented here.

Resampling. When dealing with medical images and especially MRI images, different scanners and protocols typically result in an anisotropic voxel spacing across the dataset. This can negatively affect the learning process of CNNs due to inconsistent voxel geometries across the dataset. Thus, in order to allow our architecture to learn about the spatial dimensions of anatomical structures, we resample our images and ground-truth segmentation masks to the median voxel spacing of the used training dataset. Resampling is performed as a nearest-neighbor interpolation for the one-hot encoded segmentation mask, while third-order spline interpolation is carried out for the images. Additionally, when resampling the plane constructed by the high resolution axis separately, bicubic interpolation is used.

Resizing. In order to provide as much context as possible to the CNN, we did not process the images patch-based but as a whole. To allow batch-processing in the training routine, the images were resampled to a common shape. This is necessary because, similar to the voxel spacings, the original shape of the images varies across different MRI scanners. Furthermore, the difference in voxel spacings may also lead to a different shape after resampling, even when the original shape is equal. Therefore, we have decided to resize all the inputs to the median shape of the dataset that is acquired after resampling the images to the median voxel spacing. Resizing is performed by cropping and padding operations to ensure that the voxel spacing across all images remains consistent.

Intensity Normalization. The intensity ranges of the acquired images typically vary across MRI scans from different scanners and protocols. Therefore, we normalized our input images by performing a z-score normalization using the standard deviation and the mean of the respective images. However, as proposed in [4], normalization was performed only on the non-zero part if one quarter of the image is non-zero.

3.2 Augmentation

Augmentation techniques are applied to the input image and the corresponding one-hot encoded segmentation mask to increase the robustness of our model to unseen data. These include mirroring along the image plane and an elastic deformation to simulate the contraction and relaxation of the respective left- and right ventricle as well as the left ventricular myocardium. Following the proposed method for performing the elastic deformation by Simard et al. [6], we first create random displacement fields $\Delta x(x, y)$ and $\Delta y(x, y)$ in the range between -1 and +1. Subsequent convolution with a Gaussian filter then allows a conform displacement of adjacent pixels by choice of a sufficiently large standard deviation σ. In this case, σ has been chosen to equal 30. In order to control the intensity of the deformation, the resulting displacement field is additionally multiplied by a factor α, whose value has been chosen to equal 1550 in our experiments.

3.3 Architecture

Similarly to the network architecture chosen by Iseensee et al. [4] we have decided to base our architecture on the U-Net [5]. In order to allow our model to appropriately adjust to dataset-specific features, the number of pooling layers and the corresponding kernels are chosen automatically. This was done by defining a minimal edge length for the feature map in the bottleneck that should not be undershot when downsampling the image by means of a max-pooling operation. As a result, every axis may be subject to a different number of pooling layers. To ensure that axes with a relatively small dimensionality don't fall below the minimal edge length in the bottleneck, while still enforcing the previously computed number of pooling steps, no max-pooling was performed on these axes in the first layers of the network. This was achieved by assigning a value of 1 to the corresponding kernel dimension of the max-pooling layer.

On the other hand, the kernel dimension corresponding to the longest axis is set to 2, starting at the first layer of the encoder network. Throughout the course of our experiments, two different U-Net architectures have been implemented to deal with the three-dimensional input data: The first one consists of a simple 2D U-Net, where the input to the network is chosen to be the image plane, formed by the two axes with the highest resolution. The second architecture replaces every 2D layer in the 2D U-Net by a corresponding 3D layer, in order to construct a 3D network.

3.4 Hyperparameters

While most hyperparameters are set automatically by the dataset analysis explained above, we still had to tune few parameters manually. Given that we have trained our models on a GTX 1080 TI and have chosen the median size after resampling as the common size for all inputs, the batch size has been tuned to consume the remaining GPU memory. This adjustment results in a batch size of 2 in the case of our 3D U-Net and a batch size of 23 in the case of the 2D U-Net. Training has been performed using the Adam optimizer with a learning rate of $1 \cdot 10^{-3}$. The loss function used throughout our training routine is the soft dice loss.

3.5   TTA and Ensembling

In order to increase the accuracy of our model we incorporated test time augmentation (TTA) into our post-processing pipeline TTA is applied by performing horizontal flips and rotations in 90 degree steps from 0 to 270. The final model output is then a result of taking the mean over the predictions for each augmentation step. To further boost the performance, we also built an ensemble of models by means of a majority voting over the predicted labels. The models used in the ensembling are acquired by performing a 5-fold cross-validation for the 2D and 3D U-Net.

4   Results

In order to evaluate our approach and tune hyperparmeters such as early stopping, we performed a 5-fold cross validation over the 150 labelled training images. Splitting between training, validation and test set was performed at patient level to prevent information leakage. For each split, we assigned 64% of the data to the training set, 16% to the validation set and the remaining 20% to the test set for performance assessment. The results for 2D and 3D U-Nets with various combinations of post-processing are depicted in Table 1. With a 3D network

Table 1. Result metrics for all models. All values refer to medians and interquartile ranges based on the 5 folds on the annotated training data.

Predictor	Dice LV	Dice RV	Dice MYO	Hausdorff LV	Hausdorff RV	Hausdorff MYO
2D noTTA noEnsembling	0.69 [0.52, 0.77]	0.78 [0.66, 0.84]	0.43 [0.29, 0.52]	50.33 [27.57, 78.91]	29.50 [15.14, 44.84]	110.43 [92.25, 133.1]
2D TTA ensembling	0.72 [0.60, 0.80]	0.83 [0.78, 0.88]	0.60 [0.50, 0.69]	21.2 [16.37, 27.44]	13.94 [10.57, 15.42]	17.31 [14.50, 23.75]
3D+2D TTA ensembling	0.82 [0.73, 0.87]	0.90 [0.86, 0.93]	0.79 [0.73, 0.82]	15.32 [12.15, 20.04]	**9.96 [8.47, 11.53]**	12.98 [10.72, 17.19]
3D noTTA noEnsembling	0.83 [0.75, 0.88]	0.90 [0.86, 0.93]	0.81 [0.76, 0.84]	19.36 [13.64, 49.93]	10.40 [8.55, 12.63]	13.30 [10.35, 17.78]
3D TTA noEnsembling	0.83 [0.76, 0.89]	0.90 [0.85, 0.93]	0.81 [0.76, 0.85]	14.64 [12.0, 19.3]	10.04 [8.25, 11.3]	12.46 [10.27, 16.31]
3D noTTA ensembling	**0.85 [0.79, 0.89]**	**0.91 [0.87, 0.94]**	0.81 [0.77, 0.85]	14.20 [11.83, 18.46]	10.03 [7.87, 11.9]	12.43 [10.39, 15.91]
3D TTA ensembling	**0.85 [0.79, 0.89]**	**0.91 [0.87, 0.94]**	**0.82 [0.77, 0.85]**	13.72 [11.51, 18.39]	9.98 [7.2, 10.87]	**12.35 [10.2, 15.83]**

employing TTA and ensembling, the best performance is achieved. The 2D approach performs considerable worse than the 3D approach. In Fig. 1 we can see that the performance depends on the imaging centre. Example segmentations are provided in Fig. 2.

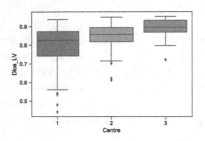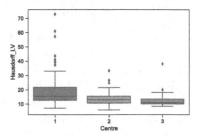

Fig. 1. Dice score (left) and Hausdorff distance (right) performance for the three imaging centres. The evaluation has been performed across the whole training set using the 3D U-Net with TTA and ensembling over all five folds.

5   Discussion

We have described an approach for automated segmentation of left ventricle, right ventricle and myocardium in cardiac MR images. Our results show that generally, a good performance is achieved for a variety of imaging protocols. The 3D segmentation approach is clearly superior to the 2D variant. Interestingly, even when 2D and 3D are combined in an ensemble, the 3D-only approach is superior. Furthermore we found that especially for the left ventricle that tends to be rather challenging to segment, TTA and ensembling strongly improve the performance in terms of both median and interquartile ranges. This indicates that TTA and ensembling are an effective tool for improving generalization. Manually inspecting the segmentations of our approach, we found that most segmentation errors arise at the boundaries of the cardiac structures along the long axis. In this area, we assume that even the ground truth is ambiguous and subject to high intra- and inter-rater variability, especially across institutions and acquisition protocols.

Moreover, the performance is still dependent on the imaging center. Baumgartner et al. achieved a higher segmentation performance on a different dataset using a 2D U-Net [1]. The same applies for [3] using an ensemble of 2D and 3D U-Nets.

In future work, our approach will be refined by further morphological postprocessing. In addition, more thorough evaluations should assess to what extends certain sequences lead to segmentations with high and low performance, respectively.

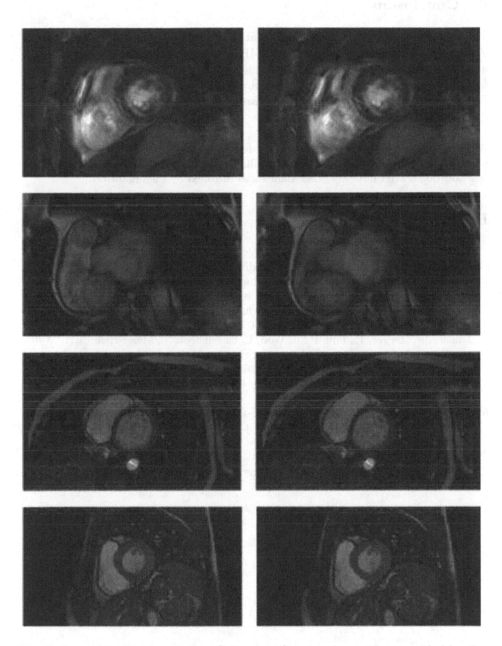

Fig. 2. Ground truth segmentations (left column) and segmentations provided by the 3D U-Net with TTA and ensembling (right column). The first and second row show poorly performing cases, whereas the third and fourth row show examples for cases with good performance.

6   Conclusion

We have shown an approach for automated segmentation of left ventricle, right ventricle and myocardium on cardiac MRI images. Our method is based on an adaptive preprocessing that takes voxel geometry and acquisition parameters for the parametrization of the U-Net architecture and preprocessing pipeline. Furthermore, we have shown that TTA and ensembling improve the performance and generalization.

Acknowledgements. The authors declare that the segmentation method implemented for participation in the M&Ms challenge has not used any pre-trained models nor additional MRI datasets other than those provided by the organizers.

References

1. Baumgartner, C.F., Koch, L.M., Pollefeys, M., Konukoglu, E.: An exploration of 2d and 3d deep learning techniques for cardiac mr image segmentation (2017)
2. Bernard, O., et al.: Deep learning techniques for automatic MRI cardiac multi-structures segmentation and diagnosis: Is the problem solved? IEEE Trans. Med. Imaging **37**(11), 2514–2525 (2018). https://doi.org/10.1109/tmi.2018.2837502, https://doi.org/10.1109/tmi.2018.2837502
3. Isensee, F., Jaeger, P., Full, P.M., Wolf, I., Engelhardt, S., Maier-Hein, K.H.: Automatic cardiac disease assessment on cine-mri via time-series segmentation and domain specific features (2017). https://doi.org/10.1007/978-3-319-75541-0
4. Isensee, F., et al.: nnu-net: self-adapting framework for u-net-based medical image segmentation (2018)
5. Ronneberger, O., Fischer, P., Brox, T.: U-net: convolutional networks for biomedical image segmentation. In: Medical Image Computing and Computer-Assisted Intervention - MICCAI, pp. 234–241. Springer International Publishing (2015)
6. Simard, P.Y., Steinkraus, D., Platt, J.C., et al.: Best practices for convolutional neural networks applied to visual document analysis. In: Icdar vol. 3 (2003)

Deidentifying MRI Data Domain
by Iterative Backpropagation

Mario Parreño[1(✉)], Roberto Paredes[1], and Alberto Albiol[2]

[1] PRHLT Research Center, Universitat Politècnica de València, València, Spain
{maparla,rparedes}@prhlt.upv.es
[2] iTeam Research Institute, Universitat Politècnica de València, València, Spain
alalbiol@iteam.upv.es

Abstract. Medical images acquired at various hospitals can differ significantly in their data distribution. We can find multiple sources of divergence evaluating images from different clinical centers, patient diseases, vendors, or even configurations on the same scanner. Typically at deployment, when we are facing real-world domain, data is collected from it and the trained model is adapted. This is not practical in all scenarios like medical imaging due to the lack of data and the strict protection to which it is subjected. We investigate this challenging problem by evaluating a novel domain adaptation procedure. First, a classifier model is trained to distinguish between which data distribution comes from. Once trained, the images from each vendor are modified iteratively using the gradients of the error obtained when the target is set arbitrarily. Finally, when we have a new sample we only have to carry out the same process of domain adaptation by error backpropagation. The experiments were performed on Multi-Centre, Multi-Vendor & Multi-Disease Cardiac Image Segmentation Challenge (M&Ms), comparing the segmentation metrics obtained for studies from vendors present in the training set and an additional studies from an unseen vendor.

Keywords: Domain adaptation · Image segmentation · MRI · M&Ms

1 Introduction

Magnetic resonance imaging (MRI) helps to diagnose a disease, an injury, or monitor how well a patient is doing with treatment. MRI is currently one of the most important medical imaging methods due to its non-invasive nature as well as does not use the damaging ionizing radiation of X-rays. However, MRI data have artifacts such as noise and inhomogeneities due to stochastic processes in the image acquisition, where using non-standardized imaging protocols is common [10]. Adjusting voltage will affect directly in the range of the MRI pixel values, changing amperage will affect image noise rate, and the selection of reconstruction algorithms will result in different image texture features. Varying these scanner parameters alter the noise distribution of the images and

© Springer Nature Switzerland AG 2021
E. Puyol Anton et al. (Eds.): STACOM 2020, LNCS 12592, pp. 277–286, 2021.
https://doi.org/10.1007/978-3-030-68107-4_28

make more difficult the task of pattern recognition [15] between vendors or even the same scanners. Because of all of this, current studies tend to use MRI data obtained from multi-centre, multi-vendor, and multi-disease sources to test algorithms with reasonable robustness to such different domains, where data is scarce and training data distribution differs from the test one.

Building generalizable models that can be deployed to real-world applications is complex when training and test data are sampled from different distributions. For the sake of diminishing these differences researchers have proposed domain adaptation methods that normally learn to align source and target data in a domain-invariant discriminative feature space [4]. The problem of these methods is that they rely on to take some knowledge of the test distribution, for example, adapting a trained system through a few labeled or unlabelled data from the new domain, but in most real-world applications, this information about the new domain is not available.

Furthermore, research has focused on model generalization to unseen domains by different techniques [16]. The most common one is data augmentation, where training data is randomly transformed trying to capture the possible variability that test distribution will present. Transformations can go from simple rotations and flips to more elaborate methods that support the training process trying to draw samples from vicinity distributions [20].

These complications appear in all kinds of problems when they are deployed but in the medical imaging field they are worsened by the lack of data, something essential in the training of neural networks. We introduce baseline results for the MRI segmentation problem on the multi-centre, multi-vendor, and multi-disease brand new dataset M&Ms. Our approximation takes state-of-the-art techniques who have proven their good performance on similar MRI segmentation problems and will establish a strong starting point for practitioners. Moreover, we face the domain adaptation problem to unseen vendors by iteratively moving their images to a reference domain through backpropagating the error of a simple discriminator.

2   Related Work

Convolutional Neural Networks (CNNs) have achieved impressive results in image segmentation, but these can drop drastically when tested on real-world applications where the distribution of the images can be different. MRI modality lack of uniformity and exhibit high variance between subjects, with intensity variations up to 40% [17].

One of the most common techniques to lower this gap between training and test distributions is domain generalization. This aims to make the model generalize to unseen domains without any knowledge about test distribution during training time. Image augmentations are included on all training procedures including geometric transformations, color space augmentations [19], random erasing [3], kernel filters, mixing images [20], etc. But these techniques are often limited and help only if represent variations in the target domain.

On the other hand, other techniques try to adapt test images to the training distribution or narrow the domain shift via a domain-invariant discriminative feature space. This is done usually with adversarial learning [21] or using maximum mean discrepancy [6]. More classical methods focus on the domain shift problem as an intensity transformation between two fixed image settings. For eliminating this intensity difference, histogram-matching and joint histogram registration methods are used [9,12]. The problem of these methods is that require paired source-target MRI obtained from the same person in a short time to extract the necessary features, and the major one, they are not able to process new images that are not from any of the MRIs groups of the training data. Neural networks can solve this problem but most of the algorithms need paired input-output samples. A method that lacks supervision in the form of paired samples is Cyclegan [21]. Modanwal et al. [11] modified the original Cyclegan discriminator architecture and forced the penalization to focus more on features of breast tissue.

These methods evaluate the success of their approaches qualitatively by visual inspections of whether the organ shape and tissue region shapes were altered or not. Meanwhile, quantitative results are calculated measuring mean intensity values between vendors [11], structural similarity index measure (SSIM) and the peak signal-to-noise ratio (PSNR) [2], or mean square error (MAE) [18]. The problem of these metrics is that they do not help us to know if another machine learning algorithm can take advantage when organ structures should be preserved as in segmentation.

To address this issue, we propose a novel approach for multi-centre, multi-vendor & multi-disease cardiac studies based on domain classification and error backpropagation. Our first approach uses a state-of-the-art segmentation architecture with accordingly data augmentations for our dataset, M&Ms, with cardiovascular magnetic resonance (CMR) volumes. Furthermore, following the guidelines of the challenge where the dataset was released, we have not used any pre-trained models nor additional MRI datasets other than those provided by the organizers. Next, we tackle the domain adaptation problem with the use of a simple discriminator by iteratively backpropagating the error of the classification to an arbitrary target.

3   Methods

Our initial approach tackles the problem of image segmentation via CNNs. Next, we introduce the data augmentation techniques that were used. Then, we specify the training details and a model combination method that fits properly on segmentation problems. Finally, we present a simple domain adaptation procedure as a baseline method for future researchers.

3.1   Model

Our model follows the U-Net [13] architecture idea, an encoder that tries to capture the low-level features and a decoder that uses this information to build the prediction. The encoder is composed by the feature extractor from Resnet-34 model [7] with intermediate spatial and channel Squeeze and Excitation blocks (scSE) [14]. These blocks are intended to recalibrate the feature maps separately along channel and space by finally combining their output. When the images are forwarded through the encoder, a feature pyramid attention (FPA) module [14] for pixel-precise attention is applied. This uses different pyramid scales varying convolutional sizes to extract more context. Next, the decoder takes the FPA output and gradually reconstructs the segmentation of the input image. For that, as in the original U-Net architecture, intermediate outputs from the encoder blocks are merged with enlarged features from the decoder using transposed convolutions. These combined maps at the decoder phases are stored and finally convolved obtaining the output segmentation map.

3.2 Data Augmentation

Data Augmentation can improve the performance of neural networks and expand limited datasets to take advantage of the capabilities of big data. The augmented data will represent a more comprehensive set of possible data points, thus trying to minimize the distance between the training and validation or test sets. We can see usually the use of this technique in all Machine Learning problems, so we incorporate Data Augmentation to extract more information from the training dataset through image augmentations. The selected augmentations preserve the coherence between input MRI images and their corresponding output masks, so the same geometrical transformations parameters are used for both images and masks. The chosen augmentations are flips, rotations, image shift, elastic transforms, and optical distortions. Finally, we discarded some common data augmentations that not make sense to apply to this problem but are common in Computer Vision as Cutout [3]. This is because neither the training images nor the test images will present the variations that these transformations introduce.

3.3 Training Details, Weakly Labelling, and Model Combination

Our experiments followed a five fold cross-validation method folding by patients to select the best model and hyperparameters. Models were trained for 100 epochs with a batch size of 32 images. All images are preprocessed by first padding the image if needed to 224×224 pixels, this allows us to do a center crop without any kind of resizing and makes prediction easier by adding or removing background pixels to them as a border. Adam is chosen as optimizer with a learning rate of 0.001 decayed by 0.1 at epoch 55 and 90. Two binary losses are computed, one based on the Dice Score with weight 0.4, and another one penalizing border regions by using an average pool operator to weight a binary cross-entropy loss function, with a weight of 0.1. Additionally, cross-entropy loss is added with a weight of 0.5.

When the initial training using the labeled data from vendor A and B finish, unannotated data from vendor C is labeled using the last checkpoint. Then we retrain the model with a lower learning rate a few epochs.

Once previous models are trained, Stochastic Weight Averaging (SWA) [8] is used to reduce generalization error simulating the combination of multiple model predictions, also known as model ensemble. This technique is simpler to use than other ensemble techniques, as test time augmentations, because it can be done in a single forward pass and we do not have to average multiple mask predictions. Then, starting from our last checkpoint, we continue training using a constant learning rate of 0.00256 for 50 epochs. At the end of each epoch, an intermediate checkpoint is stored for finally averaging the weights of all networks to get the final model.

3.4   Iterative Backpropagation

If we want to capture the differences between domains through Machine Learning, we first think about how to make a model distinguish between these domains. The first idea that comes to mind is a simple classifier. Through a classifier we can use all available data without having paired samples of how domain adaptation should be done. Then, first we take all the CMR images and train a Resnet-34 classifier model using their corresponding vendor label. Once the classifier is trained, if we want to adapt an image to one of the known domains we need only to iteratively backpropagate the error we get with an image when setting an arbitrary target using categorical cross-entropy loss. As an initial constraint L1 loss was added as in [21] to do not distort the images and preserve the organ structures to segment, but we observed that with minor modifications the image is classified as the desired target.

This approach is completely different from previous ones as [5] where the model backbone is trained to learn to distinguish between domains and the desired task, e.g. segmentation. Meanwhile, we train separately the segmentation and classification models, using them independently.

4   Experiments

All the experiments were implemented using Pytorch framework and trained on two NVIDIA RTX 2080 GPU. Data augmentations were made using albumentations library[1]. The Pytorch code, models, and full results can be found at our GitHub.

4.1   M&Ms Dataset

M&Ms [1] is a CMR segmentation dataset that aims to contribute to the effort of building generalizable models that could be applied consistently across clinical centres. It is part of the MICCAI Conference as an Image Segmentation Challenge at the Statistical Atlases and Computational Modeling of the Heart (STACOM) 2020 workshop. Is composed of 350 patients with hypertrophic and

---

[1] https://albumentations.readthedocs.io/.

Table 1. Comparison of the validation results obtained adding techniques incrementally on M&Ms dataset. Mean values for all vendors.

Method	Jaccard	Dice	Hausdorff	ASSD
Resnet-34 U-Net	0.733 ± 0.164	0.821 ± 0.110	19.327 ± 20.007	1.544 ± 1.642
+ Data Augmentations	0.776 ± 0.133	0.867 ± 0.105	16.522 ± 19.429	1.222 ± 1.422
++ Weakly Labelling	0.787 ± 0.125	0.875 ± 0.099	12.190 ± 14.223	1.042 ± 1.157
+++ SWA	0.791 ± 0.120	0.877 ± 0.096	**12.154** ± 11.429	**1.008** ± 1.028
++++ Iterative Backpropagation	**0.793** ± 0.122	**0.879** ± 0.086	14.134 ± 19.195	1.127 ± 1.335

dilated cardiomyopathies as well as healthy subjects. Four different magnetic resonance scanner vendors (Siemens, General Electric, Philips, and Canon) were used among six hospitals. The training set contains 150 annotated images from two different MRI vendors, A and B with 75 cases each one, and 25 unannotated from a third one, vendor C. Testing set has 200 cases equally distributed by vendors, where a fourth unseen vendor D is introduced. 20% of this testing dataset is used for validation and the rest is reserved for testing and the final ranking. All the data is annotated following a Standard Operating Procedure (SOP) where clinical contours were corrected by two in-house annotators that had to agree on the final result following a set of rules.

4.2   Results

Table 1 shows results obtained for validation set. We can see the biggest gap when data augmentation is used. At this point, qualitatively we observe how most of the predictions are very accurate, limiting the range of improvement for the following techniques. For that reason we argue that the next techniques do not introduce such a big impact on final results. Meanwhile, Table 2 shows the results for the test set divided by each vendor and the average metrics. We can see that, despite vendor B, there is no big difference between different vendor metrics. Then, we should analyze the reasons that there is this difference even with known vendors. Finally, we establish a baseline where both known and unknown vendors can be further improved with better domain adaptation techniques.

For qualitative results comparison, Fig. 1 shows test prediction overlays for different vendors. We can see how the predictions fit very well the desired structures. We can observe variations in the zoom and rotation of the images, as well as in the captured noise especially with vendor A.

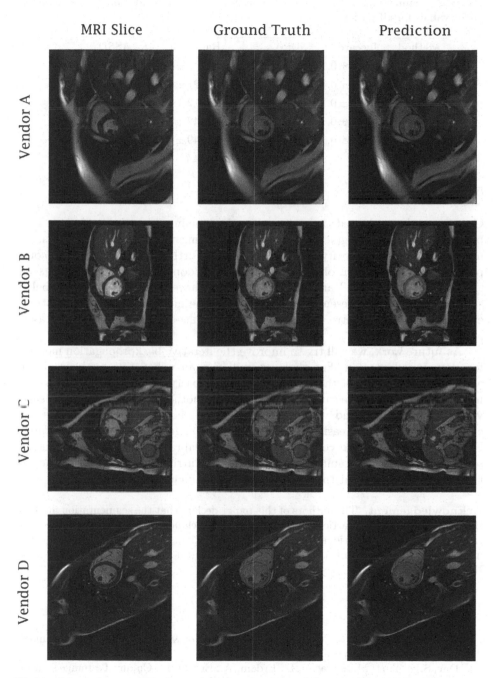

Fig. 1. Selected test slices from different vendors. Columns show MRI slice, mask ground truth, and mask prediction per vendor.

Table 2. Comparison of the test results obtained for each vendor on M&Ms dataset. Mean values for all patients.

Method	Jaccard	Dice	Hausdorff	ASSD
Vendor A	0.772 ± 0.108	0.867 ± 0.071	17.342 ± 18.346	1.161 ± 1.106
Vendor B	0.802 ± 0.105	0.886 ± 0.069	10.627 ± 6.061	0.953 ± 1.016
Vendor C	0.776 ± 0.110	0.869 ± 0.075	13.012 ± 12.623	1.129 ± 1.239
Vendor D	0.761 ± 0.137	0.856 ± 0.113	14.442 ± 18.923	1.185 ± 1.666
Average	0.778 ± 0.117	0.869 ± 0.085	13.849 ± 15.119	1.107 ± 1.285

5   Conclusions

Our experiments showed that with a simple domain adaptation technique it is possible to improve segmentation metrics, putting the focus on segmentation metrics and not in intensity or noise indicators. Furthermore, based on the comparative results, we can observe a gap between domains that have never seen the segmentation model and those for that we have labeled or weakly-labeled samples. Finally, we presented a powerful baseline model for future practiotioners on a new multi-centre, multi-vendor, and multi-disease brand new dataset, M&Ms.

As future work, we will try to improve the iterative backpropagation method for domain adaptation. For that, we will run an exhaustive analysis of the adapted images. One of the problems that we observed is that with very small modifications one image can be transformed to another vendor without, at least, visual modifications and very low values shift. Then, we should improve the classifier to do not be easily fooled by minor changes or add some constraint. Finally, another way to conduct the domain adaptation could be introducing some parameters as a gamma, blurring, and unblurring matrix with parameters that could be learned at the same time that the image is modified.

Acknowledgement. The authors of this paper declare that the segmentation method they implemented for participation in the M&Ms challenge has not used any pretrained models nor additional MRI datasets other than those provided by the organizers. The authors thank the EU-FEDER Comunitat Valenciana 2014-2020 grant IDIFEDER/2018/025.

References

1. Campello, V.M., et al.: Multi-centre, multi-vendor & multi-disease cardiac image segmentation. in preparation (2020)
2. Dar, S.U., Yurt, M., Karacan, L., Erdem, A., Erdem, E., Çukur, T.: Image synthesis in multi-contrast MRI with conditional generative adversarial networks. IEEE Trans. Med. Imaging **38**(10), 2375–2388 (2019)
3. DeVries, T., Taylor, G.W.: Improved regularization of convolutional neural networks with cutout (2017). http://arxiv.org/abs/1708.04552

4. Fan, C., Liu, P., Xiao, T., Zhao, W., Tang, X.: Domain adaptation based on domain-invariant and class-distinguishable feature learning using multiple adversarial networks. Neurocomputing **411**, 178–192 (2020). https://doi.org/10.1016/j.neucom.2020.06.044, http://www.sciencedirect.com/science/article/pii/S0925231220310158

5. Ganin, Y., Ustinova, E., Ajakan, H., Germain, P., Larochelle, H., Laviolette, F., Marchand, M., Lempitsky, V.: Domain-adversarial training of neural networks. J. Mach. Learn. Res. **17**(1), 2096–2030 (2016)

6. Gretton, A., Sejdinovic, D., Strathmann, H., Balakrishnan, S., Pontil, M., Fukumizu, K., Sriperumbudur, B.K.: Optimal kernel choice for large-scale two-sample tests. In: Pereira, F., Burges, C.J.C., Bottou, L., Weinberger, K.Q. (eds.) Advances in Neural Information Processing Systems 25, pp. 1205–1213. Curran Associates, Inc. (2012). http://papers.nips.cc/paper/4727-optimal-kernel-choice-for-large-scale-two-sample-tests.pdf

7. He, K., Zhang, X., Ren, S., Sun, J.: Deep residual learning for image recognition. In: 2016 IEEE Conference on Computer Vision and Pattern Recognition (CVPR), pp. 770–778 (2015)

8. Izmailov, P., Podoprikhin, D., Garipov, T., Vetrov, D., Wilson, A.: Averaging weights leads to wider optima and better generalization (2018)

9. Jäger, F., Deuerling-Zheng, Y., Frericks, B., Wacker, F., Hornegger, J.: A new method for MRI intensity standardization with application to lesion detection in the brain. Vision Modeling and Visualization, January 2006

10. Midya, A., Chakraborty, J., Gönen, M., Do, R., Simpson, A.: Influence of CT acquisition and reconstruction parameters on radiomic feature reproducibility. J. Med. Imaging **5**, 1 (2018). https://doi.org/10.1117/1.JMI.5.1.011020

11. Modanwal, G., Vellal, A., Buda, M., Mazurowski, M.: MRI image harmonization using cycle-consistent generative adversarial network, p. 36, March 2020. https://doi.org/10.1117/12.2551301

12. Nyul, L.G., Udupa, J.K., Zhang, X.: New variants of a method of MRI scale standardization. IEEE Trans. Med. Imaging **19**(2), 143–150 (2000)

13. Ronneberger, O., Fischer, P., Brox, T.: U-Net: convolutional networks for biomedical image segmentation. In: Navab, N., Hornegger, J., Wells, W.M., Frangi, A.F. (eds.) MICCAI 2015. LNCS, vol. 9351, pp. 234–241. Springer, Cham (2015). https://doi.org/10.1007/978-3-319-24574-4_28

14. Roy, A.G., Navab, N., Wachinger, C.: Recalibrating fully convolutional networks with spatial and channel "squeeze and excitation" blocks. IEEE Trans. Med. Imaging **38**(2), 540–549 (2019). https://doi.org/10.1109/TMI.2018.2867261

15. Saha, A., Yu, X., Sahoo, D., Mazurowski, M.A.: Effects of MRI scanner parameters on breast cancer radiomics. Expert Syst. Appl. **87**, 384–391 (2017). https://doi.org/10.1016/j.eswa.2017.06.029, http://www.sciencedirect.com/science/article/pii/S0957417417304463

16. Shorten, C., Khoshgoftaar, T.M.: A survey on Image Data Augmentation for Deep Learning. J. Big Data **6**(1), 60 (2019). https://doi.org/10.1186/s40537-019-0197-0, https://doi.org/10.1186/s40537-019-0197-0

17. Simkó, A., Löfstedt, T., Garpebring, A., Nyholm, T., Jonsson, J.: A generalized network for MRI intensity normalization. In: International Conference on Medical Imaging with Deep Learning - Extended Abstract Track, London, United Kingdom (2019). https://openreview.net/forum?id=HyeL2iQRYE

18. Welander, P., Karlsson, S., Eklund, A.: Generative adversarial networks for image-to-image translation on multi-contrast MR images - a comparison of cyclegan and unit. ArXiv abs/1806.07777 (2018)

19. Xiao, Y., Decencière, E., Velasco-Forero, S., Burdin, H., Bornschlögl, T., Bernerd, F., Warrick, E., Baldeweck, T.: A new color augmentation method for deep learning segmentation of histological images. In: 2019 IEEE 16th International Symposium on Biomedical Imaging (ISBI 2019), pp. 886–890 (2019)
20. Zhang, H., Cisse, M., Dauphin, Y., Lopez-Paz, D.: mixup: beyond Empirical Risk Minimization (2017)
21. Zhu, J.Y., Park, T., Isola, P., Efros, A.A.: Unpaired image-to-image translation using cycle-consistent adversarial networks. In: 2017 IEEE International Conference on Computer Vision (ICCV) (2017)

A Generalizable Deep-Learning Approach for Cardiac Magnetic Resonance Image Segmentation Using Image Augmentation and Attention U-Net

Fanwei Kong[(✉)] and Shawn C. Shadden

Department of Mechanical Engineering, University of California,
Berkeley, CA 94720, USA
{fanwei_kong,shadden}@berkeley.edu

Abstract. Cardiac cine magnetic resonance imaging (CMRI) is the reference standard for assessing cardiac structure as well as function. However, CMRI data presents large variations among different centers, vendors, and patients with various cardiovascular diseases. Since typical deep-learning-based segmentation methods are usually trained using a limited number of ground truth annotations, they may not generalize well to unseen MR images, due to the variations between the training and testing data. In this study, we proposed an approach towards building a generalizable deep-learning-based model for cardiac structure segmentations from multi-vendor,multi-center and multi-diseases CMRI data. We used a novel combination of image augmentation and a consistency loss function to improve model robustness to typical variations in CMRI data. The proposed image augmentation strategy leverages un-labeled data by a) using CycleGAN to generate images in different styles and b) exchanging the low-frequency features of images from different vendors. Our model architecture was based on an attention-gated U-Net model that learns to focus on cardiac structures of varying shapes and sizes while suppressing irrelevant regions. The proposed augmentation and consistency training method demonstrated improved performance on CMRI images from new vendors and centers. When evaluated using CMRI data from 4 vendors and 6 clinical center, our method was generally able to produce accurate segmentations of cardiac structures.

Keywords: MRI segmentation · Generalization · Image augmentation · Deep learning

1 Introduction

Cine cardiac MRI (CMRI) is considered a reference standard for assessments of the function and morphology of the heart. Analysis of the heart from CMRI

This works was supported by the National Science Foundation, Award No. 1663747.

E. Puyol Anton et al. (Eds.): STACOM 2020, LNCS 12592, pp. 287–296, 2021.
https://doi.org/10.1007/978-3-030-68107-4_29

can be essential in disease diagnosis and treatment planning. This analysis is greatly facilitated by proper identification of the left ventricular blood pool, myocardium, and the right ventricular blood pool at both end diastolic and end systolic phases. Recent developments in deep learning (DL) are accelerating this previously time-consuming identification process. This has been accomplished by supervised learning of deep neural networks with previously annotated data [1,2]. However, it is well known that DL methods are prone to over-fitting training data and in turn under-performing on real-world data, especially when the amount of training data is limited. Prior DL-based CMRI segmentation methods were usually trained using small datasets obtained from only one or two sources [6]. However, CMRI data are sensitive to a number of factors, including differences in vendor, magnetic coil types and/or acquisition protocols. Thus, the performance of DL based methods can drop significantly when tested on images that differ from the training data [3,9]. An outstanding challenge has been to develop generalizable DL based methods that can perform consistently well across different centers, making them useful for real-world clinical applications.

Recent works have helped to improve the generalization capabilities of DL based models. Tao et al. trained a conventional U-Net model on a large multi-vendor, multi-center training set from patients with various cardiovascular diseases [9]. Chen et al. showed that training models using a single yet large data source with appropriate data normalization and augmentation could also achieve promising performance on data from other sources [3]. However, the collection and labeling of such large or diverse datasets are extremely expensive, which limits real-world applicability or adaptability to other segmentation tasks. Therefore, several studies have sought to use unsupervised domain adaptation techniques to optimize the model on an unannotated target dataset [4,5]. Such methods require images from new sources and their generalization capabilities were usually tested with only one new data source with a limited number of samples. Thus, building generalizable DL models that can be reliably and efficiently applied to data from new clinical centers and scanner vendors remains to be demonstrated.

In this study, we proposed an approach towards building a generalizable DL based model for cardiac structure segmentations from multi-vendor, multi-center and multi-diseases CMRI data. We develop a fully automated segmentation model based on an attention-gated U-Net model [8]. To improve model robustness to typical spatial and intensity variations of cardiac MR images, we propose a novel combination of image augmentation and consistency loss. The proposed image augmentation strategy leverages un-labeled data by a) using CycleGAN to generate images in different styles and b) exchanging the low-frequency features of images between different vendors. A consistency loss is introduced to coerce the model to generate consistent predictions on images with the same anatomical features but different appearances. Our framework demonstrated improved segmentation performance on CMRI images from new vendors and clinical centers.

2   Methods

2.1   Image Dataset Information and Pre-processing

Image data from the 2020 Multi-Centre, Multi-Vendor& Multi-Disease Cardiac Image Segmentation Challenge (M&Ms) was utilized. This dataset contains CMRI scans from 4 vendors and 6 clinical centers. The training set consisted of 150 annotated images from 2 vendors (Vendor A and B) and 25 unannotated images from another vendor (Vendor C). The ground truth annotations include the left and right ventricular blood pools and left ventricular myocardium. Trained models were validated using a separate validation set, which contained a collection of 80 CMRI scans from all vendors/centers. The final model was evaluated on the M&Ms test set containing 160 CMRI scans. We sliced each stacked 3D image volume into 2D short-axis images and resampled each 2D slice to the same size of 256×256 and a pixel spacing of $1.2\,\mathrm{mm}$. For each scan, we clipped the pixel intensity values between 0 and the 99th percentile to reduce bright artifacts, and normalized the pixel intensity for each slice to zero mean and unit variance.

2.2   Image Augmentation

Image augmentation was used to improve the robustness of deep neural network models to certain image variations. We considered two categories of common variations for cardiac MR images, spatial and appearance. For spatial variations of the heart, we randomly scaled the training images by a factor of 0.8 to 1.2. We also randomly rotated the images clockwise or counter-clockwise by 90° and then applied a small amount of random rotation by up to 10°. For appearance variations of the images, we used frequency-domain augmentation and Cycle GAN to change pixel intensities, as described in more detail below.

Frequency Domain Augmentation (FDA). The intensity of MR images does not have fixed meaning; tissue intensities can vary significantly across different vendors and clinical centers even after applying intensity normalization as described above. To better handle this factor, we propose perturbing the low-frequency contents of an MR image. Namely, as shown in Fig. 1, we augmented Vendor A and B images with the low-frequency features of the unlabeled Vendor C images, to introduce the low-level statistics of Vendor C images into our training data, similar to a strategy proposed in [10]. The two image slices are extracted at the same relative location from their image stacks so that they show a similar region of the hearts. Namely, for a labeled image slice x^U from Vendor A or B and its corresponding unlabeled Vendor C image slice x^L showing the similar location of the heart, the augmented image slice $x^{L \rightarrow U}$ can be obtained by Eq. 1, where $\mathcal{F}_A$ and $\mathcal{F}_P$ denote the amplitude and phase components of the Fouirer transform $\mathcal{F}$, and M is a mask with zero values except for the center square as illustrated in Fig. 1. We used a dimension ratio of 0.02 between the swapped region and the full image.

$$x^{L \to U} = \mathcal{F}^{-1} \left([M \circ \mathcal{F}_{\mathcal{A}}(x^U) + (1-M) \circ \mathcal{F}_{\mathcal{A}}(x^L), \\ M \circ \mathcal{F}_{\mathcal{P}}(x^U) + (1-M) \circ \mathcal{F}_{\mathcal{P}}(x^L)] \right) \tag{1}$$

Our method differed from [10] in that we swapped both the amplitude and phase information in the frequency domain; swapping only the amplitude, as in [10], led to numerous artifacts.

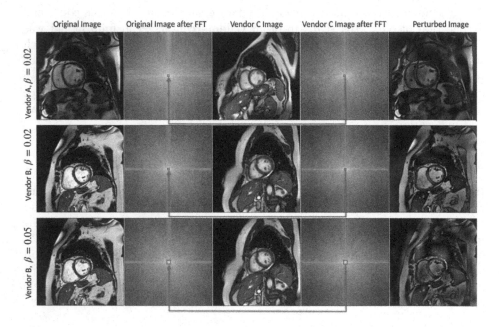

Fig. 1. Illustration of image augmentation in the frequency domain. An image slice (first column) from a labeled dataset (e.g., Vendor A in the first row or Vendor B in the second row) is selected and its magnitude spectrum from FFT is computed (second column). A corresponding image slice (third column) from the unlabeled Vendor C data is selected, and its magnitude spectrum from FFT is computed (fourth column), and subsequently used to augment the spectrum of the original image to generate a perturbed image (fifth column). The third row illustrates the influence of using a larger dimension ratio of 0.05 between the swapped region and the full image. Red boxes represent the center regions where the amplitude and phase components of fourier transformed images were swapped.

Appearance Augmentation Using Cycle GAN. Although frequency domain augmentation perturbs image intensity while preserving anatomical features, it sometimes introduces unrealistic intensity inhomogeneity. Observing that MR images from different vendors are different in appearance, we used

Cycle GAN [11] to transfer the appearance of images from Vendor B to Vendor A or C and vice versa. Briefly, Cycle GAN takes in two images from two styles and output the corresponding synthetic images that have the texture appearance in the other style, respectively. It consists of two generator networks to generate synthetic images and two discriminator networks that attempt to discriminate generated images from real images. Compared with the original implementation [11], we reduced the learning rate of the discriminator to 0.00002 to achieve a better balance between the generators and the discriminators and replaced the transpose convolution layer with bilinear upsampling followed by a convolution layer to reduce checkerboard artifacts [7]. We trained the Cycle GAN models for 30 epochs and saved the model weights at the end of each epoch after the 5th epoch. For each image in Vendor A or B, we augmented it with 10 CycleGAN models randomly picked from the saved ones. As shown in Fig. 2, the augmented images resemble the appearance of images from the other vendor and by using CycleGAN models saved at different epochs, we obtained further intensity variations among the generated images.

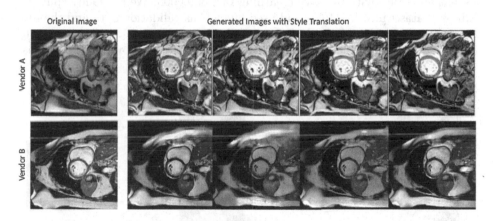

Fig. 2. Original images and the generated images after style translation.

2.3   Training with Consistency Loss

A robust segmentation model should generate consistent predictions for two images with the same anatomical features but different appearance. That is, after applying intensity augmentation to an image slice as described in the previous section, a robust model should predict similar probability maps between the augmented image and the original image. Therefore, we trained our models with two types of loss. One was a hybrid loss L_{seg}, accounting for cross entropy loss and the dice loss, which optimizes for the accuracy of the predicted segmentation. The other was a consistency loss $L_{consistency}$, which regularizes the differences between the predictions P of one image slice I and its intensity-augmented version $A(I)$. The hybrid loss and consistency loss are defined by the following equations, where G is the one-hot coded ground truth segmentation and N is the number of segmentation domains:

$$L_{seg}(I, G) = -\frac{1}{N} \sum_{i=1}^{N} \sum_{x \in I} G_i(x) \log(P_i(x)) + N - \sum_{i=1}^{N} \frac{2 \sum_{x \in I} G_i(x) P_i(x)}{\sum_{x \in I} G_i(x) + \sum_{x \in I} P_i(x)}$$
(2)

$$L_{consistency}(I, A(I)) = -\frac{1}{N} \sum_{i=1}^{N} |P_i(I) - P_i(A(I)))|.$$
(3)

As illustrated in Fig. 3, during each training iteration, our proposed pipeline takes in one image slice and its intensity-augmented version and generates predictions separately for the two inputs. The model parameters are then updated based on the sum of the hybrid loss computed for the two predictions and the consistency loss. Our model architecture is based on an attention-gated U-Net model that learns to focus on cardiac structures of varying shapes and sizes while suppressing irrelevant regions [8]. We used an Adam stochastic gradient descent algorithm with an initial learning rate of 0.0005. We randomly split the training dataset into five folds and used one fold as validation data and the rest as training data. We adopted a learning rate schedule where the learning rate was reduced by 20% if the validation dice score had not improved for 10 epochs.

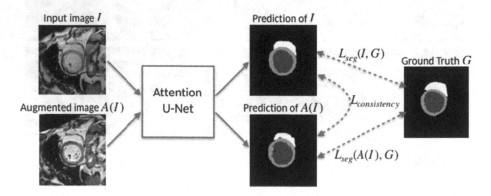

Fig. 3. Proposed training pipeline with both segmentation and consistency losses

2.4   Weighted Ensemble

Ensemble learning with deep neural networks can reduce DL model variance and thus better generalize to unseen data. We observed that our models were more likely to produce mistakes or inconsistent predictions on apical or basal slices. Indeed, the ventricles imaged in these slices usually appear very small in size or have large anatomical variation and noisy ground truth labels. Therefore, we augmented the apical and basal slices four times more than the others using the image augmentation methods described above. We computed a weighted average of the probability maps predicted by models trained with and without such extra augmentation on high-variance image slices. Namely, for apical or basal slices, we assigned a higher weight ratio of 3:2 between models trained with and without extra augmentation on high-variance image slices, respectively. While for other image slices, we assigned a lower weight ratio of 1:2. These weight ratios were experimentally determined by validation.

3 Results

We compared the performance of five segmentation models trained under different augmentation and ensemble settings 1) the baseline attention-UNet model trained without image augmentation (NoAug), 2) with only spatial augmentation(Sptl.Aug), 3) with spatial augmentation and intensity augmentation in the frequency domain (Sptl.Aug+FDA) 4) with spatial augmentation and intensity augmentation using CycleGAN (Sptl.Aug+CycleGan), and 5) ensemble of 3 sets of models trained with frequency domain augmentation (Sptl.Aug+FDA+Ens.). The segmentation predictions generated by these five models were evaluated on the validation dataset from the M&Ms challenge and Table 1 compares the segmentation accuracy for each vendor/center. Spatial augmentation consistently improved segmentation performance for all vendors and centers. Adding frequency domain and CycleGAN augmentation significantly improved the dice scores for the fourth unseen vendor, Vendor D. Compared with a single model, a weighted ensemble consistently improved the performance for all vendors/centers.

Table 1. Dice scores and average surface distance errors (ASSD) of segmentations generated by different models. The dice scores and ASSD were calculated as the averages over the three cardiac structures. The yellow-colored cells represent clinical centers with no annotated training data and the bold numbers in black are the best scores among the models without using ensemble. The bold numbers in blue are when using ensemble achieves the best performance among all models.

Metric	Dice						ASSD					
Vendor	A		B		C	D	A		B		C	D
Center	1	6	2	3	4	5	1	6	2	3	4	5
NoAug	0.867	0.822	0.883	0.928	0.848	0.837	1.303	1.746	0.894	0.358	1.249	1.421
Sptl.Aug	0.874	0.851	0.889	**0.929**	**0.866**	0.828	1.340	1.531	0.780	**0.300**	1.246	1.782
Sptl.Aug+FDA	**0.883**	0.839	**0.899**	0.926	0.862	0.850	**1.133**	1.615	**0.605**	0.421	1.230	1.436
Sptl.Aug+CycleGan	0.863	**0.853**	0.898	0.923	0.863	**0.859**	1.330	**1.359**	0.609	0.363	**1.230**	**1.194**
Sptl.Aug+FDA+Ens.	**0.888**	0.849	**0.904**	**0.932**	**0.868**	**0.862**	**1.016**	1.460	**0.557**	**0.290**	1.246	1.211

As ensemble learning improved the segmentation accuracy for most clinical centers, we selected the best three sets of models, with each set trained using different training/validation splits. Each set contains two models that were selected based on their performance—both on our own validation split and the M&Ms validation dataset. Specifically, the model ensemble consists of three models trained

Table 2. Dice and ASSD values of the final model evaluated on the test set.

Metric	Dice						ASSD					
Vendor	A		B		C	D	A		B		C	D
Center	1	6	2	3	4	5	1	6	2	3	4	5
LV	0.924	0.904	0.900	0.919	0.891	0.903	1.010	1.168	1.168	0.913	1.374	1.184
RV	0.876	0.864	0.872	0.869	0.819	0.882	1.183	1.323	1.201	1.324	2.064	1.347
Myo	0.826	0.839	0.843	0.873	0.817	0.820	0.725	0.821	0.835	0.658	0.974	0.986

with FDA, two models trained with CycleGAN augmentation and one model trained with FDA and extra augmented apical and basal slices. Table 2 displays the segmentation accuracy of our final submission evaluated on the M&Ms test data. Overall, our method achieved promising results for most of the vendors and clinical centers, although our method was only trained with annotated data from two vendors and three centers. Figure 4 displays example segmentations of our method on Vendor C and D, which did not have annotated training data. Generally, our method predicted segmentations that closely resemble the ground truths. For some cases, our method tends to make mistakes on apical or basal slices, while generating better predictions in the middle part of the heart.

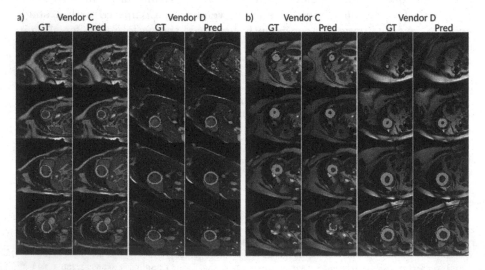

Fig. 4. Examples of predicted segmentations that were better (a) and worse (b) than average compared with predictions by other participants. Segmentations are overlaid with the corresponding image data. Predictions (Preds) are compared with ground truths (GTs) for Vendor C and D. From top row to bottom row are apical, middle and basal slices.

4 Conclusion

We presented a DL-based automatic cardiac segmentation framework that demonstrated promising performance across multi-scanner and multi-site CMRI scans. We showed that using image augmentation to simulate appearance variations of CMRI data while at the same time constraining the model to generate similar predictions on appearance-augmented images can lead to improved generalization to previously unseen samples from a new vendor or clinical center. We also explored and compared two effective appearance augmentation techniques, frequency-domain augmentation and CycleGAN based augmentation, that can leverage information from an unlabeled data source to enrich the training dataset. The proposed method can be applied not only to CMRI segmentation but may be readily adapted to other segmentation tasks. Future work will aim to combine the proposed augmentation and consistency training methods with semi-supervised learning to further leverage the unlabeled data.

Acknowledgement. This works was supported by the National Science Foundation, Award No. 1663747. The authors declare that the segmentation method implemented for participation in the M&Ms challenge has not used any pre-trained models nor additional MRI datasets other than those provided by the organizers.

References

1. Avendi, M., Kheradvar, A., Jafarkhani, H.: A combined deep-learning and deformable-model approach to fully automatic segmentation of the left ventricle in cardiac MRI. Med. Image Anal. **30**, 108–119 (2015)
2. Bai, W., et al.: Semi-supervised learning for network-based cardiac MR image segmentation, pp. 253–260 (2017)
3. Chen, C., et al.: Improving the generalizability of convolutional neural network-based segmentation on CMR images. Front. Cardiovasc. Med. **7**, 105 (2020)
4. Chen, C., Dou, Q., Chen, H., Qin, J., Heng, P.A.: Synergistic image and feature adaptation: towards cross-modality domain adaptation for medical image segmentation. AAAI **33**, 865–872 (2019)
5. Dou, Q., Ouyang, C., Chen, C., Chen, H., Heng, P.A.: Unsupervised cross-modality domain adaptation of convnets for biomedical image segmentations with adversarial loss, pp. 691–697 (2018)
6. Isensee, F., Jaeger, P.F., Full, P.M., Wolf, I., Engelhardt, S., Maier-Hein, K.H.: Automatic cardiac disease assessment on cine-MRI via time-series segmentation and domain specific features. In: Pop, M., et al. (eds.) STACOM 2017. LNCS, vol. 10663, pp. 120–129. Springer, Cham (2018). https://doi.org/10.1007/978-3-319-75541-0_13
7. Odena, A., Dumoulin, V., Olah, C.: Deconvolution and checkerboard artifacts. Distill (2016). https://doi.org/10.23915/distill.00003, http://distill.pub/2016/deconv-checkerboard
8. Oktay, O., et al.: Attention u-net: Learning where to look for the pancreas. ArXiv abs/1804.03999 (2018)
9. Tao, Q., et al.: Deep learning-based method for fully automatic quantification of left ventricle function from cine MR images: a multivendor, multicenter study. Radiology **290**, 180513 (2018)

10. Yang, Y., Soatto, S.: Fda: fourier domain adaptation for semantic segmentation. In: CVPR (2020)
11. Zhu, J.Y., Park, T., Isola, P., Efros, A.A.: Unpaired image-to-image translation using cycle-consistent adversarial networks. In: ICCV (2017)

Generalisable Cardiac Structure Segmentation via Attentional and Stacked Image Adaptation

Hongwei Li[1], Jianguo Zhang[2](✉), and Bjoern Menze[1]

[1] Department of Computer Science, Technical University of Munich, Munich, Germany
{hongwei.li,bjoern.menze}@tum.de

[2] Department of Computer Science and Engineering, Southern University of Science and Technology, Shenzhen, China
zhangjg@sustech.edu.cn

Abstract. Tackling domain shifts in multi-centre and multi-vendor data sets remains challenging for cardiac image segmentation. In this paper, we propose a generalisable segmentation framework for cardiac image segmentation in which multi-centre, multi-vendor, multi-disease datasets are involved. A generative adversarial networks with an attention loss was proposed to translate the images from existing source domains to a target domain, thus to generate good-quality synthetic cardiac structure and enlarge the training set. A stack of data augmentation techniques was further used to simulate real-world transformation to boost the segmentation performance for unseen domains. We achieved an average Dice score of **90.3%** for the left ventricle, **85.9%** for the myocardium, and **86.5%** for the right ventricle on the hidden validation set across four vendors. We show that the domain shifts in heterogeneous cardiac imaging datasets can be drastically reduced by two aspects: 1) good-quality synthetic data by learning the underlying target domain distribution, and 2) stacked classical image processing techniques for data augmentation.

Keywords: Model generalisability · Image segmentation · GANs

1 Introduction

Fully automatic cardiac segmentation methods can help clinicians to quantify the heart structure (e.g. left ventricle (LV), myocardium (MYO), and right ventricle (RV)) from cardiac magnetic resonance (CMR) images for diagnosis of multiple heart diseases [1]. Deep learning-based methods have shown promising avenues for cardiac image segmentation [3]. However, existing work [7] have shown that the segmentation performance of such methods may drop in when they are directly tested to scans acquired from different centres or vendors. The degradation of performance is not only caused by the varying cardiac morphology

© Springer Nature Switzerland AG 2021
E. Puyol Anton et al. (Eds.): STACOM 2020, LNCS 12592, pp. 297–304, 2021.
https://doi.org/10.1007/978-3-030-68107-4_30

but also the differences of acquisition parameter, resolution, intensity distribution, etc. [6] as shown in Fig. 1. All these factors pose obstacles for deploying deep learning-based segmentation algorithms in real-world clinical practice.

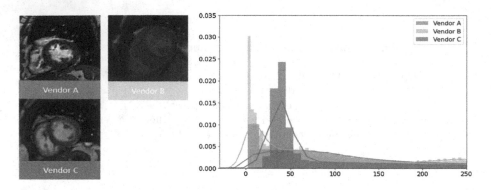

Fig. 1. The differences of image appearance, resolution and intensity distributions in the cardiac structures (i.e. LV, MYO, RV) of three vendors.

One simple way to address the above issues is to fine-tune a model learned from some datasets (source domains) with extra labelled data from another dataset (target domain). However, collecting sufficient pixel-level labelled medical data for all centres and vendors is extremely difficult which require fully clinical studies. To mitigate these issues, domain adaptation methods have been proposed to generalise one algorithm trained on some datasets (source domain) with additional data (either labelled or not labelled) from another dataset (target domain) [4,8,11]. Data augmentation-based methods are further proposed to enhance the generalisability of cardiac image segmentation models [2,9]. Generally, it is clinically relevant to explore how to learn a generalisable model that can be successfully applied to other datasets without additional model tuning.

In this work, we present a fully automatic segmentation framework to segment three cardiac structures (i.e. LV, MYO, and RV) and to mitigate the above issues caused by domain shifts. It is achieved by leveraging generative adversarial networks to transfer image style and stacked image processing techniques to augment the training samples and thus to generalise the segmentation model to unseen domains. Specifically, our approach mainly consists of three modules:

(i) A target domain transfer network. This module is used for learning the underlying intensity distribution of the unlabeled vendor and translate the labeled vendor to the target vendor, to augment the training set with **synthetic vendor-C-like** images and annotation from vendor A and B. Specifically we develop an attention-GANs with a focus on the cardiac structure.

(ii) A stacked image transformation function. To simulate real-world testing conditions and to increase data variations, we apply a stack of six spatial and intensity transformations to overcome the domain shifts.

(iii) A segmentation model. A residual U-shape convolutional neural network with dilated convolutions [5] with a larger receptive field compared with U-Net is used to perform segmentation. It can capture richer context information with less parameters and is trained with all the data simulated above.

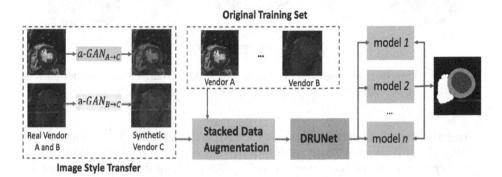

Fig. 2. Schematic view of our approach. It consists of three modules: a) image style transfer involving an attention GAN with a focus on cardiac structure, b) stacked data augmentation of several image transformations, and c) a segmentation network called *DRUNET* with lighter number of weight but capture rich context information.

2   Methodology

Given the challenge setting[1], the proposed method aims at learning an generalisable segmentation model for labelled vendor A and labelled vendor B:$\{(x_A, y_A), (x_B, y_B)\}$, unlabelled vendor C: $\{x_C\}$ where 25 scans were given but without labels, and the unseen vendor D, where the data was hidden.

As shown in Fig. 2, our framework mainly included three modules as mentioned above. Specifically, the image style transfer module translated the images from vendor A and vendor B to unlabeled vendor C and further augmented the training set with synthetic vendor-C-like images and the annotations from vendor A and B. We enhanced the image quality by introducing an attention loss. The stacked data augmentation module aimed to further increase the data variation by employing several intensity and spatial transformation on the original and synthetic data after the first module. Last, the segmentation models were trained with all the original and synthetic data and performed inference with an ensemble model.

---

[1] https://www.ub.edu/mnms/.

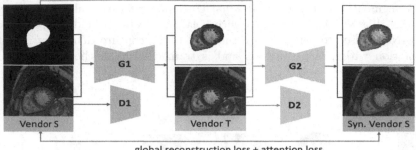

Fig. 3. Overview of the image translation network. An attention- reconstruction loss was introduced to enhance the image quality of cardiac region.

Generative Adversarial Network with Attention. We adopted the CycleGAN [10] as the basic architecture for one-to-one image translation. This included a pair of generators $\{G_1, G_2\}$ and a pair of domain discriminators $\{D_1, D_2\}$. As illustrated in Fig. 3, generator G_1 aimed to translate images from a source vendor to a target vendor while generator G_2 aimed to translate from the target one to the source one. Discriminators $\{D_1, D_2\}$ are trained to distinguish if the generated images are real or fake in the two domains respectively. In a min-max game, the generators try to fool the discriminators by good image translation. However, since the cardiac structures were of interests, we further introduced a attention-reconstruction loss with a focus on cardiac structure:

$$\mathcal{L}_{a\_rec} = \mathbb{E}_{x \sim \mathcal{S}}[||(x - G_2(G_1(x))) \odot m||_1] + \mathbb{E}_{y \sim \mathcal{T}}[||y - G_1(G_2(y))||_1] \quad (1)$$

Notably the mask m was only applied to the source domains where the labels are available. Thus, in addition to the original global reconstruction loss $\mathcal{L}_{g\_rec}$, we used a composite loss function which consists of two loss terms: $\mathcal{L}_{rec} = \mathcal{L}_{g\_rec} + \lambda \mathcal{L}_{a\_rec}$. In our experiments, we set $\lambda = 0.5$ to balance the contribution of the two losses.

Stacked Augmentation. We proposed a sequence of k stacked image transformations $f(\cdot)$ to simulate image distributions for unseen domains. Given training data x_t and the corresponding label y_t, augmented data $\hat{x}_S$ and the paired label $\hat{y}_t$ can be generated after k transformations.

$$(\hat{x}_t, \hat{y}_t) = f^k(f^{k-1}(...f^1(x_t, y_t))) \quad (2)$$

We selected three intensity transformations and three spatial transformations:

(1) image blurring and sharpening. Gaussian filtering was utilised to smooth the image to simulate blurry produced by motions. The standard deviation of a Gaussian kernel is ranged between $[0.1, 2]$. Meanwhile, we sharpened the image by using a unsharp masking.

(2) intensity perturbation. The intensity range was shifted with a magnitude range between [−0.05, 0.05].

(3) gamma correction. This was used to adjust the contrast of the image with a range between [0.6, 1.7].

(4) shearing. This was aimed to equip the network with variance to deformations with a magnitude range between [−0.1, 0.1] was used for both images and masks.

(5) rotation. A range between [−15°, 15°] was used for both images and masks.

(6) scaling. This enforces the network to be scale-invariant to resolution. A magnitude range between [−0.1, 0.1] was used for both images and masks.

Segmentation Network. We adopted a top-perform 2D architecture named Dilated Residual U-Net (DRUNet) [5], which was used for both brain and cardiac segmentation. DRUNet exploited the inherent advantages of the skip connections, residual learning and dilated convolutions to capture rich context information with a minimal number of trainable parameters. The network was trained with a weight cross entropy loss function (Table 1).

3   Experiments

Table 1. Multi-vendor datasets. Resolutions of scans from the same vendor are even different.

Vendor	A	B	C	D
Numbers (training/test)	75/50	75/50	25/50	0/50
Annotation Availability (%)	Yes	Yes	No	No

3.1   Experimental Setting

Datasets. The released training set consists of 150 annotated scans from two different MRI vendors (75 for each) and 25 unannotated scans from a third vendor as shown in Table 1. The CMR scans have been segmented by experienced clinicians, with contours for the left (LV) and right ventricle (RV) blood pools, and the left ventricular myocardium (MYO). The segmentation pipeline was evaluated on it and the results on the hidden validation set provided by the challenge organisers were presented. To optimise the segmentation model We use four labeled scans (2 from vendor A and 2 from vendor B) as a validation set, and the remains as a training set. For the final submission, we used all the whole released training set.

Pre- and Post- processing. The pre-processing of all the images (including the hidden test set) were performed in a slice-wise manner by three steps. First, non-local means denoising was performed for each slice to reduce the noise level considering that the image quality from multiple centres are diverse; second, the intensity range was normalised to [0, 1] to facilitate the model training; third, the images and masks are cropped or padded to [256, 256]. For post-processing, we performed connected component analysis and removed small structures with less than 30 voxels.

Network Training. (1) For the image translation network, we used the Cycle-GAN implementation for the one-to-one mappings: $A \rightarrow C$, and $B \rightarrow C$. Network configuration and hyper-parameters were kept the same as in [10] except the input and output images are single-channel 2D images. It was trained for 100 epochs with a batch size of 5 involving around 3500 images for each vendor including scans from multiple time points. (2) For training the segmentation model, the weights for background, LV, MYO, and RV in the weighted cross entropy loss were empirically set to 0.19 : 0.24 : 0.31 : 0.26 based on the performance on the validation set. The algorithm was implemented using python and Tensorflow and was trained for 100 epochs in total on an NVIDIA® Titan V GPU. The training of the segmentation model took around 5 h.

3.2 Results

We conducted three experiments to illustrate the effectiveness of our approach. First, we trained the DRUNet on solely the labeled datasets: vendor A and B, referred as *baseline* in Table 3. Second, we used the attention-GANs to generate good-quality vendor C-like images and include those synthetic images and their corresponding labels for training, referred as *baseline+a-GANs* in Table 3. Lastly, we further incorporated the stacked image transformation and train the model from scratch, referred as *ours* in the table. We found that after including the stacked image transformation, we drastically improved the segmentation performance on the hidden vendor D, e.g. Dice for RV is improved from 14.5% to 72.7%. On the unlabeled vendor C, we achieved average Dice score of 86.8% for the left ventricle, 83.4% for the myocardium, and 83.2% for the right ventricle; on the hidden vendor, we achieved average Dice score of **89.3%** for the left ventricle, **83.4%** for the myocardium, and 72.7% for the right ventricle. Qualitative segmentation result from vendor C is shown in Fig. 4 (Table 2).

Table 2. Average Dice scores of all vendors, the highest performance in each class is highlighted.

Method	$Dice_A$ (%)			$Dice_B$ (%)			$Dice_C$ (%)			$Dice_D$ (%)		
	LV	MYO	RV	LV	MYO	RV	LV	MYO	RV	LV	MYO	RV
Baseline	85.7	77.1	66.6	92.2	83.9	87.7	86.0	81.0	76.5	72.3	51.7	14.5
Baseline+a-GANs	88.5	81.6	71.8	94.2	86.6	91.5	87.7	84.5	80.1	65.9	58.0	13.3
Ours (a-GANs+Stacked)	90.5	84.1	85.1	93.6	87.5	91.1	86.8	83.4	83.2	89.3	81.4	72.7

Table 3. Average Hausdorff distance (HD) of all vendors, the highest performance in each class is highlighted.

Method	HD_A (mm)			HD_B (mm)			HD_C (mm)			HD_D (%)		
	LV	MYO	RV	LV	MYO	RV	LV)	MYO	RV	LV	MYO	RV
Baseline	23.7	37.0	44.11	14.0	20.7	23.4	17.7	19.4	31.5	27.5	35.8	61.8
Baseline+a-GANs	21.4	31.2	21.1	**7.5**	11.7	12.4	14.4	16.3	**16.7**	22.3	30.3	42.0
Ours (a-GANs+Stacked)	**15.8**	**16.7**	**16.2**	7.9	**11.0**	**11.5**	**10.9**	15.0	23.6	**17.3**	**24.6**	**17.6**

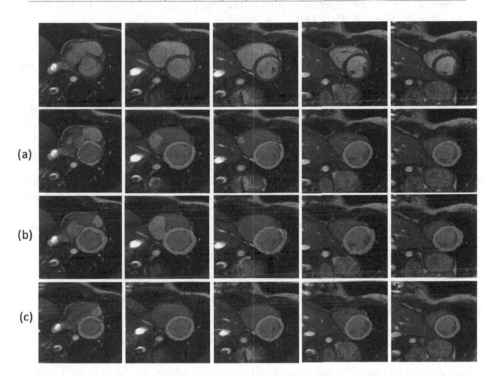

Fig. 4. Qualitative segmentation result of one subject from vendor C. a) baseline results; b) results of baseline+a-GANs; c) results of our method. Please note that the ground truth for vendor C is not available.

4   Conclusion

We proposed a cardiac structure segmentation framework and segmented three structures: LV, MYO, and RV. We demonstrated that domain shifts in heterogeneous cardiac imaging datasets can be drastically reduced by two aspects: 1) good-quality synthetic data by learning the underlying target domain distribution, and 2) stacked classical image transformation techniques for data augmentation.

References

1. Bernard, O., et al.: Deep learning techniques for automatic MRI cardiac multi-structures segmentation and diagnosis: is the problem solved? IEEE Trans. Med. Imaging **37**(11), 2514–2525 (2018)
2. Chen, C., et al.: Improving the generalizability of convolutional neural network-based segmentation on CMR images. arXiv preprint arXiv:1907.01268 (2019)
3. Chen, C., et al.: Deep learning for cardiac image segmentation: a review. Front. Cardiovasc. Med. **7**, 25 (2020)
4. Chen, C., Dou, Q., Chen, H., Qin, J., Heng, P.A.: Synergistic image and feature adaptation: Towards cross-modality domain adaptation for medical image segmentation. In: Proceedings of the AAAI Conference on Artificial Intelligence, vol. 33, pp. 865–872 (2019)
5. Li, H., Zhygallo, A., Menze, B.: Automatic brain structures segmentation using deep residual dilated U-Net. In: Crimi, A., Bakas, S., Kuijf, H., Keyvan, F., Reyes, M., van Walsum, T. (eds.) BrainLes 2018. LNCS, vol. 11383, pp. 385–393. Springer, Cham (2019). https://doi.org/10.1007/978-3-030-11723-8_39
6. Petitjean, C., Dacher, J.N.: A review of segmentation methods in short axis cardiac MR images. Med. Image Anal. **15**(2), 169–184 (2011)
7. Yan, W., et al.: MRI manufacturer shift and adaptation: increasing the generalizability of deep learning segmentation for MR images acquired with different scanners. Radiol. Artif. Intell. **2**(4), e190195 (2020)
8. Yan, W., et al.: The domain shift problem of medical image segmentation and vendor-adaptation by Unet-GAN. In: Shen, D., et al. (eds.) MICCAI 2019. LNCS, vol. 11765, pp. 623–631. Springer, Cham (2019). https://doi.org/10.1007/978-3-030-32245-8_69
9. Zhang, L., et al.: Generalizing deep learning for medical image segmentation to unseen domains via deep stacked transformation. IEEE Trans. Med. Imaging **39**(7), 2531–2540 (2020)
10. Zhu, J.Y., Park, T., Isola, P., Efros, A.A.: Unpaired image-to-image translation using cycle-consistent adversarial networks. In: Proceedings of the IEEE International Conference on Computer Vision, pp. 2223–2232 (2017)
11. Zhuang, X., et al.: Cardiac segmentation on late gadolinium enhancement MRI: a benchmark study from multi-sequence cardiac MR segmentation challenge. arXiv preprint arXiv:2006.12434 (2020)

Style-Invariant Cardiac Image Segmentation with Test-Time Augmentation

Xiaoqiong Huang[1,2], Zejian Chen[1,2], Xin Yang[1,2], Zhendong Liu[1,2], Yuxin Zou[1,2], Mingyuan Luo[1,2], Wufeng Xue[1,2], and Dong Ni[1,2(✉)]

[1] School of Biomedical Engineering, Shenzhen University, Shenzhen, China
nidong@szu.edu.cn
[2] Medical UltraSound Image Computing (MUSIC) Laboratory, Shenzhen University, Shenzhen, China

Abstract. Deep models often suffer from severe performance drop due to the appearance shift in the real clinical setting. Most of the existing learning-based methods rely on images from multiple sites/vendors or even corresponding labels. However, collecting enough unknown data to robustly model segmentation cannot always hold since the complex appearance shift caused by imaging factors in daily application. In this paper, we propose a novel style-invariant method for cardiac image segmentation. Based on the zero-shot style transfer to remove appearance shift and test-time augmentation to explore diverse underlying anatomy, our proposed method is effective in combating the appearance shift. Our contribution is three-fold. First, inspired by the spirit of universal style transfer, we develop a zero-shot stylization for content images to generate stylized images that appearance similarity to the style images. Second, we build up a robust cardiac segmentation model based on the U-Net structure. Our framework mainly consists of two networks during testing: the ST network for removing appearance shift and the segmentation network. Third, we investigate test-time augmentation to explore transformed versions of the stylized image for prediction and the results are merged. Notably, our proposed framework is fully test-time adaptation. Experiment results demonstrate that our methods are promising and generic for generalizing deep segmentation models.

Keywords: Style transfer · Cardiac image segmentation · Test-time augmentation

1 Introduction

Delineation of the left ventricular cavity (LV), myocardium (MYO), and right ventricle (RV) from cardiac magnetic resonance (CMR) images (multi-slice 2D cine MRI) is a common clinical task to establish the diagnosis. It is of great

---

X. Huang and Z. Chen – Equally contribute to this work.

© Springer Nature Switzerland AG 2021
E. Puyol Anton et al. (Eds.): STACOM 2020, LNCS 12592, pp. 305–315, 2021.
https://doi.org/10.1007/978-3-030-68107-4_31

interest to develop an accurate automated segmentation method since manual segmentation is tedious and likely to suffer from inter-observer variability. Deep learning cardiac segmentation models have achieved remarkable success based on a large amount of labeled data. However, as shown in Fig. 1, learning-based models often subject to severe performance drop due to testing data that has different distributions from the training data. This is a highly desirable but challenging task that makes deep models robust against the complex appearance shift of testing images [1,13] caused by different sites, scanner vendors, imaging protocols, etc.

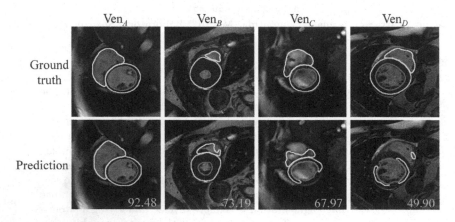

Fig. 1. Illustration of segmentation degradation on cardiac images from four vendors (Ven$_A$, ..., Ven$_D$). Green, red and yellow curves represent LV, MYO and RV, respectively. Orange digits denote the average Dice index over three structures. The model trained on the images of Ven$_A$ performs a notable drop on images of other verdors. (Color figure online)

To mitigate the performance degradation, one straightforward choice is data augmentation [10,11]. It can help suppress overfitting but cannot guarantee the generalization ability of deep models. Recently, Domain Adaptation (DA) [4] and Domain Generalization (DG) [5] have been common methods for coping with the appearance shift. As main branches of DA/DG, aligning appearance level or feature level among different domains via adversarial learning were explored. Although DA/DG is attractive, it depends heavily on sufficient data from the target domain or requires enough multiple labeled source data. It is also confined by its domain mapping and may not extend to images from unknown domains. By revisiting the basic definition of appearance shift, style transfer [6] (ST) inspires a new and intuitive way for the problem. ST removes appearance shift by rendering the appearance of the content image as the style image [3,8,9]. Compared to DA, ST is independent on the target domain, retraining-free and suitable for images with unknown appearance shifts. In [9], Ma *et al.* made the early attempt to exploit an online ST to reduce the appearance variation for better cardiac MR segmentation. But such optimization-based ST has high latency and restrains real-time applications. Liu *et al.* [8] proposed an Adaptive Instance Normalization (AdaIN) [7] based ST module for vendor adaption to achieve real-time arbitrary style transfer. However, it directly utilized the pre-trained VGG-16 as the ST backbone, which may be unadaptable for the medical image to retain a more realistic semantic content structure.

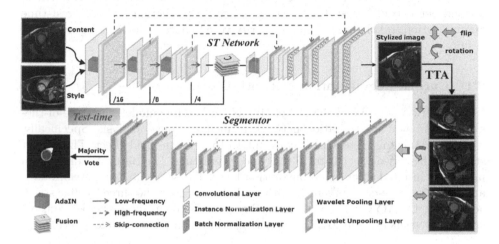

Fig. 2. Schematic view of our proposed framework.

In this paper, based on Wavelet Corrected Transfer network (WaveCT) [14] and WaveCT-AIN [8], we propose an improved ST network to generate style-invariant images for removing appearance shift and test-time augmentation to enhance the segmentation results. Our contribution is three-fold. First, inspired by the spirit of universal style transfer, we develop a zero-shot stylization for testing (content) images to generate stylized images that appearance similarity to the source (style) images. Second, we build up a robust model based on the U-Net structure for cardiac segmentation. Our framework is a two-stage system during testing: we utilize the ST network to generate the stylized image, then feed it into the segmentation model. Third, we investigate test-time augmentation to explore transformed versions of the stylized image for inference, followed by inverse transformation and predictions mergence to get the final segmentation result. In particular, we make two experiments to verify our proposed framework. 1) segmentation model trained on the original dataset and 2) segmentation model trained on the style-unified dataset generated by our zero-shot ST network from the original dataset.

2   Methodology

Figure 2 is the schematic view of our proposed method. The universal 2D U-Net and VGG-16 networks serve as the backbones for segmentation and ST, respectively. We first train the segmentation model on the source data, then develop a ST network to generate stylized images that suitable for the segmentation model. Specifically, the proposed framework is a two-stage system for segmenting images with appearance shifts. In the first stage, the testing image is transferred into a stylized image refer to the source image appearance. In the second stage, the segmentation model trained on source data is performed on the stylized image and get the segmentation result. Moreover, we explore transformed versions of the stylized testing image for prediction by using test-time augmentation and then perform a majority vote to obtain the final segmentation result.

2.1   Cardiac Segmentation Network Design

In this work, we modified the U-Net [12] as our baseline model, the state-of-the-art 2D semantic segmentation network in medical image analysis. Specifically, we use upsampling instead of deconvolution to avoid the grid effect. The output stride of the network is cut to 16 to reduce overfitting. The Batch Normalization layers are inserted after each convolution layer. The segmentation network aims at predicting four-class pixel-wise probabilistic maps for the three cardiac structures (i.e., LV, MYO, RV) and the background. To train the network, we use a composite segmentation loss function L_{seg} which consists of two loss terms:

$$L_{ce} = -\sum_c y^c log(p^c), \quad L_{Dice} = \sum_c 1 - \frac{2|X^c \cap Y^c|}{|X^c| + |Y^c|}$$

$$L_{seg} = L_{ce} + \lambda L_{Dice}$$

(1)

The first term L_{ce} is a categorical cross entropy loss, where p^c denotes the corresponding predicted probability map of different classes. The second term is a Dice loss to measure the similarity between probability map X^c and ground truth Y^c. We set $\lambda = 0.5$ to balance the contribution of the two losses.

2.2   Zero-Shot Style Transfer

ST enables us to transfer the style of an image called style image to that of an image called the content image, rendering the low-level visual style while preserving its high-level semantic content structure. Inspired by the spirit of ST can remove appearance shift to approach generalize image analysis, we develop a ST network to generate style-invariant images for generalizing segmentation model. Different from the optimization-based or feed-forward approximate stylization, we utilize zero-shot ST to achieve real-time arbitrary stylization without training on any pre-defined styles.

To meet the requirement of universal and stable transfer between any content-style pairs, we adopt the WaveCT network recently used in WCT2 [14] and make improvements to preserve image structure details and render the style features. Different from previous online ST methods [9] used to remove appearance shift which may distort image details, the WaveCT network replaces vanilla max-pooling/unpooling with the Haar wavelet pooling/unpooling layers that maintain the content structure to the great extent. In particular, WaveCT splits the features into low-frequency and high-frequency components via Haar wavelet pooling, then low-frequency information passes the main network and high-frequency information skips to connect between encoder and decoder.

The ST network proposed in this paper is a significant extension of our prior conference paper proposed WaveCT-AIN [8], regarding the following highlighted points. As the ST network depicted in Fig. 2, first, we design a multi-scale feature fusion layer after the encoder, in return mitigate the variation of background area without information in the image. Second, we keep the ST module in the encoder and add an extra AdaIN after the feature fusion. Besides, we enhance the style-invariance by introducing the Instance Normalization (IN) layer into the decoder. In this respect, we focus on rendering the style texture representations in the low-level features and preserves its invariance in the high-level patterns. Third, we simplify the case-specific style image selection strategy more concisely and effectively, which is directly considered selecting the reference style image that as close as possible to the mean and standard deviation of the testing image. Especially, the ST network utilizes the pre-trained VGG-16 as backbone while feature fusion layer and IN layers are embedded, thus it needs to be fine-tuned with image reconstruction task.

2.3   Test-Time Augmentation

Data augmentation significantly improves robustness to appearance shift and can be used as a simple strategy for generalizing model performance. Data augmentation at training time has been commonly used to increase the amount of data for improving performance [11]. Recent works also demonstrated the usefulness of data augmentation directly at test time, for achieving more robust predictions [10]. For the point of data acquisition, a testing image is only one of many possible observations of the underlying anatomy. Therefore, we explore multiple transformed versions of the testing image for robust segmentation. Test-time augmentation includes four procedures: augmentation, prediction, inverse-augmentation, and merging. We firstly consider different transformations on the testing image. For our case, we have already remove appearance shift through the ST network, thus we apply flip and rotation transformations for stylized cardiac images instead of complicated contrast or brightness change. In particular, we make three different transformations on the stylized testing image and inference each version of the testing image, thus four predictions are obtained by the inverse transformation. Then we perform a majority vote to obtain the final segmentation result, that is, once the pixel is predicted twice or more, it will be regarded as the target area.

3   Experimental Results

3.1   Dataset and Implementation Details

Notably, we make two experiments to verify our proposed framework. *Exp.1*) segmentation model trained on the original training dataset, denoted as *SegO* and *Exp.2*) segmentation model trained on the style-unified dataset generated by our zero-shot ST network from the original dataset, denoted as *SegST*.

Table 1. Quantitative comparison results of the *Exp.1*.

Metrics	SegO				STSegO				STSegO-TTA			
	Ven_A	Ven_B	Ven_C	Ven_D	Ven_A	Ven_B	Ven_C	Ven_D	Ven_A	Ven_B	Ven_C	Ven_D
Dice_{AVG}	80.72	86.82	81.20	62.23	82.92	**89.72**	85.27	64.87	**84.70**	89.57	**85.56**	68.01
Jac_{AVG}	68.93	77.98	69.53	49.81	71.82	**82.04**	75.01	53.49	**74.29**	81.98	**75.45**	55.80
HDB_{AVG}	19.52	14.24	18.30	44.88	17.54	9.93	13.27	44.53	**14.88**	9.04	**12.94**	38.46
ASSD_{AVG}	1.96	0.89	1.84	12.44	1.85	0.65	1.46	13.17	**1.51**	0.65	**1.44**	7.38
Dice_{LV}	89.14	93.07	85.12	72.97	88.71	**92.90**	86.46	78.21	89.75	92.45	86.39	74.91
Jac_{LV}	81.01	87.56	75.56	62.83	80.27	**87.39**	77.34	68.09	81.95	86.92	77.26	64.94
HDB_{LV}	13.58	9.55	13.38	33.68	13.28	6.72	12.05	28.23	11.59	6.03	10.90	31.42
ASSD_{LV}	1.53	0.61	1.91	10.86	1.54	**0.63**	1.74	5.76	1.34	0.68	1.76	8.28
Dice_{MYO}	72.71	76.63	73.60	51.76	81.84	**85.69**	83.83	60.76	**83.13**	85.68	**83.92**	63.19
Jac_{MYO}	57.47	62.64	58.72	36.92	69.49	75.19	72.52	47.01	**71.32**	75.21	72.63	48.53
HDB_{MYO}	20.27	21.56	19.15	34.00	16.54	11.36	**14.77**	27.22	14.02	10.74	14.85	30.40
ASSD_{MYO}	1.92	1.35	2.00	7.13	1.29	0.59	1.29	**2.67**	1.09	**0.55**	1.28	3.04
Dice_{RV}	80.31	90.77	84.88	61.96	78.21	90.58	85.51	55.66	**81.22**	**90.60**	**86.39**	65.94
Jac_{RV}	68.32	83.74	74.30	49.69	65.71	83.55	75.18	45.38	**69.60**	**83.79**	**76.47**	53.94
HDB_{RV}	24.72	11.61	22.38	66.95	22.79	11.71	**12.98**	78.14	19.03	10.35	13.07	**53.55**
ASSD_{RV}	2.43	0.70	1.61	19.34	2.72	0.74	1.35	31.08	**2.09**	**0.72**	1.27	10.83

Dataset. The framework was trained and evaluated on the Multi-Centre, Multi-Vendor & Multi-Disease Cardiac Image Segmentation Challenge (M&Ms 2020) dataset [2]. Two subsets of 75 CMR images from vendor A and vendor B (denoted as Ven_A and Ven_B) with only ES & ED annotated are provided as training data, respectively. Additionally, 25 unannotated images are also given from Ven_C. However, our methods do not use it because we are concentrate on generalizing the model to other more unknown data, not just Ven_C. For evaluation, the results are evaluated on not only 50 new studies from each of $\text{Ven}_{A,B,C}$, but also 50 else studies from Ven_D.

Implementation Details. We obtain 3284 training slices along the anatomical plane from the ES & ED images of Ven_A and Ven_B. All slices are resized to 256×256. For training the *SegO*, we apply elastic deformations to the available training slices (i.e., random expand, flip, rotation, mirror, contrast change, and brightness change). Whereas the training of *SegST* is not necessary to make contrast or brightness change because its training data already have a particular appearance distribution. We use 3000 cardiac slices to fine-tune the ST Network as the extra feature fusion layer and IN layers are embedded into the pre-trained VGG-16. For the segmentation network, it was trained for 60k iterations with a batch size of 24 and was optimized using the composite loss L_{seg} where Adam optimizer with a learning rate of 10^{-3} initially then decreased to 10^{-5}. We implement all experiments with PyTorch on two GeForce® RTX 2080 Ti GPU.

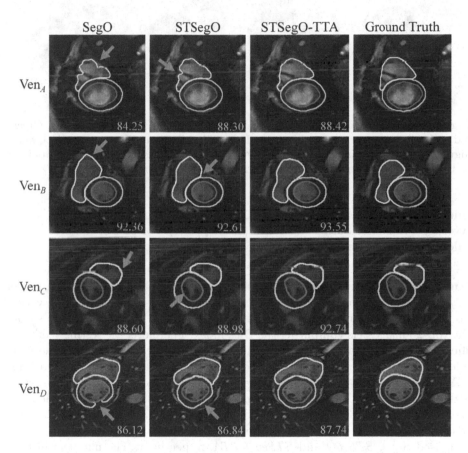

Fig. 3. Visualization of our better 3D segmentation results. From left to right are cases from Ven_B, Ven_C, Ven_D and Ven_D, respectively. Green, red and yellow areas represent LV, MYO and RV, respectively. (Color figure online)

3.2 Quantitative and Qualitative Evaluation

Metrics. To evaluate the accuracy of segmentation performance, we adopt in total 4 indicators including the Dice similarity index (Dice, %), Jaccard

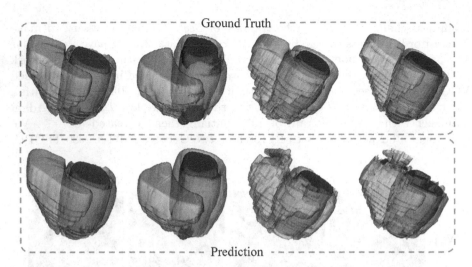

Fig. 4. Visualization of the 2D segmentation results of our proposed methods. Green, red and yellow curves represent the boundary of LV, MYO and RV, respectively. Orange digits denote the average Dice index. The performance is gradually improved from left to right methods, especially in the boundaries of RV. (Color figure online)

similarity index (Jac, %), Hausdorff Distance of Boundaries (HDB, pixel), and Average Symmetric Surface Distance (ASSD, pixel). For ease of comparison, we calculate the average (AVG) per indicator over the three structures (LV, MYO, RV).

Quantitative Results of the *Exp.1.* We first train the segmentation model *SegO* based on the available labeled images from Ven_A and Ven_B. Then we set up a style image library from the training data, which contains 221 slices from the top-20 images via computing the Dice index. Therefore, the testing (content) slice selects the reference style slice from the library through our simple style selection strategy. Subsequently, the content-style pairs feed into our zero-shot ST network to generate the stylized slice for segmentation. This two-stage system is denoted as *STSegO*. Likewise, *STSegO* with test-time augmentation is denoted as *STSegO-TTA*. Table 1 shows the quantitative results based on the 200 test cases correspond to four different vendors. We compare three versions of our proposed *SegO*, *STSegO* and *STSegO-TTA*, respectively. The numbers in bold indicate the best results of multiple vendors among different methods. Both *STSegO* and *STSegO-TTA* get consistent improvements over the pure *SegO*, in which the best results are achieved by the *STSegO-TTA*. It almost improves the Dice index by 5% and the Jaccard index by 6% on average for each vendor. The HDB and ASSD also improve about 5.5 pixels and 1.5 pixels, respectively. Obviously, the segmentation model can be well generalized to the images of Ven_C, but the performance on the Ven_D shows relatively poor. This may be due to the large difference in the data distribution and anatomical structure between images of Ven_D and the source data. Figure 4 visualizes the 2D segmentation results of our proposed methods on unseen cases from four vendors. *STSegO-TTA* produces more anatomically plausible results on the images. Figure 3 visualizes the 3D segmentation results of the *STSegO-TTA* on four unseen cases.

Quantitative Results of the *Exp.2*. Different from the *SegO* trained on the original dataset, *SegST* utilized the style-unified dataset for training, which is generated by our zero-shot ST network from the original dataset. Notably, we randomly select a slice serve as the style slice to generate stylized data, thus the style-unified training data has a particular appearance distribution. Consequently, the ST network only takes over this style slice to achieve stylization during testing. Likewise, *SegST* with test-time augmentation is denoted as *SegST-TTA*. As can be seen from Table 2, *STSeg-TTA* shows improvements over *STSeg*, the Dice index and Jaccard index are both raised about 2% on average for each vendor, and the HDB is improved about 1 pixel. However, the performance of Ven_D shows worse compared with *Exp.1*, which may be caused by the target style slice is not suitable for the images of Ven_D. Thus it is crucial to choose a universal style slice to generate the style-unified dataset, which will be our further study.

Table 2. Quantitative comparison results of the *Exp.2*.

Metrics	SegST				SegST-TTA			
	Ven_A	Ven_B	Ven_C	Ven_D	Ven_A	Ven_B	Ven_C	Ven_D
$Dice_{AVG}$	84.19	89.63	**85.74**	53.84	**85.99**	**90.28**	85.23	**58.86**
Jac_{AVG}	73.31	82.06	**75.71**	42.64	**76.03**	**82.74**	74.96	**47.90**
HDB_{AVG}	16.89	15.46	20.22	**35.56**	13.62	9.28	13.39	47.86
$ASSD_{AVG}$	1.71	0.73	1.44	**14.55**	1.34	0.58	1.37	17.52
$Dice_{LV}$	87.82	92.54	87.04	63.71	**89.64**	**93.83**	**87.50**	**68.34**
Jac_{LV}	78.84	86.95	78.23	52.92	**81.85**	**88.68**	78.87	**58.54**
HDB_{LV}	16.78	14.45	17.34	**48.06**	10.96	6.22	11.52	49.84
$ASSD_{LV}$	1.90	0.76	1.74	**20.87**	1.33	0.47	1.49	21.27
$Dice_{MYO}$	81.04	**86.61**	**84.81**	53.49	82.92	86.54	84.70	**56.97**
Jac_{MYO}	68.37	**76.81**	**73.96**	40.61	71.00	76.44	73.74	**44.49**
HDB_{MYO}	16.42	20.60	22.42	**24.78**	13.46	10.11	14.08	30.12
$ASSD_{MYO}$	1.39	0.59	1.28	12.26	1.07	0.49	1.01	7.68
$Dice_{RV}$	83.70	89.73	**85.37**	44.31	**85.41**	**90.46**	83.49	**51.26**
Jac_{RV}	72.71	82.42	**74.94**	34.40	**75.25**	**83.09**	72.28	**40.66**
HDB_{RV}	17.45	**11.32**	20.90	**32.93**	16.45	11.49	14.57	63.62
$ASSD_{RV}$	1.85	0.84	1.30	**10.52**	1.64	0.79	1.62	23.62

4   Conclusion

In this paper, we proposed a zero-shot ST network to generate style-invariant images for removing appearance shift and test-time augmentation to enhance the segmentation results. By investigating the two experiments *Exp.1* and *Exp.2*, we showed that *SegO* and *STSeg* with their variants present promising performance in segmenting cardiac images across the multi-vendor and multi-cencre dataset.

Acknowledgement. This work was supported by the grant from National Key R &D Program of China (No. 2019YFC0118300), Shenzhen Peacock Plan (No. KQTD2016053112051497, KQJSCX20180328095606003) and National Natural Science Foundation of China (No. 61801296).

References

1. Abràmoff, M.D., Lavin, P.T., Birch, M., Shah, N., Folk, J.C.: Pivotal trial of an autonomous AI-based diagnostic system for detection of diabetic retinopathy in primary care offices. NPJ Digit. Med. **1**(1), 1–8 (2018)
2. Campello, V.M., et al.: Multi-centre, multi-vendor & multi-disease cardiac image segmentation. (in preparation)
3. Chen, C., et al.: Unsupervised multi-modal style transfer for cardiac MR segmentation. In: Pop, M., et al. (eds.) STACOM 2019. LNCS, vol. 12009, pp. 209–219. Springer, Cham (2020). https://doi.org/10.1007/978-3-030-39074-7_22
4. Chen, C., Dou, Q., Chen, H., Qin, J., Heng, P.A.: Unsupervised bidirectional cross-modality adaptation via deeply synergistic image and feature alignment for medical image segmentation. IEEE Trans. Med. Imaging (2020)
5. Dou, Q., de Castro, D.C., Kamnitsas, K., Glocker, B.: Domain generalization via model-agnostic learning of semantic features. In: Advances in Neural Information Processing Systems, pp. 6450–6461 (2019)
6. Gatys, L.A., Ecker, A.S., Bethge, M.: Image style transfer using convolutional neural networks. In: Proceedings of the IEEE Conference on Computer Vision and Pattern Recognition, pp. 2414–2423 (2016)
7. Huang, X., Belongie, S.: Arbitrary style transfer in real-time with adaptive instance normalization. In: Proceedings of the IEEE International Conference on Computer Vision, pp. 1501–1510 (2017)
8. Liu, Z., et al.: Remove appearance shift for ultrasound image segmentation via fast and universal style transfer. In: 2020 IEEE 17th International Symposium on Biomedical Imaging (ISBI), pp. 1824–1828. IEEE (2020)
9. Ma, C., Ji, Z., Gao, M.: Neural style transfer improves 3D cardiovascular MR image segmentation on inconsistent data. In: Shen, D., et al. (eds.) MICCAI 2019. LNCS, vol. 11765, pp. 128–136. Springer, Cham (2019). https://doi.org/10.1007/978-3-030-32245-8_15
10. Moshkov, N., Mathe, B., Kertesz-Farkas, A., Hollandi, R., Horvath, P.: Test-time augmentation for deep learning-based cell segmentation on microscopy images. Sci. Rep. **10**(1), 1–7 (2020)
11. Perez, L., Wang, J.: The effectiveness of data augmentation in image classification using deep learning. arXiv preprint arXiv:1712.04621 (2017)

12. Ronneberger, O., Fischer, P., Brox, T.: U-Net: convolutional networks for biomedical image segmentation. In: Navab, N., Hornegger, J., Wells, W.M., Frangi, A.F. (eds.) MICCAI 2015. LNCS, vol. 9351, pp. 234–241. Springer, Cham (2015). https://doi.org/10.1007/978-3-319-24574-4_28
13. Yang, X., et al.: Generalizing deep models for ultrasound image segmentation. In: Frangi, A.F., Schnabel, J.A., Davatzikos, C., Alberola-López, C., Fichtinger, G. (eds.) MICCAI 2018. LNCS, vol. 11073, pp. 497–505. Springer, Cham (2018). https://doi.org/10.1007/978-3-030-00937-3_57
14. Yoo, J., Uh, Y., Chun, S., Kang, B., Ha, J.W.: Photorealistic style transfer via wavelet transforms. In: Proceedings of the IEEE International Conference on Computer Vision, pp. 9036–9045 (2019)

Automatic Evaluation of Myocardial Infarction from Delayed-Enhancement Cardiac MRI Challenge (EMIDEC)

Comparison of a Hybrid Mixture Model and a CNN for the Segmentation of Myocardial Pathologies in Delayed Enhancement MRI

Markus Huellebrand[1,2(✉)], Matthias Ivantsits[1], Hannu Zhang[1],
Peter Kohlmann[2], Jan-Martin Kuhnigk[2], Titus Kuehne[1,3,4],
Stefan Schönberg[5], and Anja Hennemuth[1,2,4]

[1] Charité – Universitätsmedizin Berlin, Augustenburger Pl. 1, 13353 Berlin, Germany
markus.huellebrand@charite.de
[2] Fraunhofer MEVIS, Am Fallturm 1, 28359 Bremen, Germany
[3] Department of Radiology and Nuclear Medicine,
University Medical Center Mannheim, 68167 Mannheim, Germany
[4] German Heart Institute Berlin, Augustenburger Pl. 1, 13353 Berlin, Germany
[5] DZHK (German Centre for Cardiovascular Research), Berlin, Germany

Abstract. DE-MRI provides a reliable and accurate imaging technique for the assessment of pathological alterations in myocardial tissue. The clinically applied thresholding techniques enable the assessment of the amount of diseased tissue. To also assess distribution patterns, transmurality and micro-vascular obstruction, more accurate segmentation methods are needed. We compare a hybrid CNN and mixture model approach with a two single-stage U-Net segmentation: one based on the EMIDEC challenge data set, one with additional training data, and could achieve DICE coefficients of 84.8%, 84.08%, and 82.95%, respectively. We hope to further improve the promising results through an extension of the training set.

Keywords: Delayed enhancement MRI · Mixture model · CNN · U-Net

1 Introduction

The analysis of delayed-enhancement magnetic resonance imaging provides an effective technique to analyze the state of the myocardial tissue after myocardial infarction. The analysis helps to select treatment, i.e., can give insight if a revascularization therapy will be successful. An automated, standardized way of segmenting the different areas such as the myocardium, the infarcted tissue, and the permanent microvascular obstruction will improve the diagnosis and therapeutic decision. Over the last decades, several segmentation approaches have been developed. However, the analysis of these DE-MRI is still a challenging task due to image contrast, image and motion artifacts (Fig. 1). In this

© Springer Nature Switzerland AG 2021
E. Puyol Anton et al. (Eds.): STACOM 2020, LNCS 12592, pp. 319–327, 2021.
https://doi.org/10.1007/978-3-030-68107-4_32

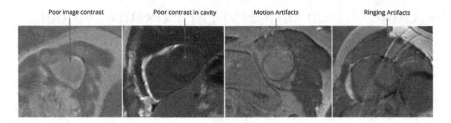

Fig. 1. Example cases with difficult image quality. From left to right: The first image shows a low contrast between myocardium and blood-pool in apical slice. The second example shows a low signal inside the left and right cavity. In the third example, there are motion artifacts due to myocardial motion. The last image shows ringing artifacts.

challenge, we tested two different approaches: one traditional mixture-model based approach and one based on CNN for the differentiation between normal myocardium, regions with late enhancement, and no-reflow areas where neither wash-in nor wash-out of contrast agent can be observed because of microvascular obstructions.

2   Materials and Methods

2.1   Image Data

The image data for the EMIDEC challenge [6] consists of 150 cases 100 diseased patients and 50 normal cases. The data set is split into a training set with 100 cases and a testing set with 50 cases. The training set as well as the testing set contains 1/3 normal and 2/3 of pathological cases, which roughly corresponds to real life observations in clinical settings. The data was acquired on Siemens MRI scanners on 1.5T (Area) and 3T (Skyra); the in-plane resolution was 1.25 × 1.25 mm^2 and 2 × 2 mm^2, a slice thickness of 8 mm and a distance between slices of 8 to 13 mm. In a post-processing step the image slices were realigned to prevent any drawbacks resulting from breathing motions.

Furthermore, the EMIDEC challenge organizers provided ground-truth segmentations for the training set, containing labels for the **cavity**, **myocardium**, **myocardial infarction**, and **no-reflow** areas. Figure 2 illustrates an example of a normal case and two cases with myocardial infarction and no-reflow areas.

2.2   Methods

Background. There have been many attempts to provide segmentation methods for a reproducible quantitative assessment of myocardial fibrosis based on late gadolinium enhancement imaging. The overview paper from the STA-COM challenge by Karim et al. presented many conventional voxel classification approaches based on intensity distributions [5]. Most approaches are organized

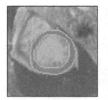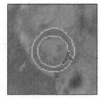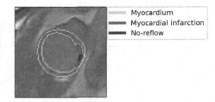

Fig. 2. Left: an MRI acquisition of a patient without myocardial infarction. **Center**: a pathological case with myocardial infarction. **Right**: a pathological case with myocardial infarction and no-reflow regions.

in two steps: first the myocardium is segmented to reduce the problem and provide anatomical context information such as the relative position of a voxel with regard to the endocardial border of the myocardium, then the myocardium is analyzed. More recent approaches make use of modern machine learning techniques such as CNNs [11]. Because of the lack of large training cohorts most successful approaches still use a multi-step approach, which combines different types of segmentation and classification methods. Zabihollaly et al. suggest the combination of a myocardium segmentation U-Net with an ensuing CNN-based classification of the myocardial voxels [12]. The approach by de la Rosa et al. applies a CNN for the preselection of image slices to analyze with regard to myocardial pathologies [9]. The reported DICE-values of these multi-step approaches are 88% and 77% respectively. The approaches that only used a single CNN obviously depend more on the training set. The reported DICE coefficients in published approaches are minimum 48% [2] and can be improved through additional data sets generated with Generative Adversarial Networks (GAN) [7].

CNN Segmentation. Over the last years, CNN-based segmentation approaches often outperformed traditional approaches in medical image processing challenges. In former work [10], we already applied a 2D 4-layer U-Net [8] successfully to segment the left and right ventricle on cine MRI. In this work, we tried to use that prior knowledge for the challenge.

We chose the same architecture (Fig. 3) as in our previous work and tested if transfer learning can improve the robustness and quality of the results. We used 100 additional DE-MRI images for which only blood-pool and myocardial labels were provided. Based on these data, we trained two CNNs. One U-Net that was only trained with the challenge data. The other U-Net was first trained on the additional DE-MRI data. As these images only contain labels for the left ventricular cavity and myocardium, we could not directly train our model on both data sets together, or start the training on the additional data and resume it on the challenge data directly. To support the five classes, we had to reset the last layer and restart the training in order to incorporate the additional labels. As the number of feature maps in the up-sampling path of U-Nets is large in contrast to the basic fully-connected-network architecture, in which the number

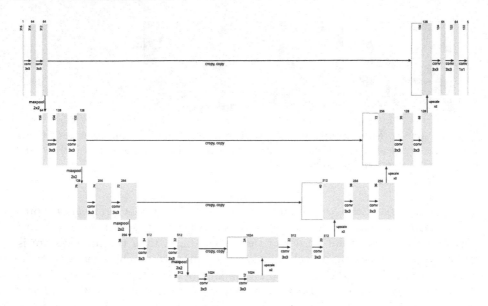

Fig. 3. Architecture of the proposed U-Nets.

of feature map equals the number of classes, we did not reset the other layers on the up-sampling path.

For training of both experiments, we chose a learning rate of 0.005, categorical cross-entropy as loss function, and used dropout and batch normalization.

Hybrid Mixture Model-Based Segmentation. Our mixture-model app-roach is based on [3] but is adapted to the current data, definition of the micro-vascular obstructions, and the definition of thresholds to generate binary masks. The approach uses the intensity distribution in the myocardium and the expected location and size of myocardial infarcts. In the first step, the myocardium has to be delineated. We used the same 4-layer U-Net architecture for this task. As we only have to differentiate background, cavity and myocardium started with a U-Net trained on cine MRI from the ACDC challenge data set [1] and in-house cine MRI data sets. In a second step, we used the additional DE-MRI data as well as the challenge data and resumed the training with that data. For the challenge data, we replaced the labels for infarction and no-reflow areas by the label for the myocardium. In a second step, we performed a mixture model fit of an expected distribution model to the myocardial histogram. For MRI data, a mixture of Rice and a Gaussian distribution or a mixture of Rayleigh and Gaussian can be assumed. During the training phase, we achieved better results using a mixture of Rayleigh and Gaussian (**MM-RG**). As a third step, the infarcted tissue is segmented, incorporating expected position and spatial connectivity. In [4], a watershed segmentation based on automatically extracted seed points in high-intensity voxels, located at the inner part of the myocardium, was used.

Table 1. Internal quantitative evaluation on 5-fold cross-validation

Metric	MM-RG	Transfer-learning	EMIDEC
Dice myocardium (%)	92.07	82.43	81.00
Volume difference myocardium (mm^3)	3863.89	11766.66	13655.55
Hausdorff myocardium (mm)	9.30	15.75	16.72
Dice MI (%)	57.25	34.09	36.08
Volume difference MI (mm^3)	6048.54	5994.9	8980.5
MI volume difference ratio (%)	4.99	4.94	7.07
Dice no-reflow (%)	39.29	40.26	54.15
Volume difference no-reflow (mm^3)	1301.80	1520	1501.73
No-reflow volume difference ratio (%)	0.97	1.11	1.08

As a binary segmentation is needed in this challenge, we chose the threshold such that resulting probabilities were optimal on the training set using a brute force optimization approach. No re-flow areas are detected by morphological closing because we assumed that they are surrounded by blood-pool and/or enhanced fibrotic tissue.

3   Results

We evaluated all models based on the metrics provided for the challenge.

For the myocardial region the **DICE** index, **Hausdorff distance**, and **volume difference** are calculated. For infarcted tissue and micro-vascular obstructions the **DICE** index, **volume difference**, and **volume difference ratio, according to the myocardium,** are used.

For the pure CNN based approaches, we used 5-fold cross-validation during our experiments.

Table 1 gives an extensive overview of the conducted experiments' metrics on the internal validation. The CNN for the hybrid mixture model shows superior DICE-values on the myocardium (see Fig. 4). Notably, the mixture of Rayleigh and Gaussian fails to accurately infer a segmentation on the no-reflow areas within the myocardial tissue. Remarkable is the results of the randomly initialized models, which produces the top performance in the no-reflow area.

Again the mixture model of Rayleigh and Gaussian demonstrates preferable results on the Hausdorff distance (Fig. 4). With a 9.3 × 5.4 4 mm distance to the ground-truth surface, the results are considerably exceeding the performance of the other tested methods. Moreover, the margin of error is explicitly lower with even fewer outliers compared to the transfer-learning and randomly initialized method. Figure 5 shows the volume differences and volume difference ratios of the proposed methods for the different structures. The results for infarctedion and no-reflow areas is quite similar for the different methods. For the myocardium the volume difference is the smallest. Table 2 shows the final result of the different approaches on the test set. The performance of the CNN based approaches

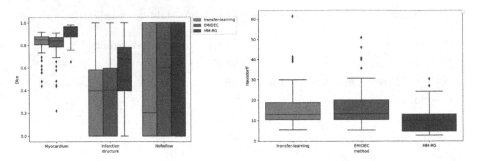

Fig. 4. An illustration of the models results. **Left**: DICE index for the myocardium infarction and no re-flow area. For the myocardium and the no re-flow areas, value zero and one in the box plots result from cases, in which bright regions were classified as infarct in healthy subjects. **Right**: Hausdorff distance for the different segmentation approaches. The CNN that was only trained to segment the myocardium, on additional DE-MRI data as well as on cine MRI data, outperformed the CNN approaches that also tried to detect infarcted tissue.

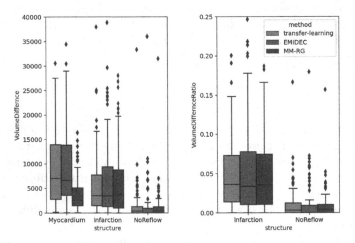

Fig. 5. Left: illustration of the volume difference between proposed methods and the ground truth. For the myocardium, the mixture model approach achieved the best results. In the infarction and the micro-vascular obstruction, the three approaches achieved similar results. **Right**: illustration of the volume difference ratio of infarction and micro-vascular obstructions in relation to the complete myocardial volume.

showed a better performance on the test set compared to the internal tests. The drop of performance of the mixture model results come from the fact that we did not perform a cross validation and used the final model to generate the myocardium segmentation. However, the performance of the myocardium segmentation by this method still achieved the best results for DICE index and Hausdorff distance.

Table 2. Quantitative comparison of the final results on testing set.

Metric	MM-RG	Transfer-learning	EMIDEC
Dice myocardium (%)	84.8	84.08	82.95
Volume difference myocardium (mm^3)	12582.76	10874.47	9060.61
Hausdorff myocardium (mm)	15.93	18.3	16.16
Dice MI (%)	35.83	37.87	37.33
Volume difference MI (mm^3)	6100.92	6166.01	6021.72
MI volume difference ratio (%)	5.17	4.93	4.87
Dice no-reflow (%)	41.25	52.25	48.39
Volume difference no-reflow (mm^3)	1090.04	953.47	990.29
No-reflow volume difference ratio (%)	0.76	0.64	0.64

4   Discussion

We have demonstrated three different approaches to segment the myocardium, infarcted tissue, and micro-vascular obstructions from DE-MRI data. For pure CNN based segmentation approaches, the results on the actual testing set were better than in our internal evaluation using 5-fold cross-validation. As we retrained our models for the submission using all data, we think that the improvement comes from the additional cases. The slightly better results for the transfer learning approach also back this hypothesis. The U-Net that was additionally trained on cine data showed the best results in the segmentation of the myocardium according to the Dice index and Hausdorff distance. This could also be due to the better generalization thanks to the additional data. The improvement could also come from the reduction of the number of classes to be classified. The drop in the performance of the results by the hybrid approach can be explained by using the final CNN to extract the myocardial segmentation. Here we can see that the quality of the segmentation of the infarction and no-reflow areas is highly dependent on a robust segmentation of the myocardium.

Figure 6 shows several segmentation results. In the top row, healthy tissue is segmented correctly by all approaches. Row two shows an example of infarction with no re-flow area that was also correctly depicted by all methods and shows a good agreement with the ground truth. In row three, both pure CNN based approaches misclassified brighter regions in the basal slices as infarction. In row four, an example is provided where the mixture model approach could not correctly segment the no re-flow area.

We saw in several cases that the segmentation of the infarction was underestimated by the mixture model approach or the no-reflow areas were not surrounded by the infarction mask. Thus the no-reflow areas could not be segmented using simple morphological closing operations. Here one could try using radial closing approaches to try to overcome this limitation. To reduce the number of cases without infarction in which voxels were misclassified as infarcted tissue, one could additionally analyze the fitted distributions. That means analyze whether the mean of the distributions are close together.

In future work, it makes sense to investigate the performance of 2-step U-Net approaches, also incorporating the results of our mixture model analysis. To achieve better results for the smaller regions infarction and no-reflow areas we could try to use weighted categorical cross-entropy as a loss function. Additionally, we could try if self-learning on the test set can further improve the overall performance of the proposed method.

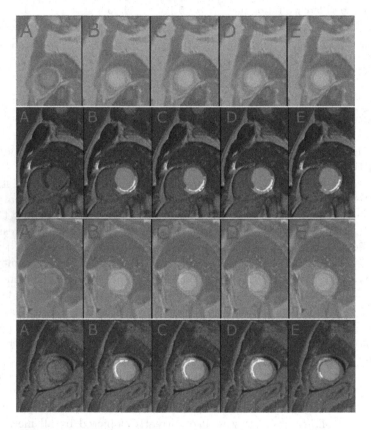

Fig. 6. Example results from the training data set. Each row shows a case on different locations in the heart. On the left, the original image is shown (A), followed by the ground truth (B), transfer learning (C), EMIDEC training (D) and the Mixture model of Rayleigh and Gaussian distribution on the right (E).

References

1. Bernard, O., Lalande, A., Zotti, C., Cervenansky, F., Yang, X., Heng, P., et al.: Deep learning techniques for automatic MRI cardiac multi-structures segmentation and diagnosis: Is the problem solved? IEEE Trans. Med. Imaging **37**(11), 2514–2525 (2018)

2. Fahmy, A.S., et al.: Three-dimensional deep convolutional neural networks for auto-mated myocardial scar quantification in hypertrophic cardiomyopathy: a multicen-ter multivendor study. Radiology **294**(1), 52–60 (2020)
3. Hennemuth, A., Friman, O., Huellebrand, M., Peitgen, H.-O.: Mixture-Model-based segmentation of myocardial delayed enhancement MRI. In: Camara, O., Mansi, T., Pop, M., Rhode, K., Sermesant, M., Young, A. (eds.) STACOM 2012. LNCS, vol. 7746, pp. 87–96. Springer, Heidelberg (2013). https://doi.org/10.1007/978-3-642-36961-2_11
4. Hennemuth, A., et al.: A comprehensive approach to the analysis of contrast enhanced cardiac MR images. IEEE Trans. Med. Imaging **27**(11), 1592–1610 (2008)
5. Karim, R., et al.: Evaluation of state-of-the-art segmentation algorithms for left ventricle infarct from late Gadolinium enhancement MR images. Med. Image Anal. **30**, 95–107 (2016)
6. Lalande, A., et al.: Emidec: a database usable for the automatic evaluation of myocardial infarction from delayed-enhancement cardiac MRI. Data **5**(4), 89 (2020)
7. Lau, F., Hendriks, T., Lieman-Sifry, J., Sall, S., Golden, D.: ScarGAN: chained generative adversarial networks to simulate pathological tissue on cardiovascular MR scans. In: Stoyanov, D., et al. (eds.) DLMIA/ML-CDS -2018. LNCS, vol. 11045, pp. 343–350. Springer, Cham (2018). https://doi.org/10.1007/978-3-030-00889-5_39
8. Ronneberger, O., Fischer, P., Brox, T.: U-Net: convolutional networks for biomed-ical image segmentation. In: Navab, N., Hornegger, J., Wells, W.M., Frangi, A.F. (eds.) MICCAI 2015. LNCS, vol. 9351, pp. 234–241. Springer, Cham (2015). https://doi.org/10.1007/978-3-319-24574-4_28
9. de la Rosa, E., Sidibé, D., Decourselle, T., Leclercq, T., Cochet, A., Lalande, A.: Myocardial Infarction Quantification From Late Gadolinium Enhancement MRI Using Top-hat Transforms and Neural Networks (2019)
10. Tautz, L., et al.: Cardiac radiomics an interactive approach for 4d data exploration. Curr. Directions Biomed. Eng. (2020). https://doi.org/10.1515/cdbme-2020-0008
11. Zabihollahy, F., Rajan, S., Ukwatta, E.: Machine learning-based segmentation of left ventricular myocardial fibrosis from magnetic resonance imaging. Curr. Car-diol. Rep. **22**(8), 1–8 (2020). https://doi.org/10.1007/s11886-020-01321-1
12. Zabihollahy, F., Rajchl, M., White, J.A., Ukwatta, E.: Fully automated segmen-tation of left ventricular scar from 3D late gadolinium enhancement magnetic res-onance imaging using a cascaded multi-planar U-Net (CMPU-Net). Med. Phys. **47**(4), 1645–1655 (2020)

Cascaded Convolutional Neural Network for Automatic Myocardial Infarction Segmentation from Delayed-Enhancement Cardiac MRI

Yichi Zhang[✉]

School of Biological Science and Medical Engineering,
Beihang University, Beijing, China
coda1998@buaa.edu.cn

Abstract. Automatic segmentation of myocardial contours and relevant areas like infraction and no-reflow is an important step for the quantitative evaluation of myocardial infarction. In this work, we propose a cascaded convolutional neural network for automatic myocardial infarction segmentation from delayed-enhancement cardiac MRI. We first use a 2D U-Net to focus on the intra-slice information to perform a preliminary segmentation. After that, we use a 3D U-Net to utilize the volumetric spatial information for a subtle segmentation. Our method is evaluated on the MICCAI 2020 EMIDEC challenge dataset and achieves average Dice score of 0.8786, 0.7124 and 0.7851 for myocardium, infarction and no-reflow respectively, outperforms all the other teams of the segmentation contest.

Keywords: Magnetic resonance imaging · Myocardial infarction · Segmentation · Convolutional neural network

1 Introduction

Myocardial infarction (MI) is a myocardial ischemic necrosis caused by coronary artery complications that cannot provide enough blood and has become one of the leading causes of death and disability worldwide [8]. The viability of the cardiac segment is an important parameter to assess the cardiac status after MI, such as whether the segment is functional after the revascularization. Delayed-enhancement MRI (DE-MRI) performed several minutes after the injection is a method to evaluate the extent of MI and assess viable tissues after the injury. According to the World Health Organization (WHO), cardiovascular diseases are the first one cause of death worldwide and 85% of the deaths are due to heart attacks and strokes.

Automatic segmentation of the different relevant areas from DE-MRI, such as myocardial contours, the infarcted area and the permanent microvascular obstruction area (no-reflow area) could provide useful information like the absolute value (mm^3) or percentage of the myocardium, which can provide useful

© Springer Nature Switzerland AG 2021
E. Puyol Anton et al. (Eds.): STACOM 2020, LNCS 12592, pp. 328–333, 2021.
https://doi.org/10.1007/978-3-030-68107-4_33

information for quantitative evaluation of MI. However, myocardial infarction segmentation is still a challenging task due to the morphological similarity. Recently, deep learning-based methods have achieved state-of-the-art results for various image segmentation tasks and shown great potential in medical image analysis and clinical applications.

In this paper, we propose a cascaded convolutional neural network for automatic myocardial infarction segmentation from delayed-enhancement cardiac MRI. We first use a 2D U-Net to focus on the intra-slice information and perform a prelimi-nary segmentation. After that, a 3D U-Net is applied to utilize the volumetric spatial information for a subtle segmentation. All the training procedure of our network are based on MICCAI 2020 EMIDEC Challenge dataset[1] [5].

2   Method

Network Architecture. As the most well-known network structure for medical image segmentation, U-Net [7] is a classical encoder-decoder segmentation network and achieves state-of-the-art results on many segmentation challenges [3,4]. The encoder is similar with the typical classification network and uses convolution-pooling module to extract more high-level semantic features layer by layer. Then the decoder recovers the localization for every voxel and utilizes the extracted feature information for the classification of each pixel. To incorporate multi-scale features and employ the position information, skip connections are constructed between the encoder and decoder in the same stage.

For the segmentation of 3D biomedical images, many 3D segmentation networks like [2,6] are proposed to extract volumetric spatial information using 3D convolutions instead of just focusing on intra-slice information. However, for some volumes with highly anisotropic voxel spacings, 3D networks may not always outperform 2D networks when the inter-slice correlation information is not rich [1]. For example, Case N042 is a 3D MRI volume with an image shape of 166 * 270 * 7 and voxel spacing of 1.667mm * 1.667mm * 10mm on x, y, and z-axis, respectively. This means x and y-axis preserve much higher resolution and richer information than the z-axis. Under this circumstance, using pure 3D network that treats the three axes equally may not be the best choice.

To issue this problem, we propose a cascaded convolutional neural network for automatic myocardial infarction segmentation from delayed-enhancement cardiac MRI. As illustrated in Fig. 1, our network can be mainly divided into two stages. Firstly, after the preprocessing of input data, the MRI volume is divided into a sequence of slices for input of 2D U-Net to obtain a preliminary segmentation based on the intra-slice information. However, the results of 2D network ignore the correlation between slices, which would lead to limited segmentation accuracy, especially for challenging pathological areas that are difficult to distinguish only based on intra-slice information. Therefore, in the second stage, we use a 3D U-Net to utilize the volumetric spatial information and make a subtle segmentation. Specifically, we concatenate the 2D coarse results in the first stage

---

[1] http://emidec.com.

with the input volume for the input of 3D U-Net in the second stage as a spatial prior. In the end, after the postprocessing like removing the scattered voxels, we get the final segmentation results.

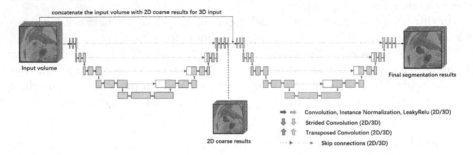

Fig. 1. The overall architecture of our cascaded convolutional neural network. The blocks and ar-rows viewed in blue and red denote the corresponding structure for 2D and 3D networks.

Implementation Details. All the training procedure of our network is performed on NVIDIA Tesla V100 GPUs using the Pytorch framework based on the nnU-Net implementation [4]. During training, we use Adam optimizer with an initial learning rate of 0.01. Instead of patch-based methods, we use the whole short-axis slice and whole volume for the input of the 2D and 3D networks. To enhance the attention of foreground voxels, we use the summation of cross-entropy (CE) loss and dice loss [6] as the loss function for the training of our network.

Dataset and Evaluation Metrics. The EMIDEC Challenge dataset consists of delayed-enhancement cardiac MRI with a training set of 100 patients including 67 pathological cases and 33 normal cases, and a testing set of another 50 patients including 33 pathological cases and 17 normal cases. For training cases, manual annotations are provided with 0 for background, 1 for cavity, 2 for normal myocardium, 3 for myocardial infarction, and 4 for no-reflow.

For evaluation of segmentation results, clinical metrics include the average errors for the volume of the myocardium of the left ventricle, the volume and the percentage of MI and no-reflow and geometrical metrics include the average Dice coefficient for the different areas and Hausdorff distance for the myocardium.

3   Experiments and Results

There are totally 100 scans with published labels to train our network, while the other 50 scans remained for final evaluation. We make random 5-fold cross validation by randomly shuffling the sequence of cases and splitting the training dataset into 5 fixed folds with 20 MR scans in each fold, using 4 folds for training and the other one for testing. In this way, we can make a more comprehensive evaluation of our method.

Table 1 and Table 2 respectively represent the cross-validation results of our 2D coarse segmentation output and final segmentation output. The evaluation of clinical and geometrical metrics is based on the official code[2]. From the result, we can see that the application of 3D U-Net can make use of the volumetric spatial information and improve the segmentation result.

Table 1. Quantitative 5-fold cross-validation results of 2D coarse segmentation output.

Targets	Metrics	fold 0	fold 1	fold 2	fold 3	fold 4
Myocardium	Dice(%)	83.98	85.29	85.94	85.83	85.59
	VolDif(mm^3)	10906.43	6384.88	6012.93	5423.57	11629.66
	HSD(mm)	17.01	13.77	13.68	12.60	12.65
Infarction	Dice(%)	44.39	53.12	48.10	50.55	66.34
	VolDif(mm^3)	9883.17	4821.30	3449.09	4986.66	6621.52
	Ratio(%)	7.10	4.39	2.96	4.68	5.00
NoReflow	Dice(%)	65.26	63.84	70.61	60.24	66.67
	VolDif(mm^3)	2703.34	775.07	480.32	703.7	443.86
	Ratio(%)	1.68	0.68	0.37	0.67	0.32

Table 2. Quantitative 5-fold cross-validation results of our final segmentation output.

Targets	Metrics	fold 0	fold 1	fold 2	fold 3	fold 4
Myocardium	Dice(%)	86.66	86.46	87.87	87.61	87.13
	VolDif(mm^3)	8680.23	5405.52	6087.88	4880.5	7317.76
	HSD(mm)	15.88	14.12	12.96	13.43	13.79
Infarction	Dice(%)	61.44	72.08	81.51	68.48	76.87
	VolDif(mm^3)	6536.55	3233.94	3514.97	4091.74	3520.3
	Ratio(%)	4.67	2.91	2.85	3.96	2.64
NoReflow	Dice(%)	68.47	68.33	79.67	65.12	73.48
	VolDif(mm^3)	2158.34	712.36	451.84	620.93	649.98
	Ratio(%)	1.37	0.65	0.35	0.61	0.46

For our final segmentation results, the network performs well on myocardium segmentation, with an average dice score of 0.8715. However, for more challenging segmentation of pathological areas, the average dice score is only 0.7208 and 0.7101 for infarction and no-reflow. Also, the performance variance is very small,

---

[2] https://github.com/EMIDEC-Challenge/Evaluation-metrics.

which indicates the robustness of our method. Figure 2 illustrates two samples of our segmentation results and corresponding ground truth in the validation set of our own split. We can see that the segmentation results closely approximate the ground truth.

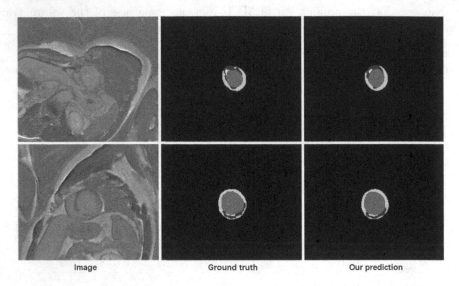

Fig. 2. Two samples of our segmentation results. The columns from left to right are the image, ground truth, and prediction. (cavity in red, myocardium in green, infarction in blue, no-reflow in yellow) (Color figure online)

In the inference stage, we obtain the final prediction of the testing set by ensembling the segmentation results of each fold using majority voting. The evaluation results of our method on the testing set of EMIDEC dataset is presented in Table 3. The average Dice score is very similar to our cross-validation results (even higher on some metrics), which indicates that our method is stable for the myocardial infarction segmentation task.

Table 3. The evaluation results of our method on the EMIDEC test set.

Targets	Dice (%)	VolDif (mm^3)	HSD (mm)	ratio (%)
Myocardium	87.86	9258.24	13.01	–
Infarction	71.24	3117.88	–	2.38
NoReflow	78.51	634.69	–	0.38

4 Conclusion

In this paper, we propose a cascaded convolutional neural network for automatic myocardial infarction segmentation from delayed-enhancement cardiac MRI. The network consists of a 2D U-Net to focus on the intra-slice information to perform a preliminary segmentation and a 3D U-Net to utilize the volumetric spatial information to make a subtle segmentation. Our method is trained and validated on MICCAI 2020 EMIDEC challenge dataset. For the testing stage, our ensembled model has achieved an average Dice score of 0.8786, 0.7124 and 0.7851 for myocardium, ranking first of the segmentation challenge.

References

1. Abulnaga, S.M., Rubin, J.: Ischemic stroke lesion segmentation in CT perfusion scans using pyramid pooling and focal loss. In: Crimi, A., Bakas, S., Kuijf, H., Keyvan, F., Reyes, M., van Walsum, T. (eds.) BrainLes 2018. LNCS, vol. 11383, pp. 352–363. Springer, Cham (2019). https://doi.org/10.1007/978-3-030-11723-8_36
2. Çiçek, Ö., Abdulkadir, A., Lienkamp, S.S., Brox, T., Ronneberger, O.: 3D U-Net: learning dense volumetric segmentation from sparse annotation. In: Ourselin, S., Joskowicz, L., Sabuncu, M.R., Unal, G., Wells, W. (eds.) MICCAI 2016. LNCS, vol. 9901, pp. 424–432. Springer, Cham (2016). https://doi.org/10.1007/978-3-319-46723-8_49
3. Heller, N., et al.: The state of the art in kidney and kidney tumor segmentation in contrast-enhanced CT imaging: results of the KiTS19 challenge. Med. Image Anal. **67**, 101821 (2020)
4. Isensee, F., Petersen, J., Kohl, S.A., Jäger, P.F., Maier-Hein, K.H.: nnU-Net: breaking the spell on successful medical image segmentation **1**, 1–8. arXiv preprint arXiv:1904.08128 (2019)
5. Lalande, A., et al.: Emidec: a database usable for the automatic evaluation of myocardial infarction from delayed-enhancement cardiac MRI. Data 5(4), 89 (2020)
6. Milletari, F., Navab, N., Ahmadi, S.A.: V-net: fully convolutional neural networks for volumetric medical image segmentation. In: Fourth International Conference on 3D Vision (3DV), pp. 565–571. IEEE (2016)
7. Ronneberger, O., Fischer, P., Brox, T.: U-Net: convolutional networks for biomedical image segmentation. In: Navab, N., Hornegger, J., Wells, W.M., Frangi, A.F. (eds.) MICCAI 2015. LNCS, vol. 9351, pp. 234–241. Springer, Cham (2015). https://doi.org/10.1007/978-3-319-24574-4_28
8. Thygesen, K., Alpert, J.S., White, H.D., et al.: Universal definition of myocardial infarction. J. Am. Coll. Cardiol. **50**(22), 2173–2195 (2007)

Automatic Myocardial Disease Prediction from Delayed-Enhancement Cardiac MRI and Clinical Information

Ana Lourenço[1,2], Eric Kerfoot[1], Irina Grigorescu[1], Cian M. Scannell[1], Marta Varela[1,3(✉)], and Teresa M. Correia[1(✉)]

[1] School of Biomedical Engineering and Imaging Sciences, King's College London, London, UK
[2] Faculty of Sciences, University of Lisbon, Lisbon, Portugal
[3] National Heart and Lung Institute, Imperial College London, London, UK
marta.varela@imperial.ac.uk, teresa.correia@kcl.ac.uk

Abstract. Delayed-enhancement cardiac magnetic resonance (DE-CMR) provides important diagnostic and prognostic information on myocardial viability. The presence and extent of late gadolinium enhancement (LGE) in DE-CMR is negatively associated with the probability of improvement in left ventricular function after revascularization. Moreover, LGE findings can support the diagnosis of several other cardiomyopathies, but their absence does not rule them out, making disease classification by visual assessment difficult. In this work, we propose deep learning neural networks that can automatically predict myocardial disease from patient clinical information and DE-CMR. All the proposed networks achieved very good classification accuracy (>85%). Including information from DE-CMR (directly as images or as metadata following DE-CMR segmentation) is valuable in this classification task, improving the accuracy to 95–100%.

Keywords: Cardiac MRI · Late gadolinium enhancement · Myocardial infarction · Classification

1 Introduction

Delayed-enhancement cardiac magnetic resonance (DE-CMR) is considered the non-invasive gold standard for assessing myocardial infarction and viability in coronary artery disease [3,5,19] and can help differentiate ischemic from non-ischemic myocardial diseases [10]. DE-CMR images are typically acquired 10–15 min. after intravenous injection of a gadolinium-based contrast agent. The contrast agent is washed out by normal tissue, whereas contrast washout is delayed in regions with scar tissue, making these regions appear bright in T_1-weighted

---

M. Varela and T.M. Correia—Contributed equally.

© Springer Nature Switzerland AG 2021
E. Puyol Anton et al. (Eds.): STACOM 2020, LNCS 12592, pp. 334–341, 2021.
https://doi.org/10.1007/978-3-030-68107-4_34

images. The presence and extent of late gadolinium enhancement (LGE) within the left ventricular (LV) myocardium provides important diagnostic and prognostic information, including the risk of an adverse cardiac event and response to therapeutic strategies such as revascularization [2,6,7,9]. Moreover, no-reflow regions (or microvascular obstruction), which are dark in T1-weighted images, have been associated with worse clinical outcomes.

The absence of LGE does not, however, rule out the presence of myocardial disease, since patients with, for example, extensive hibernating myocardium, hypertrophic cardiomyopathy, sarcoidosis or myocarditis may not show contrast uptake [10,12,17]. This makes disease classification from DE-CMR a complex task. Therefore, DE-CMR is often combined with other CMR sequences, such as T_1 and T_2 maps, to better characterize myocardial tissue alterations in various cardiomyopathies [10].

Machine learning classification algorithms, such as support vector machines [16], random forests [4], and K-nearest neighbor [15], have been used to predict the presence/absence of cardiovascular disease. However, these techniques require complex feature extraction procedures and domain expertise to create good inputs for the classifier. On the other hand, deep learning architectures have the ability to learn features directly from the data and hence reduce the need for domain expertise and dedicated feature extraction [13].

In this work, we propose fully automatic neural networks (NNs) that perform binary classification for predicting normal vs pathological cases considering: 1) patient clinical information only (Clinic-NET), 2) clinical information and DE-CMR images (DOC-NET). Additionally, given that cardiomyopathies can cause disturbances in LV volume and wall thickness, we hypothesize that using this information as additional metadata can aid the classification task. Thus, we estimated, using a stand-alone segmentation NN, the volumes across the DE-CMR images, of three relevant cardiac regions: LV myocardium, LV blood pool and (when present) LGE uptake region. This volumetric information was used as enhanced metadata in two additional classification networks: Clinic-Net+ (classification from enhanced metadata only) and DOC-NET+ (classification from enhanced metadata and DE-CMR images).

2   Methods

Clinical Images and Metadata. The networks were trained and tested on the EMIDEC STACOM 2020 challenge dataset [11], comprising 100 cases: 33 cases with normal CMR and 67 pathological cases. For each case, Phase Sensitive Inversion Recovery (PSIR) DE-CMR images (consisting of 5–10 short-axis (SA) slices covering the left ventricle) and 12 clinical discrete or continuous variables were provided. Clinical information included sex, age, tobacco (Y/N/Former smoker), overweight (BMI > 25), arterial hypertension (Y/N), diabetes (Y/N), familial history of coronary artery disease (Y/N), ECG (ST+ (STEMI) or not), troponin (value), Killip max (1 - 4), ejection fraction of the left ventricle from echography (value), and NTproBNP (value). More details can be found in [1,11].

Image Preprocessing. DE-CMR images had variable dimensions and were zero-padded along the z-direction, when necessary, to obtain 10 slices. To remove anatomical structures unrelated to the left ventricle, they were further cropped in-plane to a matrix of 128 × 128, whose center was the centroid of the the LV blood pool segmentation label.

In the absence of ground truth segmentation labels, we propose an NN-based method to automatically perform image cropping and segmentation, as detailed below.

DE-CMR Segmentation. A two-step approach based on NNs is proposed to automatically segment DE-CMR images into 3 classes (LV, healthy myocardium and LGE uptake area) and extract their volumes as additional inputs to aid classification. The segmentation NNs are based on the 2D U-Net architecture [8,14] and were trained separately on the EMIDEC dataset. The first NN was trained with the Dice loss function to identify the LV center by segmenting the LV blood pool region and calculating the LV centroid coordinates. This information was used to crop each slice. The cropped images were then sent to a second NN, which was trained with the generalized Dice loss function [18] to automatically segment: LV myocardium, LV blood pool and, where detected, LGE uptake regions.

Data Augmentation. Several randomly-chosen data augmentation functions were applied to each DE-CMR volume. These replicate some of the expected variations in real images without modifying the cardiac features of interest and include: 1) small image rotations and image flips; 2) additional cropping of the images; 3) additive stochastic noise; 4) k-space corruption; 5) smooth non-rigid deformations using free-form deformations and 6) intensity and contrast scaling.

Neural Networks for Myocardial Disease Prediction. Two classification methods are proposed: 1) Clinic-NET, classification based on clinical information only and 2) DOC-NET, classification based on DE-CMR and Other Clinical information. As explained below, both of these NNs are further compared to other two proposed networks that use volumetric information from previously segmented DE-CMR as further metadata inputs: Clinic-NET+ and DOC-NET+.

The classification networks were trained using a cross-entropy loss function and the Adam optimizer with a learning rate of 0.00001. We randomly divided the 100 labeled cases provided by the EMIDEC challenge into 3 datasets: training (70 cases), validation (10 cases) and test (20 cases). Training was performed for 170 iterations. Hyperparameters were chosen after careful empirical testing.

To assess the quality of each network, we calculated the classification's accuracy, specificity and sensitivity on the 20-case test dataset.

Clinic-NET. Clinic-NET takes the 12 provided metadata variables as inputs to a classification NN with 3 fully connected (fc) layers, which sequentially encode information in fc1 = 20, fc2 = 30, fc3 = 10 and 2 units, as shown in Fig. 1b. Parametric Rectified Linear Units (PReLu) are applied to the outputs of the first three layers.

DOC-NET. DOC-NET combines features extracted from DE-CMR images and features calculated from metadata to perform the final classification. The image feature extraction network consists of seven layers: 1) 3D convolutions with 3 × 3 × 3 kernels, a stride of 2 and a variable number of channels (4, 8, 16, 32, 64, 16, 8); 2) instance layer-normalization; 3) dropout (20% probability of being dropped out) and 4) PReLU activation (see Fig. 1a). The image feature vector was then flattened into an 8-element array and concatenated with the 12-variable metadata. This combined vector was then the input to a fully connected NN similar to Clinic-NET (see Fig. 1b). The sizes of the 3 fc layers in DOC-NET were rescaled to match the new input size, such that fc1 = 33, fc2 = 50, fc3 = 16. In the image feature extraction network, the number of feature maps decreases in the last convolutional layers to improve computational efficiency and guarantee a balanced contribution from both metadata and DE-CMR images (features) to the classification task.

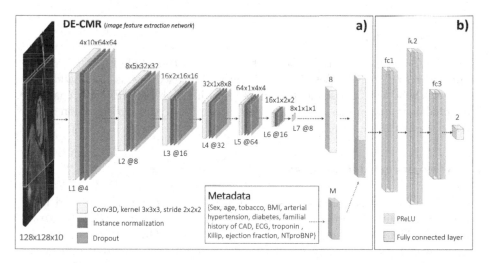

Fig. 1. DOC-NET classification network: a) image feature vectors, obtained from the last convolutional layer of an image feature extraction network, are concatenated with the metadata vector (M) and b) sent through four fully connected (fc) layers.

Clinic-NET+ and DOC-NET+. To further explore the classification task, we created additional metadata variables with the volumes of each of the segmentation labels of the DE-CMR. These variables were concatenated with the existing metadata and used as enhanced metadata inputs to the previously described

networks to create Clinic-NET+ and DOC-NET+. For this, we used the volumes
of the labels provided by the ground truth segmentations: 1) LV blood pool, 2)
healthy myocardium, 3) LGE uptake area and 4) no-reflow area.

We performed additional experiments to gauge whether Clinic-NET+ and
DOC-NET+ could still be deployed in more general circumstances in which
expert manual segmentation is not available. In these experiments, we used the
segmentation networks detailed above to automatically segment DE-CMR into
labels 1–3 and used the volume of each of these categories as enhanced metadata
for the classification NNs.

Table 1. Confusion matrix (actual vs predicted counts of pathological and normal
cases), **accuracy, sensitivity and specificity** obtained with the different classification networks. The additional metadata used in Clinic-NET+ and DOC-NET+ was
extracted from the ground truth segmentations.

		Clinic-NET		DOC-NET		Clinic-NET+		DOC-NET+	
		Pathology	Normal	Pathology	Normal	Pathology	Normal	Pathology	Normal
Actual	Pathology	12	1	12	1	13	0	13	0
	Normal	2	5	0	7	0	7	0	7
	Accuracy	0.85		0.95		1		1	
	Sensitivity	0.92		0.92		1		1	
	Specificity	0.71		1		1		1	

2.1 Live Challenge

Clinic-NET and Clinic-NET+ were retrained using the whole EMIDEC dataset
comprising 100 labeled cases. These NNs were then used to classify 50 additional
test cases that were subsequently released during the EMIDEC live challenge.
The challenge organizers calculated the accuracy of the classification results
during the challenge.

3   Results and Discussion

The best overall performance was jointly achieved by Clinic-NET+ and DOC-
NET+, both with a test accuracy of 100%, followed by DOC-NET (accuracy:
95%) and Clinic-NET (accuracy: 85%) - see Table 1. Our results suggest that
the clinical metadata information already includes very valuable information
that can be leveraged by our proposed network, Clinic-NET, to classify subjects
with an accuracy of 85%. The accuracy is greatly increased, however, when
information from DE-CMR is also provided to the network. In the live challenge,
Clinic-NET and Clinic-NET+ also performed well, achieving an accuracy of 72%
and 82%, respectively.

DOC-NET+ and Clinic-NET+ both rely on information from existing high-
quality segmentations of DE-CMR performed manually by an expert or automatically by a suitable segmentation approach, such as the one proposed here

(or from the EMIDEC DE-CMR segmentation challenge). We found that information about the size of potential infarct areas (and also LV dimensions) is most useful for the classification task. The excellent performance of Clinic-NET+ and DOC-NET+ is likely due to the very high predictive value of the LGE zone segmentation label, which was not present in any normal cases. This observation was confirmed in an additional test performed on the 20-patient test dataset: a classification NN that takes the volumetric metadata as its sole input showed a comparable performance to Clinic-Net+ (100% accuracy).

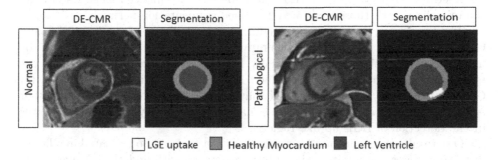

Fig. 2. DE-CMR images and segmentation of the left ventricle, normal myocardium, and region of LGE uptake (if present) obtained with the proposed automatic segmentation method for two slices from two representative subjects.

Table 2. Confusion matrix (actual vs predicted counts of pathological and normal cases), **accuracy, sensitivity, and specificity** obtained with the Clinic-NET+ and DOC-NET+ classification networks. The additional metadata was extracted from the automatic segmentation of DE-CMR images.

		Clinic-NET+		DOC-NET+	
		Pathology	Normal	Pathology	Normal
Actual	Pathology	13	0	13	0
	Normal	0	7	0	7
Accuracy		1		1	
Sensitivity		1		1	
Specificity		1		1	

We also investigated how the performance of Clinic-NET+ and DOC-NET+ was affected by using volumetric information from our proposed automatic segmentation method, which did not segment the no-reflow area (label 4). The classification performance was not affected, maintaining a 100% accuracy, as can be inferred from comparing the last two columns of Table 1 with Table 2.

The proposed DE-CMR segmentation method can be particularly useful when ground truth segmentations are not available, allowing to automatically crop the region of interest and determine LV, healthy myocardium and LGE enhancement volumes (Fig. 2). However, the proposed method does not currently segment the no-reflow area. For this particular classification task, the absence of this information did not affect the accuracy of the results. However, including information about the presence and/or volume of the no-reflow area in classification NNs may be particularly useful when predicting clinical outcomes in patients with known or potential heart problems.

The excellent results obtained when incorporating information from DE-CMR segmentations suggest that the performance of the image feature extractor included in DOC-NET/DOC-NET+ may be further enhanced when its weights are initialized with those from a well-trained segmentation network.

The lower accuracy values obtained in the live challenge with Clinic-NET and Clinic-NET+ indicate that the NNs only generalized to a certain extent to the new data. Overfitting could potentially have been reduced by augmenting the metadata in addition to the performed image augmentation. Furthermore, for Clinic-NET+, the segmentation NN may have underperformed in images that exhibited novel image artefacts. These may have potentially led to false positives in the detection of LGE uptake regions.

Our results were calculated in a small dataset and will be validated in a larger number of cases in the future, including in patients with undetected myocardial infarction on DE-MRI.

4 Conclusions

For the EMIDEC classification challenge, we propose Clinic-NET, a 4-layer fully-connected NN which uses 12 clinical variables as an input and shows a very good classification performance. Improved performance is obtained with DOC-NET, which additionally includes DE-CMR images as inputs, which are processed using 3D convolutional layers. Further performance improvements can be obtained when providing DE-CMR information distilled as the volume of segmentation labels, either from expert manual segmentation or from a proposed segmentation NN.

Acknowledgments. This work was supported by the Wellcome/EPSRC Centre for Medical Engineering [WT 203148/Z/16/Z] and the British Heart Foundation Centre of Research Excellence at Imperial College London [RE/18/4/34215].

References

1. EMIDEC classification challenge (2020). http://emidec.com/classification-contest
2. Allman, K., et al.: Myocardial viability testing and impact of revascularization on prognosis in patients with coronary artery disease and left ventricular dysfunction: a meta-analysis. J. Am. Coll. Cardiol. **39**(7), 1151–8 (2002)
3. Arai, A.: The cardiac magnetic resonance (CMR) approach to assessing myocardial viability. J. Nucl. Cardiol. **18**(6), 1095–1102 (2011)

4. Baeßler, B., et al.: Mapping tissue inhomogeneity in acute myocarditis: a novel analytical approach to quantitative myocardial edema imaging by T2-mapping. J. Cardiovasc. Magn. Reson. **17**(1), 115 (2015)
5. Bettencourt, N., Chiribiri, A., Schuster, A., Nagel, E.: Assessment of myocardial ischemia and viability using cardiac magnetic resonance. Curr. Heart Fail Rep. **6**(3), 142–153 (2009)
6. Bonow, R., et al.: Myocardial viability and survival in ischemic left ventricular dysfunction. N. Engl. J. Med. **364**(17), 1617–25 (2011)
7. Gerber, B., et al.: Prognostic value of myocardial viability by delayed-enhanced magnetic resonance in patients with coronary artery disease and low ejection fraction: impact of revascularization therapy. J. Am. Coll. Cardiol. **59**(9), 825–35 (2012)
8. Kerfoot, E., Puyol Anton, E., Ruijsink, B., Clough, J., King, A.P., Schnabel, J.A.: Automated CNN-based reconstruction of short-axis cardiac MR sequence from real-time image data. In: Stoyanov, D., et al. (eds.) RAMBO/BIA/TIA -2018. LNCS, vol. 11040, pp. 32–41. Springer, Cham (2018). https://doi.org/10.1007/978-3-030-00946-5_4
9. Kim, R., et al.: The use of contrast-enhanced magnetic resonance imaging to identify reversible myocardial dysfunction. N. Engl. J. Med. **343**(20), 1445–53 (2000)
10. Kramer, C., et al.: Role of cardiac MR imaging in cardiomyopathies. J. Nucl. Med. **56**, 39S–45S (2015)
11. Lalande, A., et al.: Emidec: a database usable for the automatic evaluation of myocardial infarction from delayed-enhancement cardiac MRI. Data **5**(4), 89 (2020)
12. Lee, E., et al.: Practical guide to evaluating myocardial disease by cardiac MRI. Am. J. Roentgenol. **214**(3), 546–556 (2020)
13. Leiner, T., et al.: Machine learning in cardiovascular magnetic resonance: basic concepts and applications. J. Cardiovasc. Magn. Reson. **21**(1), 61 (2019)
14. Lourenço, A., et al.: Left atrial ejection fraction estimation using SEGANet for fully automated segmentation of CINE MRI (2020)
15. Mantilla, J., et al.: Detection of fibrosis in late gadolinium enhancement cardiac MRI using kernel dictionary learning-based clustering. In: Computing in Cardiology Conference (CinC), pp. 357–360 (2015)
16. Narula, S., et al.: Machine-learning algorithms to automate morphological and functional assessments in 2D echocardiography. J. Am. Coll. Cardiol. **68**(21), 2287–95 (2016)
17. Soriano, C., et al.: Noninvasive diagnosis of coronary artery disease in patients with heart failure and systolic dysfunction of uncertain etiology using late gadolinium-enhanced cardiovascular magnetic resonance. J. Am. Coll. Cardiol. **45**(5), 743–48 (2005)
18. Sudre, C.H., Li, W., Vercauteren, T., Ourselin, S., Jorge Cardoso, M.: Generalised dice overlap as a deep learning loss function for highly unbalanced segmentations. In: Cardoso, M.J., et al. (eds.) DLMIA/ML-CDS -2017. LNCS, vol. 10553, pp. 240–248. Springer, Cham (2017). https://doi.org/10.1007/978-3-319-67558-9_28
19. Weinsaft, J., Klem, I., Judd, R.: MRI for the assessment of myocardial viability. Cardiol. Clin. **25**(1), 35–36 (2007)

SM2N2: A Stacked Architecture for Multimodal Data and Its Application to Myocardial Infarction Detection

Rishabh Sharma[1,2(✉)], Christoph F. Eick[2], and Nikolaos V. Tsekos[1]

[1] MRI Lab, Department of Computer Science, University of Houston,
Houston, TX 77004, USA
rsharma26@uh.edu, nvtsekos@central.uh.edu
[2] DAIS Lab, Department of Computer Science, University of Houston,
Houston, TX 77004, USA
CEick@central.uh.edu
https://www.uh.edu/nsm/computer-science/

Abstract. This work introduces a novel Stacked Multimodal (SM2N2) architecture and assess its performance in classifying whether a patient have or not Myocardial Infarction. Central to this SM2N2 architecture is the use of images and clinical data as input. Comparison studies of Multimodal Neural Network (M2N2) component of SM2N2 with AlexNet3D model demonstrated that on small size dataset, M2N2 is faster, has less trainable parameters and results higher accuracy in this binary classification. In addition to M2N2 we also identify clinical features that are sufficient to classify normal vs pathological cases. We also train statistical models on identified clinical features and use stacking to combine outputs from statistical models and M2N2. Stacking generalizes the results and the new model learns how to best combine the results of the individual base models. One of the potential application of the M2N2 is that because of less parameters the network can be deployed on mobile devices for inference.

Keywords: MRI · Heart · Myocardial infarction · Normal case · Delayed-enhancement · Classification

1 Introduction

According to an article by CDC, every year about 647,000 deaths are caused by heart attack in the USA only [2]. Secondary to compromised coronary arteries blood flow, myocardial infarction (MI) may develop and progress into the oxygen starving myocardium. Timely diagnosis of myocardial infarction is required to identify the area affected, perform an intervention, and remove the blockage

---

This work was supported by the National Science Foundation award CNS-1646566. All opinions, findings, conclusions or recommendations expressed in this work are those of the authors and do not necessarily reflect the views of our sponsors.

E. Puyol Anton et al. (Eds.): STACOM 2020, LNCS 12592, pp. 342–350, 2021.
https://doi.org/10.1007/978-3-030-68107-4_35

from the artery. It is often observed that years of domain expertise is required to classify patients with myocardial infarction from regular patients. Hence, making it relatively important to develop innovative methods that can quickly and accurately identify the patients who are suffering from myocardial infarction. Traditionally, physicians have used DE-MRI images and clinical information together to identify the cases. However, there is very limited research in machine learning and intelligent systems that can combine multiple inputs to make predictions. In this work, we propose a novel method to evaluate whether we can identify myocardial infarction cases from normal cases by combining DE-MRI images and clinical information automatically. We also propose a stack block in this paper where we combine outputs of multiple independent models to make final decision on the data set. Our results and analysis show that our technique(M2N2) of combining multiple inputs to make classification is better than AlexNet3D that only takes single image input. Additionally, we also modified inputs for AlexNet3D to take multiple inputs and observed that M2N2 is still giving better accuracy than AlexNet3D with modified inputs on the limited low number of samples that were provided in the dataset. The challenge dataset [1] consists of 100 patients with clinical observations and DE-MRI images provided for training and testing the model.

2   Methods

2.1   System Overview

Figure 1 shows the graphical overview of our system. Future subsections will discuss the individual components in detail.

2.2   Data Preprocessing

There are twelve clinical features and we need to identify the features that are important for classification. We normalize the continuous variables (4 clinical attributes) using Z-Score normalization to change the values of columns to bring these features at a common scale, such that all the variables fall in the same range. A Pair-Plot is created on all the features replacing the continuous features with Z-Score Normalized features to identify the attributes that are able to divide labels linearly. We also fit a linear model (Ridge Regression) on all the features replacing the continuous features with Z-Score Normalized continuous features. Ridge regression shrinks slope asymptotically close to zero but will never become absolute zero. Beta/Coefficient for features that are a less important start to shrink when the penalty factor is increased and, after a certain number of iteration, the variables that do not contribute to the model shrink very close to zero. We remove all the variables for which we have betas/coefficients that are in the range of 0.1 to -0.1. After filtering we were finally left with six clinical features (Sex ($\beta = 0.315$), Overweight ($\beta = 0.105$), CorArtDiseaseHist ($\beta = -0.482$), ECG($\beta = -0.253$), ZScoreNormalized_Troponin($\beta = -0.123$), and

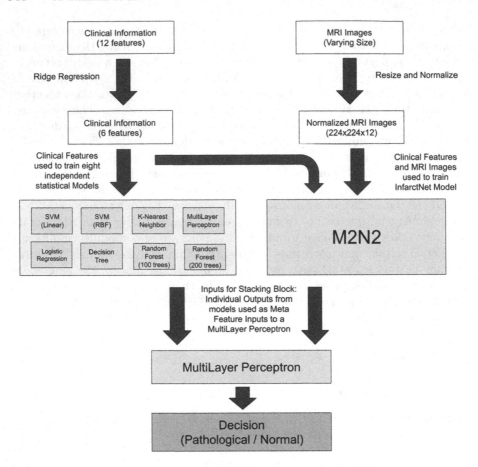

Fig. 1. SM2N2 system overview.

ZscoreNormalized_Age ($\beta = -0.124$)) to use in our network and with statistical models.

Minimization term used in the ridge regression is shown in the Eq. 1. β represent the slope for the line or coefficients of variables and λ represents the penalty factor.

$$\sum (y - y^i)^2 + \lambda \times \sum \beta^2 \tag{1}$$

Neural Networks are designed to take inputs of a constant shape, however, DE-MRI images in the challenge have inconsistent shape. DE-MRI images have a minimum of 4 slices with few having a maximum of 10 slices. To overcome the challenge of shape mismatch, we reshape our images to 224 × 224 and increase the total number of channels to 12 by adding zero padding. All the images are reshaped to 224 × 224 × 12 and are divided with 4096 to scale the pixels in the range of 0 to 1.

Finally, we split our data into 80% training data and 20% testing data. Training data is further split into 90% to train and 10% to validate the models.

2.3   M2N2 Architecture

We propose a novel neural network architecture described in Fig. 2, that combines DE-MRI images with the clinical information. The network is a multi input, multi model neural architecture which uses Depthwise Separable Convolutional layers [3] to extract three dimensional features from images and combines it with a multi layer perceptron.

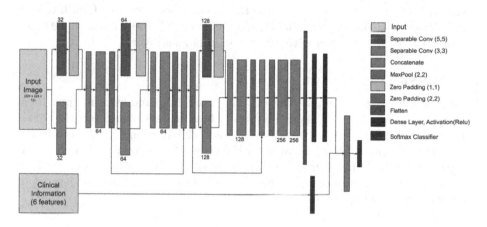

Fig. 2. M2N2 architecture: numbers below and above the layers represent the number of filters in each separable convolutional layer.

The first input consists of 3D images with a shape of 224 × 224 × 12. Since the images are reshaped, few of the local features are stretched, and few are shrunk down due to the original shape mismatch. Reshaping makes it difficult to use constant kernel size across the network to identify and extract infarction features. To overcome the above challenge and to capture infarction features, our architecture has separable convolutional layers with two kernel sizes. The first kernel is a 3 × 3 matrix, and the second kernel in parallel is a 5 × 5 matrix, capturing features of different sizes. The two kernels are followed with a ReLU [4] activation and batch normalization [5]. However, having deep convolution network raises the complication of vanishing gradients [6]. To overcome this, we use residual connections where the outputs of max pool layers and separable convolutions are concatenated to use the features from previous layers by skipping the intermediate layers. This enables the network to collect signals from max pool layer. He et al. [9] show that using residuals can overcome the issue of vanishing gradients. The shape of the feature map at the end of the CNN is 23 × 23 × 256. The later part of the network consists of two dense networks with 1024 and 1024 neurons respectively that follow Relu Activation.

The second input consists of prepossessed clinical information with only six chosen features connected to a dense layer with 12 neurons in the hidden layer, followed with a Relu Activation.

Dense features from both the outputs are concatenated together, and the concatenated features are sent to a softmax classifier to obtain the probability for target classes for the given sample. Our architecture uses Adam [7] optimizer with a categorical cross entropy [8] loss function. We train our network for 2000 epochs with a batch size of 8.

We also train nine different statistical models on the six chosen clinical attributes to identify the possibility to classify patients without MRI scans. The statistical models are SVM with RBF kernel and linear Kernel, distance weighted K-Nearest Neighbor, Logistic Regression, Decision Tree Gini Impurity as a criterion to measure the split, Random Forest with 100 trees and 200 tress, and Multi-Layer Perceptron with 12 neuron in hidden layer.

2.4   Stacking the Models

At this stage we concatenate the outputs from base models and put them together to create a set of meta features. These meta features are then used to train a multilayer perceptron with a sigmoid classifier and 18 neurons in the hidden layer to make a final decision for the class. Original class label is used as the ground truth. Multilayer perceptron used for stacking was trained for 2000 epochs with a batch size of 8. Figure 3, shows the base models and their input along with the stacking block.

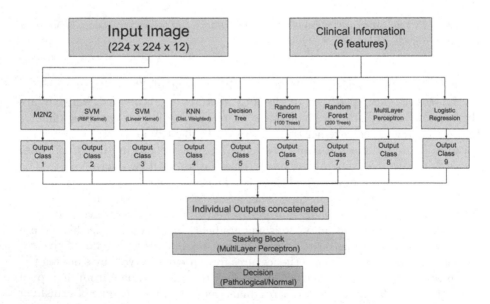

Fig. 3. Base models stacked together to choose the final class for the sample

3   Results

We compare M2N2 with AlexNet3D shown in the paper by Polat et al. [12]. We first train AlexNet3D with image inputs only and report the performance. However, the original AlexNet3D architecture is not suitable to take multiple inputs and can only take 3D images as input, because of which a comparison with M2N2 is not possible. To overcome this challenge, we modify the inputs of AlexNet3D and add a multilayer perceptron with clinical features as input. After this modification, AlexNet3D is comparable with M2N2. Figure 4, shows the modified AlexNet3D architecture with multiple inputs which is uniform with respect to the architecture of M2N2 making them comparable.

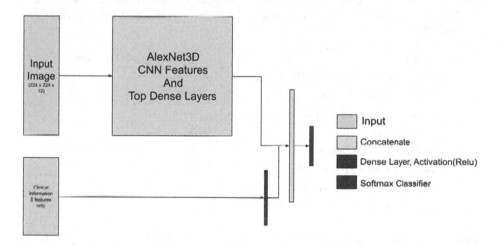

Fig. 4. AlexNet3D architecture with multiple inputs.

Table 1 represents the test and validation accuracy on the limited low number of data between AlexNet3D, AlexNet3D with multi inputs and M2N2. Table 2 compares the number of parameters for M2N2 and multi input AlexNet3D.

Table 3 represents the test accuracy after training statistical base models on clinical inputs for a limited low number of data samples. We use 10-fold cross validation on the samples and report the average test accuracy of the 10 folds. The results on statistical models show that with the limited low number of samples that are given in the dataset, it is possible to classify patients using statistical methods and clinical data alone but the M2N2 architecture has better performance then the stand alone base models. However, more research is needed on this topic which is beyond the scope of this paper.

Final classification accuracy after stacking and using base model outputs as meta feature inputs to a multilayer perceptron was 95%. Table 4 represents the confusion matrix on the limited test samples after stacking the models.

Table 1. Accuracy of models

Model	Validation accuracy (percent)	Test accuracy (percent)
M2N2	**87.25**	**90**
AlexNet3D (Multi Input)	83.33	80
AlexNet3D (Image Input Only)	83.33	75

Table 2. Evaluation of trainable parameters of models

Model	Total trainable parameters	Training time per epoch (seconds)
Our model (M2N2)	**139,970,736**	**4**
AlexNet3D	318,022,232	8

Table 3. Confusion matrix score for 20 test samples on statistical models with the six clinical inputs only. Pathological cases are 'Negative' class and Normal cases are 'Positive' class

Model	Test accuracy (percent)
Support vector machine (RBF kernel)	91
Support vector machine (Linear kernel)	90
Distance weighted K-nearest neighbors	90
Logistic regression	93
Decision tree classifier	87
Random forest (100 trees)	91
Random forest (200 trees)	92
Multi-layer perceptron (12 hidden neurons)	87

Table 4. Confusion Matrix for 20 test samples after applying the Stacking: positive is normal patient and negative are pathological patients

Model	True positive	True negative	False positive	False negative
SM2N2	6	13	0	1

4   Conclusion

We create a novel stack multimodal architecture called SM2N2, which combines 3D DE-MRI images and clinical information and allows multiple inputs to a neural network. On the limited low number of samples used for training we observed that SM2N2 has an accuracy of 95%. M2N2 component of SM2N2 serves as the center for combining images and clinical information together. On the limited low number of samples used for training we observed that M2N2 has better performance compared to multi input AlexNet3D by 10%, while reducing the trainable parameters by more than 50%. Reduction in parameters has improved the training and inference time for M2N2 and, made it possible to deploy this model on mobile devices. M2N2 is inspired by Resnet [9], Inception [10] and

MobileNet [11]. We also identified that it is possible to classify pathological patients from normal patients by using clinical information alone, however we can not make a conclusive statement about this finding with such a small dataset. Finally we use stacking on the meta features to generalize base models. We train a multilayer perceptron on concatenated outputs of base models to make the final decisions. We observe that SM2N2 gives the highest accuracy. We will use SM2N2 on the final Emidec Challenge dataset.

Feature selection technique at this stage is based on statistical analysis and needs to be verified with a physician to show the clinical impact of the chosen attributes. This finding is beyond the scope of this paper and will be explored in future work.

The current performance of our network is based on just 100 patients. The dataset is extremely small and it is a challenge to analyze the performance of a network with this small dataset. Hence, we believe that more research and a large dataset is required to analyze the performance of M2N2, to determine benefits of combining images and clinical data, and compare it with other relevant architectures for performance ranking.

5   Future Work

This is an ongoing research and future work will answer some of the questions below:

- What is the advantage of techniques that combine image and clinical data versus the techniques that only take a single input?
- What is the performance of M2N2 compared to other deep learning and statistical models?
- What is the clinical influence of attributes given in the dataset and how they impact the analysis?

References

1. Lalande, A., et al.: Emidec: a database usable for the automatic evaluation of myocardial infarction from delayed-enhancement cardiac MRI. Data **5**(4), 89 (2020)
2. CDC. https://www.cdc.gov/heartdisease/facts.htm. Last accessed 17 Aug 2020
3. Chollet, F.: Xception: deep learning with depthwise separable convolutions. In Proceedings of the IEEE conference on Computer Vision and Pattern Recognition, pp. 1251–1258 (2017)
4. Agarap, A.F.: Deep learning using rectified linear units (ReLU) (2018). arXiv preprint arXiv:1803.08375
5. Ioffe, S., Szegedy, C.: Batch normalization: accelerating deep network training by reducing internal covariate shift (2015). arXiv preprint arXiv:1502.03167
6. Hochreiter, S.: The vanishing gradient problem during learning recurrent neural nets and problem solutions. Int. J. Uncertainty Fuzziness Knowl.-Based Syst. **6**(02), 107–116 (1998)

7. Kingma, D.P., Ba, J.: Adam: a method for stochastic optimization. arXiv preprint arXiv:1412.6980 (2014)
8. Zhang, Z., Mert S.: Generalized cross entropy loss for training deep neural networks with noisy labels. In: Advances in Neural Information Processing Systems, pp. 8778–8788 (2018)
9. He, K., Zhang, X., Ren, S., Sun, J.: Deep residual learning for image recognition. In: Proceedings of the IEEE conference on Computer Vision and Pattern Recognition, pp. 770–778 (2016)
10. Szegedy, C., et al.: Going deeper with convolutions. In: Proceedings of the IEEE Conference on Computer Vision and Pattern Recognition, pp. 1–9 (2015)
11. Howard, A.G., et al.: Mobilenets: efficient convolutional neural networks for mobile vision applications. arXiv preprint arXiv:1704.04861 (2017)
12. Polat, H., Danaei Mehr, H.: Classification of pulmonary CT images by using hybrid 3D-deep convolutional neural network architecture. Appl. Sci. 9(5), 940 (2019)

A Hybrid Network for Automatic Myocardial Infarction Segmentation in Delayed Enhancement-MRI

Sen Yang[1] and Xiyue Wang[2]($\boxtimes$)

[1] College of Biomedical Engineering, Sichuan University, Chengdu 610065, China
[2] College of Computer Science, Sichuan University, Chengdu 610065, China
sen.yang.scu@gmail.com

Abstract. Delayed enhancement (DE)-MRI plays an important role in the diagnosis of various myocardial damages (such as myocardial infarct and no-reflow phenomenon). This paper proposes a hybrid U-net network to achieve the simultaneous segmentation of the background, left ventricle, left myocardium, myocardial infarction, and no-reflow regions in DE-MRI. The hybrid U-net architecture introduces the squeeze-and-excitation residual (SE-Res) module and selective kernel (SK) block in the encoder and decoder parts, respectively. The SE-Res module can address the dependencies of all feature channels, and increase the weight value on the more informative channel. The SK block can adaptively adjust the receptive field size to obtain the multi-scale feature information. Two types of labels (category label and segmentation label) and hybrid branches are used to control the whole segmentation process, which produces robust segmentation performance. The experimental result shows that the proposed model achieves high segmentation performance with the Dice score of 0.8455 for myocardium, 0.6455 for infarction, and 0.6698 for no-reflow on the validation set.

Keywords: Myocardial infarction · Segmentation · Delayed enhancement MRI · Convolution neural network

1 Introduction

In 2015, according to statistical analysis, there were about 15.9 million people suffering from myocardial infarction in the world [13]. Delayed enhancement magnetic resonance imaging (DE-MRI) is currently used for the diagnosis of myocardial involvement (myocardial infarction and no-reflow) in the current clinical environment. The no-reflow zone is characterized by persistent hypoperfusion caused by reduced blood flow [9]. For the most of patients with myocardial infarction, the myocardial infarction and no-reflow regions occupy only a fraction of the left myocardium and less area of the entire heart image. Thus, accurate delineation of the endocardial and epicardial borders of the left myocardium is a prerequisite. The manual segmentation for the left ventricle,

© Springer Nature Switzerland AG 2021
E. Puyol Anton et al. (Eds.): STACOM 2020, LNCS 12592, pp. 351–358, 2021.
https://doi.org/10.1007/978-3-030-68107-4_36

myocardium, and myocardial damage is a time-consuming (around 15–20 min per case), tedious, and experienced-dependent task. Computer-aided technology is urgently required to assist radiologists in workflow optimization.

In current years, a majority of studies have focused on the left myocardium segmentation on the late gadolinium enhancement MRI. The boundaries of ventricles are more unclear in the delayed enhancement MRI than that in the balanced-steady state free precession (BSSFP) MRI (As shown in Fig. 1). To overcome these challenges, several methods segment the myocardium region by combining the prior information from the corresponding BSSFP MRI [2–4,14,17]. These multi-modality based segmentation methods require paired MRI sequences and complex image registration operations. In addition to the segmentation of the left myocardium, 2019 MICCAI MS-CMRSeg challenge aims to simultaneously segment left ventricle, right ventricle, and left myocardium from LGE MRI with the complementary information from other MRI modalities [18]. For the small lesion myocardial infarction region, very few studies have designed automatic segmentation algorithms based on cine cardiac MRI [16] and enhancement MRI [12,15].

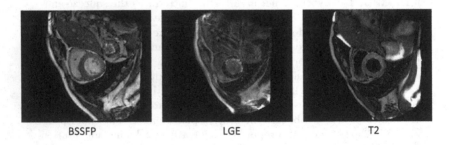

BSSFP   LGE   T2

Fig. 1. Visualization for the representation of cardiac structure in balanced-steady state free precession (BSSFP), late gadolinium enhancement (LGE), and T2-weighted cardiac MRI modalities. The variations of image intensity and contrast result in different ventricles and myocardium boundary discrimination.

However, in the current literature, the left ventricle, left myocardium, myocardial infarction, and no-reflow have not been considered together. The extent of these regions is important in the evaluation of cardiac diseases and the guide for the following treatment. This paper proposes a hybrid U-net architecture with multi-task learning to automatically segment these four regions (if have) in the DE-MRI. The encoder part in the U-net is substituted by the SE-Resnext50 [5], and the hybrid decoder part with the embedding of SK (selective kernel) block has two task branches (MI (myocardial infarction) segmentation and full segmentation) [8]. The MI segmentation branch helps target small lesion region, which is used only in the training process. The full segmentation means the simultaneous segmentation for the left ventricular cavity, left myocardium, myocardial infarction, and no-reflow (As shown in Fig. 2). Inspired by the idea

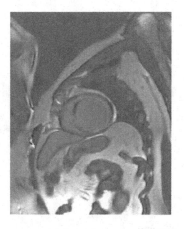 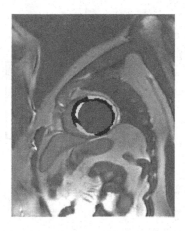

Fig. 2. An example for the left ventricular cavity, left myocardium, myocardial infarction and no-reflow region in one DE-MRI slice. The left is original image and the right is its corresponding mask. The red denotes left ventricle; green represents left myocardium; blue is the myocardial infarction region; and the yellow corresponds to the no-reflow region. (Color figure online)

of deep supervision, a classification branch is introduced to consider both the category of each image and the pixel-level classification in each image.

2   Method

In this section, the proposed cardiac structure segmentation algorithm is introduced from three parts: image preprocessing, hybrid network architecture, and image postprocessing.

2.1   Image Preprocessing

In order to remove the influence of the surrounding organ of heart in the DE-MRIs, region of interest (ROI) extraction is a crucial step in the prepossessing stage. Thus, this paper performs a statistical work to roughly locate the position of the heart. The second step in the preprocessing process is to normalize the input images as the distribution of zero mean and variance of 1. Finally, to fully utilize the dependences between slices, the neighbored three slices are stacked as the new three-channel image that has the same mechanism as the RGB channel in the color image.

2.2   Network Architecture

The modified U-net architecture adopts the SE-Resnext50 module as the encoder part, and the SK block is embodied in the decoder part. The SE-Resnext50 model helps capture the channel correlations, and the SK block with multiple kernels

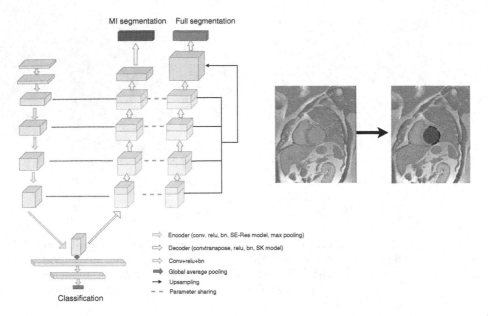

Fig. 3. The overall structure of our proposed segmentation network. The myocardial infarction (MI) segmentation branch only targets to segment the MI and no-reflow regions in the training procedure. The full segmentation means the simultaneous segmentation of the background, left ventricle, left myocardium, and MI, and no-reflow regions, which is adopted as the final segmentation. The classification is also used only in the training phase to deeply supervise the segmentation network.

and different kernel sizes can adaptively adjust the respective field size. Figure 3 illustrates the overall structure of the proposed MI detection model.

As shown in Fig. 3, the encoder part reduces the spatial resolution and extracts high-level features to obtain more precise classification results. In this architecture, the encoder sequentially includes the convolution, pooling, rectified linear unit, batch normalization, SE-Res module, and max pooling. The decoder part recovers the missing spatial information by adding the deconvolution operation. The decoder part includes the deconvolution, rectified linear unit, batch normalization, and the SK block.

Two types of labels are adopted to supervise the training procedure of the network. One label is the categorized label of each patch, which interprets whether the image is a normal or pathological slice. Another label is the segmentation mask, which includes the pixel location of the myocardial region. The two types of labels can achieve better segmentation results.

Two segmentation branches exist in this architecture. The myocardial infarction (MI) segmentation only performs the myocardial infarction region segmentation in the network training procedure, which is controlled by the category annotation provided by the radiologists. The MI segmentation branch pays more attention on the small lesion regions. It is worth noting that the MI segmentation branch is used to assist the final lesion detection, which will no longer work during the test phase. The full segmentation branch actually segments five regions (the background, left ventricle, left myocardium, MI and no-reflow), which is adopted as the final output in our segmentation network.

2.3   Image Postprocessing

The image postprocessing process aims to refine the result of the cardiac segmentation. First, the hole filling technique is applied to attain more complete segmentation. Then, a connected component analysis for all the obtained segmentation is performed [11]. Segmentations that exceed the largest connected range will be removed.

2.4   Loss Function

In the classification branch, the binary cross-entropy (BCE) loss is used as the loss function to optimize the network performance. In the two segmentation branches, Dice loss and weighted cross-entropy (WCE) loss are combined to complete the segmentation task. These loss functions are listed as follows.

$$L_{Cls} = L_{BCE} = -\sum_{l} \left[(y_l \log \hat{y}_l) + (1 - y_l) \log (1 - \hat{y}_l) \right] \tag{1}$$

where y_l and $\hat{y}_l$ denote the ground truth label and predicted output for the lth samples, respectively.

$$L_{WCE}(y, \hat{y}) = -\frac{1}{N} \sum_{i=1}^{N} \sum_{k=1}^{C} w_k y_{i,k} \log \hat{y}_{i,k}, \tag{2}$$

$$L_{DICE} = -\frac{2}{|C|} \sum_{k \in C} \frac{\sum_i y_{i,k} \hat{y}_{i,k} + \varepsilon}{\sum_i y_{i,k} + \sum_i \hat{y}_{i,k} + \varepsilon} \tag{3}$$

where C and N represent the number of classes and the number of training images, respectively. y and $\hat{y}$ denote the ground truth label and predicted output, respectively. ε is a small positive number used to ensure numerical stability. The w_k denotes the weights of each class, which are set empirically to 1, 2, 4, 4, and 4 for the background, left ventricle, left myocardium, MI and no-reflow, respectively.

$$L_{Seg} = L_{WCE} + L_{DICE} \tag{4}$$

3   Experimental Results and Discussions

3.1   Dataset and Experimental Setup

Our algorithm is evaluated on the Automatic Evaluation of Myocardial Infarction from Delayed-Enhancement Cardiac MRI (MICCAI 2020 EMIDEC challenge). The dataset consists of 150 DE-MRI cases (50 cases with normal MRI after the injection of a contrast agent and 100 cases with myocardial infarction (and then with a hyperenhanced area on DE-MRI)) [7]. The training set comprises 100 cases (67 cases with myocardial damage, 33 normal cases) and the testing set is composed of 50 cases (33 cases with myocardial damage, 17 normal cases). There are no overlapped cases in the training and testing sets.

Due to the very limited number of training images, data augmentation is applied to simulate an enlarged dataset to further improve the model generalization capability. The adopted data augmentation operations include randomized image transpose, flipping, cropping, gamma noise, and rotation. Adam optimizer is used as the optimization method for model training [6]. The initial learning rate is set to 0.0003, and reduced by a factor of 10 at the 40th and the 60th epoch, with a total of 90 training epochs. The min-batch size is set as 16.

3.2   Results

Since the labels of the test data are not opened, these experimental results are tested on the training data using the 5-fold cross-validation. Our network has three outputs, and the loss functions assigned on these outputs are cross-entropy, weight cross-entropy + Dice, and weight cross-entropy + Dice, respectively. The Dice score is adopt-ed as an indicator to evaluate the performance of the proposed model. The proposed U-net model is compared with the Linknet [1] and several U-net based models [10], which are listed in Table 1.

Table 1. The Dice score of the myocardium, infraction, and no-reflow regions segmentation using various network architectures. The cls branch means the classification branch (Fig. 3). The hybrid branch denotes the MI segmentation branch (Fig. 3).

Network architectures	Dice score		
	Myocardium	Infarction	No-reflow
Linknet	0.8023	0.5823	0.5963
U-net1(SE-Resnext50)	0.8069	0.5946	0.6038
U-net2(SE-Resnext50 + SK block)	0.8135	0.6046	0.6287
U-net3 (SE-Resnext50 + SK block) + cls branch	0.8135	0.6222	0.6464
U-net4 (SE-Resnext50 + SK block) + cls branch +hybrid branches	0.8455	0.6455	0.6698

4   Conclusion

This paper proposes an improved U-net architecture for the simultaneous segmentation of the background, left ventricle, left myocardium, myocardial infarction, and no-reflow regions in the DE-MRI. Our segmentation algorithm adopts multi-task learning to deeply supervise the final segmentation. The ablation study results have demonstrated the algorithm validity. It has the promise for clinical application in assisting radiologists for the diagnosis of cardiac diseases.

Acknowledgement. This research was funded by the National Natural Science Foundation of China, grant number 61571314.

References

1. Chaurasia, A., Culurciello, E.: Linknet: exploiting encoder representations for efficient semantic segmentation. In: 2017 IEEE Visual Communications and Image Processing (VCIP), pp. 1–4. IEEE (2017)
2. Ciofolo, C., Fradkin, M., Mory, B., Hautvast, G., Breeuwer, M.: Automatic myocardium segmentation in late-enhancement MRI. In: 2008 5th IEEE International Symposium on Biomedical Imaging: from nano to macro, pp. 225–228. IEEE (2008)
3. Dikici, E., O'Donnell, T., Setser, R., White, R.D.: Quantification of delayed enhancement MR images. In: Barillot, C., Haynor, D.R., Hellier, P. (eds.) MICCAI 2004. LNCS, vol. 3216, pp. 250–257. Springer, Heidelberg (2004). https://doi.org/10.1007/978-3-540-30135-6_31
4. El Berbari, R., Kachenoura, N., Frouin, F., Herment, A., Mousseaux, E., Bloch, I.: An automated quantification of the transmural myocardial infarct extent using cardiac DE-MR images. In: 2009 Annual International Conference of the IEEE Engineering in Medicine and Biology Society, pp. 4403–4406. IEEE (2009)
5. Hu, J., Shen, L., Sun, G.: Squeeze-and-excitation networks. In: Proceedings of the IEEE Conference on Computer Vision and Pattern Recognition, pp. 7132–7141 (2018)
6. Kingma, D.P., Ba, J.: Adam: a method for stochastic optimization. arXiv preprint arXiv:1412.6980 (2014)
7. Lalande, A., et al.: Emidec: a database usable for the automatic evaluation of myocardial infarction from delayed-enhancement cardiac mri. Data 5(4), 89 (2020)
8. Li, X., Wang, W., Hu, X., Yang, J.: Selective kernel networks. In: Proceedings of the IEEE Conference on Computer Vision and Pattern Recognition, pp. 510–519 (2019)
9. Pineda, V., Merino, X., Gispert, S., Mahía, P., Garcia, B., Domínguez-Oronoz, R.: No-reflow phenomenon in cardiac MRI: diagnosis and clinical implications. Am. J. Roentgenol. **191**(1), 73–79 (2008)
10. Ronneberger, O., Fischer, P., Brox, T.: U-Net: convolutional networks for biomedical image segmentation. In: Navab, N., Hornegger, J., Wells, W.M., Frangi, A.F. (eds.) MICCAI 2015. LNCS, vol. 9351, pp. 234–241. Springer, Cham (2015). https://doi.org/10.1007/978-3-319-24574-4_28
11. Samet, H., Tamminen, M.: Efficient component labeling of images of arbitrary dimension represented by linear bintrees. IEEE Trans. Pattern Anal. Mach. Intell. **10**(4), 579–586 (1988)
12. Tao, Q., et al.: Automated segmentation of myocardial scar in late enhancement MRI using combined intensity and spatial information. Magn. Reson. Med. **64**(2), 586–594 (2010)
13. Vos, T., et al.: Global, regional, and national incidence, prevalence, and years lived with disability for 310 diseases and injuries, 1990–2015: a systematic analysis for the global burden of disease study 2015. lancet **388**(10053), 1545–1602 (2016)

14. Wei, D., Sun, Y., Ong, S.H., Chai, P., Teo, L.L., Low, A.F.: Three-dimensional segmentation of the left ventricle in late gadolinium enhanced MR images of chronic infarction combining long-and short-axis information. Med. Image Anal. **17**(6), 685–697 (2013)
15. Xu, C., Xu, L., Brahm, G., Zhang, H., Li, S.: MuTGAN: simultaneous segmentation and quantification of myocardial infarction without contrast agents via joint adversarial learning. In: Frangi, A.F., Schnabel, J.A., Davatzikos, C., Alberola-López, C., Fichtinger, G. (eds.) MICCAI 2018. LNCS, vol. 11071, pp. 525–534. Springer, Cham (2018). https://doi.org/10.1007/978-3-030-00934-2_59
16. Xu, C., et al.: Direct delineation of myocardial infarction without contrast agents using a joint motion feature learning architecture. Med. Image Anal. **50**, 82–94 (2018)
17. Xu, R.S., Athavale, P., Lu, Y., Radau, P., Wright, G.A.: Myocardial segmentation in late-enhancement MR images via registration and propagation of cine contours. In: 2013 IEEE 10th International Symposium on Biomedical Imaging, pp. 856–859. IEEE (2013)
18. Zhuang, X.: Multivariate mixture model for myocardial segmentation combining multi-source images. IEEE Trans. Pattern Anal. Mach. Intell. **41**(12), 2933–2946 (2018)

Efficient 3D Deep Learning for Myocardial Diseases Segmentation

Khawla Brahim[1,2,3](✉), Abdul Qayyum[2](✉), Alain Lalande[2](✉),
Arnaud Boucher[2](✉), Anis Sakly[3](✉), and Fabrice Meriaudeau[2](✉)

[1] National Engineering School of Sousse, University of Sousse, Sousse, Tunisia
[2] ImViA EA 7535 Laboratory, University of Burgundy, Dijon, France
{Khawla.Brahim,Abdul.Qayyum,Alain.Lalande,Arnaud.Boucher,
Fabrice.Meriaudeau}@u-bourgogne.fr
[3] LASEE Laboratory, National Engineering School of Monastir,
University of Monastir, Monastir, Tunisia
sakly_anis@yahoo.fr

Abstract. Automated myocardial segmentation from late gadolinium
enhancement magnetic resonance images (LGE-MRI) is a critical step
in the diagnosis of cardiac pathologies such as ischemia and myocardial
infarction. This paper proposes a deep learning framework for improved
myocardial diseases segmentation. In the first step, we build an encoder-
decoder segmentation network that generates myocardium and cavity
segmentations from the whole volume, followed by a 3D U-Net based on
Shape prior to identifying myocardial infarction and myocardium ventricular obstruction (MVO) segmentations from the encoder-decoder prediction. The proposed network achieves good segmentation performance, as
computed by average Dice ratio overall predicted substructures, respectively: 'Myocardium': 96.29%, 'Infarctus': 76.56%, 'MVO': 93.12% on
our validation EMIDEC dataset consisting of LGE-MRI volumes of 16
patients extracted from the training data.

Keywords: LGE-MRI · Myocardial infarction · Deep learning

1 Introduction

According to the World Health Organization (WHO) [1], Myocardial Infarction
(MI) is one of the main cause of death globally. It essentially develops when
oxygen-rich blood flow to the myocardium is suddenly interrupted [2]. However,
when revascularization fails, MVO (also known as No-reflow) can occur in scar
regions. Efficient quantification of infarcts and MVO is essential for diagnosis
and therapy planning.

Myocardial Scar Segmentation aims to accurately recognizing myocardial
scars areas. Previous prevalent scar segmentation works were often performed
using thresholding-based methods, such as the n-standard deviations (n-SD) [3],
the full-width at half-maximum (FWHM) [4] and the region growing [5], which

© Springer Nature Switzerland AG 2021
E. Puyol Anton et al. (Eds.): STACOM 2020, LNCS 12592, pp. 359–368, 2021.
https://doi.org/10.1007/978-3-030-68107-4_37

are responsive to the regional intensity variation. However, these algorithms frequently require a prior knowledge of the expert myocardial location determined by its epicardial and endocardial annotations, to delineate the areas of interest for segmentation [6].

Manual MI delineation is time-consuming and prone to inter and intra-observer variations. Hence, there is a need for accurate and automatic segmentation models to ease the work load of medical experts. Deep learning-based methods for medical image segmentation play an important role in cardiac function analysis and follow up of different diseases due to their feature extraction effectiveness. Deep architectural betterment has been a target of several scientists for diverse works. U-Net based networks [7] have often been used. 2D U-Net network demonstrated impressive performances for valuable segmentation of myocardium structure on the ACDC 2018 challenge [8]. Fahmy et al. [9], used the U-Net based method to delineate the myocardium and the scars from LGE images obtained from subjects with hypertrophic cardiomyopathy (HCM). Fatemeh Zabihollahy et al. proposed 2D U-Net for powerful segmentation of myocardial regions from 3D LGE-MRI [10]. Applying 3D Fully Convolutional Networks (FCN), which integrate 3D context across different slices, improve estimating disease diagnosis [11]. Interestingly, frequent 3D FCN works achieve promising performances in segmenting cardiovascular volumes with robust 3D consistency [12,13]. Xu et al. [14] provided an RNN method for infarction assessment, which exploits motion patterns to segment MI regions from cine MR image sequences accurately.

2   Material and Method

2.1   Datasets and Pre-processing

The dataset [15] is supplied by the EMIDEC segmentation challenge and consists of 150 volumes, with slices number ranging from 4 to 10. Data were acquired from MRI department of the Regional University Hospital of Dijon (France). The ground truth annotation includes five labels: background (0), cavity(1), normal myocardium (2), myocardial infarction (3) and no-reflow (4). We split the labeled scans into training (68 patients), validation (16 patients) and testing (16 patients) subsets. We first cropped original volumes to a normalized set. Supplementary empty slices are added to maintain size fixed, resulting in Nifti images of shape $96 \times 96 \times 16$ for all present subjects.

2.2   3D Proposed Model

Given $V = \{V_1, V_2, ... , V_n\}$ a set of 3D LGE-MRI input volumes, our approach is trained end-to-end on each of them, and our model learns to predict 3D segmentation for the whole volume. The proposed method consists of two significant steps to reach this goal: we train the first proposed network to learn the myocardium regions. Then, the 3D pre-trained Autoencoder model's extracted attributes are transferred to 3D U-Net to segment myocardial diseases and increase the model performance. Details of each part are elucidated in the following paragraphs.

2D Myocardium Segmentation. The proposed model is designed based on the concept of encoder-decoder with skip connections. In the encoding or analysis path, the proposed Inception-Res block has been introduced with a convolutional block attention module (CBAM). The proposed EDP (expansion, depthwise, and projection layer block) module has been presented after the 2D upsampling layer in the decoding or synthesis path. The attention module has been used in skip connection that caters information at every block from the encoder to the decoder side. The number of channels is doubled at each Inception-Res block, and the input size of feature maps are reduced by half using a depthwise convolution layer in the analysis path. We have used progressive feature extraction approach at the encoder side, the number of Inception-Res block are increased progressively at each stage of the encoder side. The first encoder block used one Inception-Res block; similarly, the second, third, and fourth used 2,3,4 number of Inception-Res block, respectively. Also, in the synthesis path, the size of feature maps increases after the 2D up-sampling layer. The original size of training images will return at the output in the final layer.

We proposed a modified inception module in the encoder side of our proposed model. In the inception residual block, the features maps are aggregated from various branches using kernels of different sizes that make the network wider and having the capacity to learn more features. The residual connections provide easy learning concerning the input feature maps instead of learning an unreferenced function [16]. The proposed model is shown in Fig. 1. For reader understanding, we differentiate the layers with various colors.

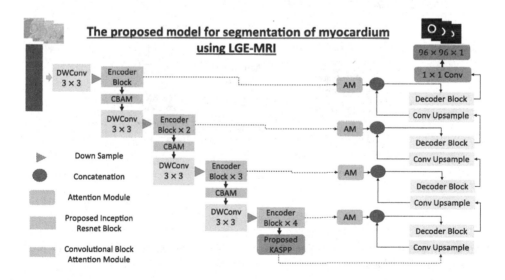

Fig. 1. Proposed model based on Inception-ResNet and EDP blocks.

The Fig. 2a shows the proposed modified Inception-Res block. As compared to the original Inception-Res architecture, batch normalization (BN) layer has been introduced after each convolutional layer except for bottleneck layers. We were using 1×1 and 3×3 kernel as a second modification and introduced a 5×5 kernel branch as inspired by the DeepLab [17]. Batch normalization layer produced smooth training and can avoid gradient vanishing while retaining convolutional layers. The feature maps are aggregated by convolving with three kernels, namely 1×1, 3×3, and 5×5. The 3×3 and 5×5 kernels are further reduced into 1×3, 3×1, 1×5 and 5×1 to minimize the number of parameters.

Assuming that x_l is the output of the lth layer, $c_{(n \times n)}(.)$ is a $n \times n$ kernel convolutional layer, $c_b(.)$ represents the batch-normalization layer and 1×1 Conv denotes the bottleneck layer. The output of each Inception-Res block module from the decoder path is given in Eq. 1.

$$x_{l+1} = c_{1 \times 1}(c_{1 \times 1}(x_l).c_b(c_{3 \times 3}(c_{1 \times 1}(x_l))).c_b(c_{3 \times 3}(c_b(c_{3 \times 3}(c_{1 \times 1}(x_l))))).c_b(c_{5 \times 5}$$
$$(c_b(c_{5 \times 5}(c_{1 \times 1}(x_l))))) + x_l$$

$$(1)$$

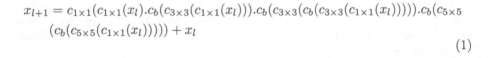

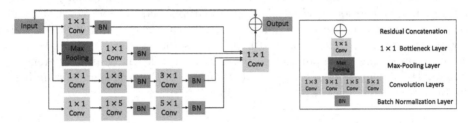

(a) Proposed Inception-ResNet Block.

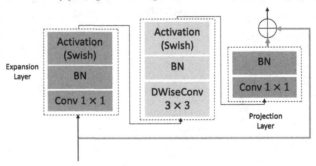

(b) Decoder Block.

Fig. 2. Proposed Inception-ResNet Block and EDP (expansion, depth-wise and projection) Decoder Block.

We have proposed EDP (expansion, depth-wise, and projection layer block) to extract useful semantic information on the decoder side. Similarly, as the encoder, the expansion layer increases the number of feature maps. The projection layer decreases the feature maps with some regularization layers such as batch normalization and activation. The complete layer structure for the decoder is shown in Fig. 2b.

The swish activation function based on the main smoothness property [18] produced a better performance than the ReLU activation function for classification based deep learning tasks. The swish activation function is given in Eq. 2.

$$f(x) = x.sigmoid(\beta x) \tag{2}$$

(Where (β is a constant or trainable parameter.)

Ye Huang et al. [19] presented a kernel-sharing atrous convolutional (KSAC) layer in the atrous spatial pyramid pooling (ASPP) module. The 3 × 3 kernel is shared with atrous convolutional layers with different dilation rates. In this paper, we have extended KSAC based ASPP (later noted KASPP) module and combined different features extracted from the down-sampling path with the various scale features (five scales) in the KASPP as shown in Fig. 3. The proposed KASPP module captures features from low-level features from various down-sampled layers to obtain texture and position information from encoder side feature maps.

In our approach, three downsampling layers and input image information are passed to the proposed KASPP module to guarantee improved cross-level feature connection and complementarity between cross-level information. The main goal of KASPP is to extract multiscale contextual information from the encoder side of the proposed model with a smaller number of parameters producing less semantic gap.

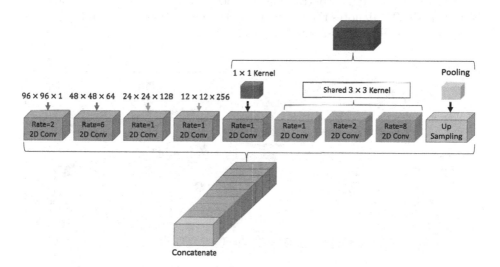

Fig. 3. The proposed K-atrous spatial pyramid pooling layer module.

3D Myocardial Diseases Segmentation. Experimental results proved that extracting volume patches of size $12 \times 12 \times 12$ pixels^3 from the training dataset attains the best results for the segmentation of diseased myocardial tissues.

The U-Net architecture is a U-shaped model that aims to catch more high-level features through an ensemble of convolutional and max-pooling layers. Then the feature maps are up-sampled to recover the segmentation maps at the original spatial dimension. Therefore, the concatenation of feature maps of the same resolution in the decoder path produces a promising medical segmentation.

As shown in Fig. 4, our 3D network incorporates 3D U-Net with a Super-Resolution (SR) module to constrain prior knowledge shape. 3D Autoencoder focuses on accurately encoding and reducing the original volume that can be rebuilt from the encoded representation. Hence a pre-trained 3D Autoencoder is effective to regularize the generated result into a realistic shape. Pre-trained 3D Autoencoder is bound to the 3D U-Net and takes the segmented scan as input. A regularization term is established for restraining the segmentation result. The final loss function is defined in Eq. 3:

$$L_{Final} = L_{Seg} + \lambda_{SR} \times L_{SR} \qquad (3)$$

(Where L_{Seg} is the cross entropy loss function, λ_{SR} is the regularization term and L_{SR} is the L2 loss function which is determined from Frobenius norm Eq. 4. We chose $\lambda_{SR} = 10^{-2}$.)

$$L_{SR} = \sum_{i=1}^{n} ||RP_i - RG_i||_F^2 \qquad (4)$$

(Where n is the number of training volumes, RG_i represents the reconstructed ground truth, RP_i denotes the reconstructed segmentation results and $||.||_F$ indicates the Frobenius norm of an $m \times n$ matrix.)

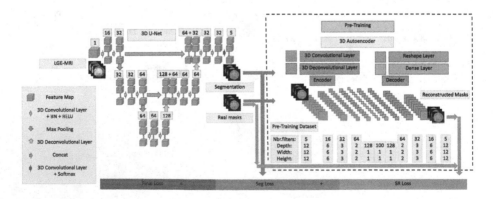

Fig. 4. Schematic representation of our approach for damaged myocardial segmentation.

Table 1. Quantitative study for myocardial segmentation. Best values are represented in bold font.

Method	Metrics	Structures		
		Myocardium	Infarctus	No-reflow
3D U-Net	DSC %	95.71	74.98	68.61
	AVD mm^3	295.50	474.12	55.75
	HD mm	4.57	–	–
	AVDR %	–	9.06	0.97
3D U-Net + 3D Autoencoder + Post-processing	DSC %	**96.29**	**76.56**	**93.12**
	AVD mm^3	**270.00**	**234.1**	**26.69**
	HD mm	**3.77**	–	–
	AVDR %	–	**4.92**	**0.59**

3   Results and Discussion

Each voxel was better determined through majority voting and morphological mathematics (erosion and dilatation) of class tissues acquired on patches. We used Dice Coefficient (DSC), Hausdorff Distance (HD), Absolute Volume Difference (AVD), and Absolute Volume Difference Rate according to the volume of the myocardium (AVDR) as evaluation metrics for all myocardial regions.

We report in Table 1 the summary of comparative evaluation on the validation set (including 16 whole volumes), showing the pertinence of 3D Autoencoder and Post-processing. Majority voting technique and morphological mathematics Post-processing are applied to increase sensitivity for quantifying scarred areas. Our proposed method successfully outperforms the baseline 3D U-Net model (average dice 88.66 % vs. 79.77 %).

To show the impact of 3D Autoencoder and Post-processing, Fig. 5, displays exemplary visualization of the segmentation myocardial structures on two validation subjects. The gold standard and segmented volumes attained using our 3D network and 3D U-Net are presented in the matching row. These results demonstrate our developed framework's performance in segmenting different interest structures by aiming attention at relevant regions.

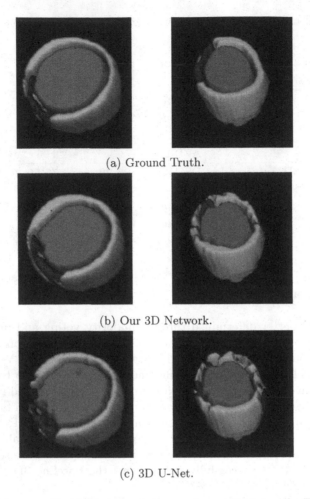

(a) Ground Truth.

(b) Our 3D Network.

(c) 3D U-Net.

Fig. 5. Exemplary results of 3D segmentation on two patients on the EMIDEC Challenge dataset. The first, second and third rows represent ground truth annotation, our 3D proposed network and 3D U-Net segmentation tissue. Our approach reaches qualitatively better segmentations than 3D U-Net.

4   Conclusion

Automated myocardial tissue segmentation is paramount for the diagnosis of cardiac diseases. In this paper, we present an end-to-end deep learning network. A modified proposed atrous convolutional layers, EDP, and Inception-Res blocks are integrated to catch more high-level features and retain finer and coarser information. The proposed myocardial segmentation adopts pre-trained 3D Autoencoder with 3D U-Net, improving the segmentation efficiency. Our experimental results prove that this approach provided the best accuracy for myocardial segmentation and can find its pertinence in different medical imaging tasks based on deep learning.

References

1. Mackay, J., Mensah, G.A.: The Atlas of Heart Disease and Stroke. World Health Organization (2004)
2. Surawicz, B., Knilans, T.: Chou's Electrocardiography in Clinical Practice E-Book: Adult and Pediatric. Elsevier Health Sciences (2008)
3. Kim, R.J., et al.: Relationship of MRI delayed contrast enhancement to irreversible injury, infarct age, and contractile function. Circulation **100**(19), 1992–2002 (1999)
4. Amado, L.C., et al.: Accurate and objective infarct sizing by contrast-enhanced magnetic resonance imaging in a canine myocardial infarction model. J. Am. Coll. Cardiol. **44**(12), 2383–2389 (2004)
5. Albà, X., Figueras i Ventura, R.M., Lekadir, K., Frangi, A.F.: Healthy and scar myocardial tissue classification in DE-MRI. In: Camara, O., Mansi, T., Pop, M., Rhode, K., Sermesant, M., Young, A. (eds.) STACOM 2012. LNCS, vol. 7746, pp. 62–70. Springer, Heidelberg (2013). https://doi.org/10.1007/978-3-642-36961-2_8
6. Carminati, M.C., et al.: Comparison of image processing techniques for nonviable tissue quantification in late gadolinium enhancement cardiac magnetic resonance images. J. Thorac. Imaging **31**(3), 168–176 (2016)
7. Ronneberger, O., Fischer, P., Brox, T.: U-Net: convolutional networks for biomedical image segmentation. In: Navab, N., Hornegger, J., Wells, W.M., Frangi, A.F. (eds.) MICCAI 2015. LNCS, vol. 9351, pp. 234–241. Springer, Cham (2015). https://doi.org/10.1007/978-3-319-24574-4_28
8. Bernard, O., et al.: Deep learning techniques for automatic MRI cardiac multi-structures segmentation and diagnosis: is the problem solved? IEEE Trans. Med. Imaging **37**(11), 2514–2525 (2018)
9. Fahmy, A.S., et al.: Automated cardiac MR scar quantification in hypertrophic cardiomyopathy using deep convolutional neural networks. JACC: Cardiovasc. Imaging **11**(12), 1917–1918 (2018)
10. Zabihollahy, F., White, J.A., Ukwatta, E.: Fully automated segmentation of left ventricular myocardium from 3D late gadolinium enhancement magnetic resonance images using a U-net convolutional neural network-based model. In: Medical Imaging 2019: Computer-Aided Diagnosis. International Society for Optics and Photonics, vol. 10950, p. 109503C, March 2019
11. Çiçek, Ö., Abdulkadir, A., Lienkamp, S.S., Brox, T., Ronneberger, O.: 3D U-Net: learning dense volumetric segmentation from sparse annotation. In: Ourselin, S., Joskowicz, L., Sabuncu, M.R., Unal, G., Wells, W. (eds.) MICCAI 2016. LNCS, vol. 9901, pp. 424–432. Springer, Cham (2016). https://doi.org/10.1007/978-3-319-46723-8_49
12. Dou, Q., et al.: 3D deeply supervised network for automated segmentation of volumetric medical images. Med. Image Anal. **41**, 40–54 (2017)
13. Yu, L., Yang, X., Qin, J., Heng, P.-A.: 3D FractalNet: dense volumetric segmentation for cardiovascular MRI volumes. In: Zuluaga, M.A., Bhatia, K., Kainz, B., Moghari, M.H., Pace, D.F. (eds.) RAMBO/HVSMR -2016. LNCS, vol. 10129, pp. 103–110. Springer, Cham (2017). https://doi.org/10.1007/978-3-319-52280-7_10
14. Xu, C., et al.: Direct delineation of myocardial infarction without contrast agents using a joint motion feature learning architecture. Med. Image Anal. **50**, 82–94 (2018)

15. Lalande, A., et al.: EMIDEC: a database usable for the automatic evaluation of myocardial infarction from delayed-enhancement cardiac MRI. Data **5**, 89 (2020). https://doi.org/10.3390/data5040089
16. Szegedy, C., et al.: Going deeper with convolutions. In: Proceedings of the IEEE Conference on Computer Vision and Pattern Recognition, pp. 1–9 (2015)
17. Chen, L.C., Zhu, Y., Papandreou, G., Schroff, F., Adam, H.: Encoder-decoder with atrous separable convolution for semantic image segmentation. In: Proceedings of the European Conference on Computer Vision (ECCV), pp. 801–818 (2018)
18. Ramachandran, P., Zoph, B., Le, Q. V.: Searching for activation functions. arXiv preprint arXiv:1710.05941 (2017)
19. Huang, Y., Wang, Q., Jia, W., He, X.: See more than once-kernel-sharing atrous convolution for semantic segmentation. arXiv preprint arXiv:1908.09443 (2019)

Deep-Learning-Based Myocardial Pathology Detection

Matthias Ivantsits[1(✉)], Markus Huellebrand[1,2], Sebastian Kelle[1,3,4],
Stefan O. Schönberg[5], Titus Kuehne[1,3,4], and Anja Hennemuth[1,2,4]

[1] Charité – Universitätsmedizin Berlin, Augustenburger Pl. 1, 13353 Berlin, Germany
matthias.ivantsits@charite.de
[2] Fraunhofer MEVIS, Am Fallturm 1, 28359 Bremen, Germany
[3] German Heart Institute Berlin, Augustenburger Pl. 1, 13353 Berlin, Germany
[4] DZHK (German Centre for Cardiovascular Research), Berlin, Germany
[5] Department of Radiology and Nuclear Medicine,
University Medical Centre Mannheim, Mannheim, Germany

Abstract. Cardiovascular diseases are the top cause of death world-wide. Commonly, physicians screen suspected pathological patients with histological examinations and blood tests. Since these clinical parameters are frequently ambiguous, they are routinely extended by delayed-enhancement magnetic resonance imaging of the myocardium.
We propose a method combining deep learning and classical machine learning to differentiate between pathological and normal cases. A convolutional neural network infers a segmentation of the left myocardium from a magnetic resonance image as a preliminary step. This segmentation is employed to determine radiomics-based features describing the morphology and texture of the myocardium. Subsequently, a multilayer perceptron deduces pathological cases from these radiomics features and clinical observations. The presented method demonstrates an accuracy of 0.96 and an F2-score of 0.98 on a nested cross-validation.

Keywords: MRI · Heart · Myocardial infarction ·
Delayed-enhancement · Machine learning · Deep learning ·
Classification

1 Introduction

Cardiac diseases, including myocardial infarction, are the world's leading cause of death [1] with approximately 31.9%. Therefore, it is highly desirable to detect heart diseases early and decide about the appropriate therapy.

The diagnosis of myocardial infarction and myocarditis can be based on histological examinations [2] and blood tests. Proteins like troponin, N-terminal pro b-type natriuretic peptide (NTproBNP), and myoglobin have shown to be very reliable indicators for pathological cases [3–6]. Clinical and patient demographic parameters are sometimes ambiguous and routinely extended with delayed enhancement magnetic resonance imaging (DE-MRI). DE-MRI image data enable the assessment of anatomy and contrast agent accumulation patterns, which are indicators for pathological alterations of the heart muscle tissue [7].

E. Puyol Anton et al. (Eds.): STACOM 2020, LNCS 12592, pp. 369–377, 2021.
https://doi.org/10.1007/978-3-030-68107-4_38

Shape-based modeling has been shown to produce decisive factors to detect heart diseases in cine MRI [8, 9], due to the structural change of the myocardium. Radiomics has proven to be a very promising toolkit for medical image processing, analysis, and interpretation. It derives vast amounts of features from imaged structures describing patterns in morphology and texture that are usually hard to differentiate with the bare eye. Initially, radiomics has been employed for oncological applications but is an emerging technique in the cardiovascular field, especially with MRI. This observation has been confirmed by studies [10–13], which are extracting shape-based radiomics features and classifying diverse heart diseases.

There is an apparent lack of the usage of features describing the texture of imaged structures in the published methods. Due to highlighted myocardial infarction areas in DE-MRI, we hypothesize that features derived from the gray level co-occurrence matrix introduce valuable information to analyze pathological cases. These parameters are used along with the clinical, demographic, and shape-based parameters to distinguish normal from pathological cases for the EMIDEC classification contest [14].

2   Method and Materials

2.1   Dataset

The EMIDEC dataset [14] consists of 150 cases in total, 100 for model training and 50 for model testing. Each observation includes a DE-MRI acquisition of the LV, covering the base to the apex. The training set with ground-truth segmentation of the LV myocardium is comprised of 100 cases (67 pathological cases, 33 normal cases). The testing includes 50 subjects (33 pathological cases, 17 normal cases), all different from those in the training set. The imbalance of normal to pathological cases roughly corresponds to real life managed exams. Along with the MRI, clinical parameters were provided: **sex, age, tobacco** (yes, no, and former), **overweight** (BMI over 25), **arterial hypertension, diabetes, family history of coronary artery disease, ECG, killip max**[1], **troponin**[2], **ejection fraction** (LV), and **NTproBNP**[3].

2.2   Method

Our proposed method consists of three major steps (1) LV segmentation, (2) radiomics feature extraction, (3) classification based on the features from the previous step plus the clinical parameters provided with the contest. For the LV segmentation we utilize a 2D U-Net variation proposed by Hüllebrand et al. [15], which was pre-trained on the ACDC dataset [16].

---

[1] A parameter based on a physical assessment, quantifying the risk of mortality.

[2] A protein released in large quantities in the event of damage to the heart muscle cells.

[3] A protein released in large quantities when the heart needs to work harder.

After the segmentation of the LV, we extract radiomics features based on the LV shape and texture. The MRI can be used as direct input to a classification via a CNN or similar architectures, but since the MRI is a very high dimensional input, it is hard to train and prone to overfit if not parameterized correctly. Therefore, we argue that using a lower-dimensional input, by deriving radiomics features, produces a more reliable classification of normal and pathological cases. Generally, radiomics features can be grouped into morphological and textural descriptors. They can be further divided into the following seven categories:

1. **First Order**, describes the voxel intensity distribution within a masked region.
2. **Shape (3D)**, describes the overall size and shape of a structure. This includes volume, surface area, sphericity, etc.
3. **Gray Level Co-occurrence Matrix (GLCM)**, describes the second-order joint probability function of an image region. This includes cluster tendency, cluster shape, contrast, etc.
4. **Gray Level Size Zone Matrix (GLSZM)** quantifies the gray level zones, which is defined as the number of connected voxels that share the same intensity. This includes gray level variance, zone entropy, zone variance, etc.
5. **Gray Level Run Length Matrix (GLRLM)** describes the length in several consecutive voxels that share the same intensity. This includes run percentage, run variance, run entropy, etc.
6. **Neighbouring Gray Tone Difference Matrix (NGTDM)** describes the difference between the intensity and the average intensity of its neighbors. This includes coarseness, contrast, busyness etc.
7. **Gray Level Dependence Matrix (GLDM)** describes the intensity dependency, which is defined as the number of connected voxels within a distance δ that depend on the center voxel. This includes dependence entropy, dependence variance, dependence non-uniformity, etc.

Shape-based features have shown to be useful for the classification of pathological cases in cine echocardiography and MRI [8–13]. We hypothesize that due to the injected contrast agent in DE-MRI, highlighted infarction areas can be described by the radiomics texture features as shown in similar approaches for T1- and T2-mapping [17,18]. The extracted radiomics features result in 108 variables for the LV, plus 12 clinical parameters per patient, summing up to 120 features. Due to this high dimensional input combined with only 100 patient observations, we propose an 8-fold nested cross-validation (CV) to select relevant radiomics feature classes. Moreover, during the CV, we perform a model selection of five classifiers and their respective hyperparameters, which are illustrated in Table 1.

Support vector machines (SVM) are particularly effective in high dimensional spaces with a clear margin between the classes and have shown to work well on pathological case detection by Cetin et al. [10]. Random forests are very robust to outliers and comparatively little impacted by noise. They have shown to be effective in medical settings, as illustrated by [11–13]. Multilayer perceptrons

Table 1. Five models and their respective hyperparameters used during the 8-fold nested cross-validation.

Model name	Hyperparameters
Support vector machine	C: 0.01, 0.05, 0.1, 1, 10, 100, 1000
	kernel: linear, poly, rbf, sigmoid
	gamma: auto, scale
Random forest	n_estimators: 100, 200, 300, 500
	max_depth: 90, 100, 110
	max_features: 2, 3
	min_samples_leaf: 3, 4, 5
Gradient boosting	loss: deviance, exponential
	n_estimators: 10, 20, 50
	max_depth: 3, 5
	max_features: log2, sqrt
	criterion': friedman_mse, mae
Multilayer perceptron	hidden_layer_sizes: (100,), (100, 50), (50,), (50, 25)
	max_iter: 200, 300, 500, 700, 1000
K-nearest neighbors	n_neighbors: 1, 3, 5, 10
	weights: uniform, distance

(MLP) have been thoroughly studied, and Isensee et al. [13] have illustrated that they can be successfully applied for pathological case detection. Gradient boosting is one of many ensemble methods, which utilizes a collection of weakly trained classifiers. It has been shown to be useful in individual treatment estimations [19]. The K-nearest neighbors (KNN) algorithm does not need any training, and therefore new data can be seamlessly added. Akhil [20] has demonstrated medical applicability by predicting various heart diseases.

3   Results

We used an 8-fold CV during all experiments, performed on an Intel(R) Core(TM) i7-8700K CPU @ 3.70 GHz with 16 GBs RAM. Since the contest is evaluated on the predictions' accuracy, we evaluate all models based on this metric. However, we argue that a false-negative classification is more harmful to the screening process, and since the dataset is imbalanced, we additionally evaluate the F2-score for each model. Where the F2-score combines sensitivity and precision and considers sensitivity twice as important as precision.

The first experiment we conducted was only using the clinical parameters to train the models described in Table 1. The results of this analysis are illustrated in Fig. 1. Based on the accuracy score, the random forest is the best performing metric with 0.89 ± 0.10, closely followed by the KNN with 0.86 ± 0.14.

When taking the F2-score into account, the random forest with 0.93 ± 0.08 is considered the best choice, followed by the MLP with 0.92 ± 0.06.

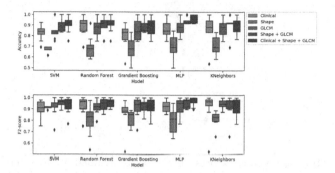

Fig. 1. An illustration of the models accuracy and F2-score produced by an 8-fold nested cross-validation. All models tested on clinical, shape, GLCM, shape+GLCM, and clinical+shape+GLCM.

In the next setup, we experimented with sole shape, GLCM, and shape+GLCM features. The results of this procedure are illustrated in Fig. 1. We observed similar accuracy compared to the clinical features when only taking the GLCM features into account. The best performing model (MLP) trained on the GLCM variables achieves an accuracy of 0.88 ± 0.07 and an F2-score of 0.92 ± 0.08. After combining the shape and GLCM features, the best performing model improves to 0.93 ± 0.05 and an F2-score of 0.94 ± 0.07.

Due to the improved accuracy with models trained on the shape and GLCM features, we combined them with the clinical observations. Clinical, morphological, and textural features sum up to 50 variables. We assume the number of features can be further compressed and subsequently improve the models' predictive power. Consequently, we performed a feature importance analysis as proposed by Breiman [21]. This analysis can be performed on any fitted model by calculating a base score produced by the model on the training or test set. This is followed by a random shuffle to one of the features and compared to the baseline's predictive power. This procedure is then repeated and applied to all features to come up with an importance score. Collinear features in the input result in lower importance scores when permuted. This can be counteracted by clustering highly correlated features and keeping only one feature per cluster.

To estimate the effect of a not perfectly segmented myocardium, we conducted experiments with different levels of structural changes to the myocardium segmentation. We performed contour distortions utilizing increasing degrees of random affine and elastic deformations [22]. An overview of the changes is illustrated in Table 2. By introducing a light structural change to the myocardial segmentation, the accuracy of the trained model stays constant (see Fig. 3). After increasing the distortion, we observed a drop in the models' performance

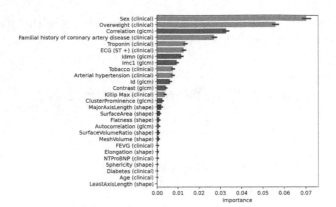

Fig. 2. The feature importance scores on the finally trained MLP, where the importance is defined by the difference of the models' baseline and the score after a feature permutation.

to 0.93. The deformed myocardium shows a dice coefficient of 0.92 and a Hausdorff distance of 1.4 mm. Further decreasing the dice coefficient to 0.87 showed similar accuracy, only after decreasing the dice to 0.75 with a Hausdorff distance of 4.3 mm shrinks the models' predictive capability to 0.87.

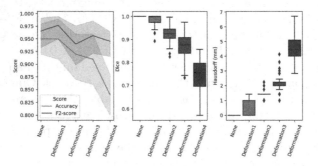

Fig. 3. An illustration of different levels of distortions performed to the myocardium segmentation. **Left**: highlights the change of performance with increasing levels of deformations. **Center**: illustrates the dice coefficient of the original myocardium segmentation and the deformed one. **Right**: shows the Hausdorff distance of the original myocardium segmentation the distorted structure in mm.

Finally, we cross-validate an MLP with optimal hyperparameters obtained in the previous experiments. The model includes one hidden layer with 100 neurons and the ReLU activation function. The network was trained with Adam (with standard parameters) and a learning rate of 0.001 for 500 iterations. This CV results in a network with an accuracy of 0.96 ± 0.05 and an F2-score of 0.98 ± 0.04. The feature importance scores of this model are illustrated in Fig. 2.

Table 2. The four levels of the distortion we performed on the myocardium segmentation. The deformations are illustrated in increasing strength.

Name	Structural changes
Mask transformation 1	Random affine transformation with scaling $[0.97, 1.03]$
Mask transformation 2	Random elastic deformation with $(7, 7, 5)$ control points and $(3, 3, 1)$ maximum displacement
Mask transformation 3	Random elastic deformation with $(7, 7, 5)$ control points and $(5, 5, 1)$ maximum displacement
Mask transformation 4	Random elastic deformation with $(15, 15, 5)$ control points and $(10, 10, 1)$ maximum displacement

This analysis shows very high importance for most clinical variables, followed by some GLCM and morphological features. For the clinical integration of the proposed method, we utilized LIME [23], which is an explanatory framework for any black-box classifier.

4   Discussion and Conclusion

We have illustrated a pathological classification pipeline, with comparable accuracy scores to state-of-the-art solutions. While clinical parameters provide an excellent baseline to distinguish normal from pathological cases, this can be further improved by including morphological and textural features extracted from DE-MRI. Our proposed method includes an automatic segmentation of the LV, an extraction of large amounts of morphological and textural radiomics features followed by classification into normal and pathological cases. A nested CV was performed to deal with the high dimensional data produced by the radiomics feature extraction. During this process, an optimal model from a pool of classifiers and hyperparameters was chosen. Moreover, the input space was further reduced by clustering highly correlated features, resulting in 27 clinical and MRI-based features. These features were ranked according to the final models' importance for clinical interpretation. The final MLP achieves an accuracy score of 0.96 and an F2-score of 0.98 on the performed nested cross-validation. We hypothesize that the utilization of GLCM features derived from DE-MRI enables the differentiation between categories of pathological cases like myocardial infarction and myocarditis. Furthermore, we illustrated the generalization capability of the proposed method by introducing distortion of the myocardium segmentation. The model exhibits invariance in performance on small to medium imperfections of the segmentation. On more intense structural changes, the model still illustrates an accuracy of 0.87 (see Fig. 3).

Interestingly, the importance of the Killip and diabetes observations are almost irrelevant to the combined model as shown in Fig. 2. A potential explanation for this phenomenon is illustrated in Fig. 4. The diabetes distribution of

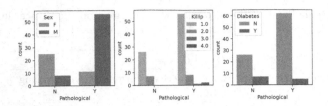

Fig. 4. Left: an illustration of pathological cases grouped by the patients' sex. **Center**: an overview of pathological cases grouped by the Killip parameter. **Right**: a plot of pathological cases grouped by diabetes.

pathological and normal cases does not show any significant difference. Moreover, the Killip classes one and two do not seem to be good predictors for pathological cases, since they mostly emerge in this class. Only classes three and four seem to be useful for predicting pathological cases, although these classes' frequency seems negligible. Notable is the high importance of the patients' sex to the prediction of pathological cases. This observation can be explained by the vast difference of men with myocardial infarction or myocarditis showing up to the emergency room compared to women. Finally, the proposed method's probable advancement is training with randomly elastically deformed segmentations of the myocardium, which should help the model to a superior generalization.

References

1. Roth, G.A., et al.: Global, regional, and national age-sex-specific mortality for 282 causes of death in 195 countries and territories, 1980–2017: a systematic analysis for the Global Burden of Disease Study 2017. Lancet **392**(10159), 1736–1788 (2017)
2. Lieberman, E.B., Hutchins, G.M., Herskowitz, A., Rose, N.R., Baughman, K.L.: Clinicopathoiogic description of myocarditis (1991)
3. Li, Y., Zhang, F., Wang, X., Wang, D.: Expression and clinical significance of serum Follistatin-like protein 1 in acute myocardial infarction (2017)
4. Kottwitz, J., et al.: Myoglobin for Detection of High-Risk Patients with Acute Myocarditis (2020)
5. Sachdeva, S., Song, X., Dham, N., Heath, D.M., DeBiasi, R.L.: Analysis of clinical parameters and cardiac magnetic resonance imaging as predictors of outcome in pediatric myocarditis (2015)
6. Shah, A.S.V., et al.: Sensitive Troponin Assay and the Classification of Myocardial Infarction (2015)
7. Jackson, E., Bellenger, N., Seddon, M., Harden, S., Peebles, C.: Ischaemic and non-ischaemic cardiomyopathies–cardiac MRI appearances with delayed enhancement (2007)
8. Tabassian, M., et al.: Machine learning of the spatio-temporal characteristics of echocardiographic deformation curves for infarct classification (2017)
9. Suinesiaputra, A., et al.: Statistical Shape Modeling of the Left Ventricle: Myocardial Infarct Classification Challenge (2018)
10. Cetin, I., et al.: A Radiomics Approach to Computer-Aided Diagnosis with Cardiac Cine-MRI (2017)

11. Wolterink, J.M., Leiner, T., Viergever, M.A., Išgum, I.: Automatic Segmentation and Disease Classification Using Cardiac Cine MR Images (2017)
12. Khened, M., Alex, V., Krishnamurthi, G.: Densely Connected Fully Convolutional Network for Short-Axis Cardiac Cine MR Image Segmentation and Heart Diagnosis Using Random Forest (2017)
13. Isensee, F., Jaeger, P.F., Full, P.M., Wolf, I., Engelhardt, S., Maier-Hein, K.H.: Automatic Cardiac Disease Assessment on cine-MRI via Time-Series Segmentation and Domain Specific Features (2017)
14. EMIDEC Classification Contest. http://emidec.com/classification-contest. Accessed 12 Sep 2020
15. Markus Hüllebrand et al.: ... (2020)
16. Automated Cardiac Diagnosis Challenge. https://www.creatis.insa-lyon.fr/Challenge/acdc/. Accessed 12 Sep 2020
17. Baessler, B., Mannil, M., Oebel, S., Maintz, D., Alkadhi, H., Manka, R.: Subacute and Chronic Left Ventricular Myocardial Scar: Accuracy of Texture Analysis on Nonenhanced Cine MR Images (2018)
18. Baessler, B., et al.: Cardiac MRI and Texture Analysis of Myocardial T1 and T2 Maps in Myocarditis with Acute versus Chronic Symptoms of Heart Failure (2019)
19. Sugasawa, S., Noma, H.: Estimating individual treatment effects by gradient boosting trees (2019)
20. Akhil, J.: Prediction of heart disease using k-nearest neighbor and particle swarm optimization (2017)
21. Breiman, L.: Random Forests (2001)
22. Pérez-García, F., Sparks, R., Ourselin, S.: TorchIO: a Python library for efficient loading, preprocessing, augmentation and patch-based sampling of medical images in deep learning (2020)
23. Ribeiro, M.T., Singh, S., Guestrin, C.: Why Should I Trust You? Explaining the Predictions of Any Classifier (2016)

Automatic Myocardial Infarction Evaluation from Delayed-Enhancement Cardiac MRI Using Deep Convolutional Networks

Kibrom Berihu Girum[1,2]([⊠]), Youssef Skandarani[1], Raabid Hussain[1], Alexis Bozorg Grayeli[1,4], Gilles Créhange[1,2,3], and Alain Lalande[1]

[1] ImViA Laboratory, University of Burgundy, Dijon, France
kibrom-berihu_girum@etu.u-bourgogne.fr
[2] Radiation Oncology Department, CGFL, Dijon, France
[3] Radiation Oncology Department, Institut Curie, Paris, France
[4] ENT Department, CHU Dijon, Dijon, France

Abstract. In this paper, we propose a new deep learning framework for an automatic myocardial infarction evaluation from clinical information and delayed enhancement-MRI (DE-MRI). The proposed framework addresses two tasks. The first task is automatic detection of myocardial contours, the infarcted area, the no-reflow area, and the left ventricular cavity from a short-axis DE-MRI series. It employs two segmentation neural networks. The first network is used to segment the anatomical structures such as the myocardium and left ventricular cavity. The second network is used to segment the pathological areas such as myocardial infarction, myocardial no-reflow, and normal myocardial region. The segmented myocardium region from the first network is further used to refine the second network's pathological segmentation results. The second task is to automatically classify a given case into normal or pathological from clinical information with or without DE-MRI. A cascaded support vector machine (SVM) is employed to classify a given case from its associated clinical information. The segmented pathological areas from DE-MRI are also used for the classification task. We evaluated our method on the 2020 EMIDEC MICCAI challenge dataset. It yielded an average Dice index of 0.93 and 0.84, respectively, for the left ventricular cavity and the myocardium. The classification from using only clinical information yielded 80% accuracy over five-fold cross-validation. Using the DE-MRI, our method can classify the cases with 93.3% accuracy. These experimental results reveal that the proposed method can automatically evaluate the myocardial infarction.

Keywords: Cardiac MRI segmentation · CNNs · Myocardial infarction · Myocardial no-reflow · Classification · Clinical information

E. Puyol Anton et al. (Eds.): STACOM 2020, LNCS 12592, pp. 378–384, 2021.
https://doi.org/10.1007/978-3-030-68107-4_39

1   Introduction

Cardiovascular diseases (CVDs) are the leading cause of death in the world [1]. Among the common CVDs, myocardial infarction (MI) is a specific cardiovascular disease. The state of the heart after myocardial infarction receiving revascularization requires careful evaluation of the myocardial segment's functionality. It requires assessment of myocardial infarction and myocardial no-reflow (persistent microvascular obstruction area) regions using magnetic resonance imaging (MRI). This is often done using a delayed enhancement-MRI (DE-MRI), which involves an MRI examination following several minutes of a contrast agent injection. Accurate and automatic patient classification is profoundly essential in this scenario. Moreover, accurate and automatic segmentation of the clinically essential regions such as myocardial infarction and myocardial no-reflow can help to assess tissue functionality and thereby make more accurate treatments based on the severity of the disease. Developing an accurate and automatic cardiac image segmentation framework is a challenging task due to low contrast boundary regions, different sizes and shapes of targets, and respiratory motion. Indeed, variations in image acquisition settings in different clinical setups are other challenges in developing accurate and robust medical image segmentation methods [2].

Recently, deep learning approaches using convolutional neural networks have provided the state of the art results in medical image analysis tasks. They have successfully been applied in classification and segmentation tasks. In this paper, we propose a deep learning framework for automatic and accurate evaluation of myocardial infarction from clinical information with and without the delayed-enhancement cardiac-MRI. The proposed framework has two main applications. The first application is to automatically detect the myocardial contours, the infarcted area, the permanent microvascular obstruction area (no-reflow area), and segments the left ventricular cavity from a short-axis DE-MRI. The second application is to automatically classify a given case into normal or pathological using the provided clinical information with or without DE-MRI. The provided clinical information includes sex, age, overweight, arterial hypertension, diabetes, familial history of coronary artery disease, Troponin, Killip Max, ejection fraction, ventricular natriuretic peptide, and the ST segment.

We evaluated our method on a MICCAI 2020 challenge dataset, named automatic Evaluation of Myocardial Infarction from the Delayed-Enhancement Cardiac MRI (EMIDEC) [3]. The experimental results show that the proposed method can be used to automatically evaluate the myocardial infarction from cardiac DE-MRI as well as from clinical information. Moreover, the proposed method can accurately segment the anatomical structures such as the myocardium and the left ventricular cavity.

2   Proposed Framework

Segmentation. To segment the anatomical and pathological (if present) structures from DE-MRI, we proposed a new deep learning framework (Fig. 1). It employed two encoder-decoder style convolutional neural networks [4]. The first network, the anatomical network, was used to segment the anatomical structures such as the myocardium and the left ventricular cavity. The second network, the pathological network, was used to segment the areas such as the myocardium infarction, the no-reflow, and the normal or healthy myocardium regions. The segmentation results from the pathological network are then masked by the segmented myocardium region from the anatomical network. It enabled us to constrain the segmentation results of the pathological network to be inside the myocardium area. However, note that the anatomical and pathological networks were trained separately. The segmented myocardium area from the anatomical network was the final myocardium segmentation. The predicted segmentation from the anatomical structure segmentation network and the masked patholog-ical areas from the pathological network are then merged to yield the final four label segmentation output.

The encoder-decoder networks' building block consists of a repeated applica-tion of 2D 3x3 convolution, exponential linear unit (ELU) activation function, batch normalization followed by squeeze-and-excitation network [5], and a 2x2 max pooling operation with stride 2 for down-sampling. Starting with 32 feature maps, after each max-pooling layer the feature channels are then doubled. The up-sampling part consists of 2x2 up-convolution and concatenation layer, fol-lowed by repeated 2D 3x3 convolution, exponential linear unit (ELU) activation function, batch normalization, and squeeze-and-excitation network [5]. The out-put 2D convolutional layer activation function was softmax for both pathological and anatomical networks. Each network was trained by calculating the average of Dice and cross-entropy loss.

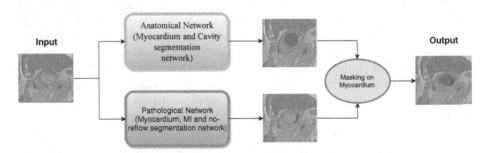

Fig. 1. The schematic representation of the proposed segmentation framework. The anatomical network is used to segment the left ventricular cavity and the myocardium. The pathological network is used to detect the pathological areas (myocardium infarc-tion (MI) and no-reflow) and the healthy or normal myocardium regions. The red, green, and blue colors, respectively, show the left ventricular cavity, the myocardium, and the myocardium infarction. (Color figure online)

Classification. The classification task revolves around the clinical information for each patient. The proposed approach used a Support Vector Machine (SVM) with a linear kernel and a hinge loss [6] as an initial feature selection step. The selected features were then passed into another SVM with a radial basis function kernel to refine the separation boundary between the two classes (i.e., normal or pathological). The cascaded SVM was designed to remove redundant features at the first step and provide robust features for final classification in the second step. Moreover, classification using the DE-MRI was performed from the segmentation results by checking if the segmented areas consist of either the myocardium infarction or the no-reflow. If a given case consists of pathological areas in two or more slices, it was considered pathological otherwise normal. The classification pipeline is shown in Fig. 2.

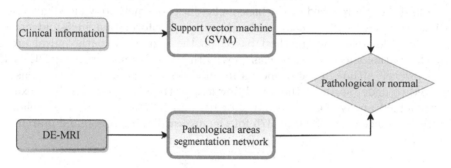

Fig. 2. The schematic representation of the proposed classification framework using the provided clinical information and cardiac DE-MRI.

Implementation Details. The proposed framework was implemented on an i7 computer with 32 GB RAM and a dedicated GPU (NVIDIA TITAN Xp, 12 GB). Python, Keras API with Tensorflow backend, was used to implement the method. We trained the model using the ADAM optimizer for 500 epochs with a learning rate of 10^{-3} and early stopping criteria of 200 [7]. All convolutional neural networks' parameters were initialized using He initialization [8]. Moreover, other hyper-parameters values, such as the ADAM optimizer parameters, were Keras' default values.

3 Experimental Setup and Results

Dataset and Evaluation Criteria. To evaluate the proposed method, we selected the EMIDEC MICCAI 2020 challenge dataset [3]. It consisted of 100 training DE-MRI images, of which 67 are with pathological and 33 normal cases. The 50 held out testing data consisted of 33 pathological conditions and 17 normal cases. The training dataset was divided randomly into 85 training and 15 validation cases. The validation cases were composed of 10 pathological and five normal cases. As the provided EMIDEC dataset was with different image resolutions, we resized it to 256x256 before feeding it to the network. Each given case was also pre-processed by zero-centering the intensity values and normalizing it by the standard deviation. Although the proposed segmentation networks are in 2D, we evaluated the results in 3D after stacking each predicted 2D images per case. Thus, the evaluation was in 3D using the Dice index (DSC), Hausdorff distance (HD), and relative volume difference (RVD) for the segmentation task. For the classification task, we used the accuracy metric.

Experimental Results. Experimental results show that the proposed method can segment the left ventricular cavity and the myocardium cardiac structures with an average Dice index of 0.93 and 0.84, and an average 3D HD value of 6.5 mm and 8.9 mm, respectively. Quantitative segmentation results are shown in Table 1. The proposed SVM-based classification method yielded a classification accuracy of 80% using only the clinical information in 5-fold cross-validation. Moreover, our method based on the segmented pathological areas from the pathological network classifies with 93.3% accuracy using DE-MRI only. However, although the proposed method detects the pathological areas, it appeared to under segment them yielding less overlapping area ratio. For example, it yielded an average Dice index of 0.4 and 0.67 and an average relative volume difference of 0.242 and 0.375 for the myocardial infarction and no-reflow, respectively.

Table 1. Quantitative segmentation results on the EMIDEC dataset (15 validation sets). Values are expressed in mean ± standard deviation (std), minimum (Min.), and maximum (Max.).

Target	Metric								
	DSC			HD(in mm)			RVD		
	Mean ± std	Min	Max	Mean ± std	Min	Max	Mean ± std	Min	Max
LV	0.93±0.018	0.90	0.96	6.5±3.7	3.9	19.2	0.06±0.05	0.003	0.20
MYO	0.84±0.03	0.75	0.88	8.9±5.3	4.9	25.9	0.08±0.08	0.004	0.27

The proposed method showed promising performance to segment the anatomical structures such as the myocardium and the left ventricular cavity from DE-MRI. To extend this approach, shape prior based deep learning methods could help constrain the segmentation of the anatomical structures [9–11]. Post processing methods using convolutional auto-encoders [12,13] could also constrain the segmentation process by guaranteeing the correct shape of the anatomical structures. Moreover, the clinical information could be used to constrain the segmentation process of the pathological areas.

Input MRI   LV/MYO Prediction   MYO/Scar Prediction   Mask Prediction   Ground Truth

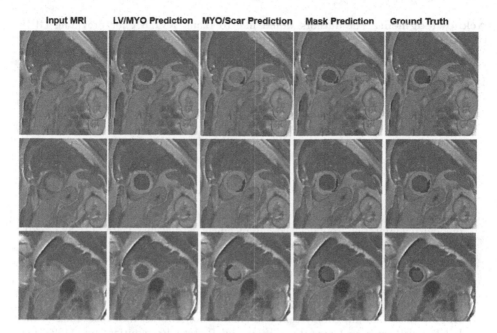

Fig. 3. Examples of segmentation results on the EMIDEC validation set. [Column 1] Input DE-MRI; [Column 2] Predicted left ventricular cavity (LV) and myocardium (MYO); [Column 3] Predicted normal MYO areas and myocardium infarction (MI) areas; [Column 4] All predicted structures including the LV, MYO, and MI; [Column 5] Ground truth level from expert. The red, green, and blue colors respectively show the LV, the MYO, and the MI. (Color figure online)

4   Discussion and Conclusion

In this paper, we proposed a deep learning framework to automatically segment the left ventricular cavity and the myocardium regions from DE-MRI. We also proposed a deep learning framework to automatically detect pathological areas, particularly the myocardium infarction and the no-reflow areas, from the DE-MRI. Moreover, the classification of a given case into pathological or normal was performed using the provided clinical information as well as using the DE-MRI.

Experimental results, obtained using the EMIDEC MICCAI 2020 challenge dataset, revealed that the proposed framework can accurately segment the heart structures. Moreover, our method can detect pathological areas. Besides, the classification of a given case can be performed using either clinical information or DE-MRI, which could benefit practitioners in assessing the state of the heart. However, the classification accuracy from DE-MRI seems to outperform the approach using only the clinical information. Next, shape prior based methods will be explored to constrain the segmentation process of anatomical structures such as the myocardium and the left ventricular cavity. The clinical information could also be used to constrain the pathological areas' segmentation process.

Acknowledgment. The authors would like to thank NVIDIA for providing GPU (NVIDIA TITAN Xp, 12 GB) through their GPU grant program.

References

1. Mozaffarian, D.: Global scourge of cardiovascular disease: time for health care systems reform and precision population health. Journal of the American College of Cardiology (2017). https://doi.org/10.1016/j.jacc.2017.05.007
2. Girum, K.B., Créhange, G., Hussain, R., Lalande, A.: Fast interactive medical image segmentation with weakly supervised deep learning method. Int. J. Comput. Assist. Radiol. Surg. **15**(9), 1437–1444 (2020). https://doi.org/10.1007/s11548-020-02223-x
3. Lalande, A., et al.: Emidec: a database usable for the automatic evaluation of myocardial infarction from delayed-enhancement cardiac mri. Data **5**(4), 89 (2020). https://doi.org/10.3390/data5040089
4. Ronneberger, O., Fischer, P., Brox, T.: U-Net: convolutional networks for biomedical image segmentation. In: Navab, N., Hornegger, J., Wells, W.M., Frangi, A.F. (eds.) MICCAI 2015. LNCS, vol. 9351, pp. 234–241. Springer, Cham (2015). https://doi.org/10.1007/978-3-319-24574-4_28
5. Hu, J., Shen, L., Sun, G.: Squeeze-and-excitation networks. In: Proceedings of the IEEE Conference on Computer Vision and Pattern Recognition, pp. 7132–7141 (2018). https://doi.org/10.1109/CVPR.2018.00745
6. Bishop, C.M.: Pattern Recognition and Machine Learning, 5th Ed. Springer (2007) ISBN: 9780387310732.
7. Kingma, D.P., Ba, J.: Adam: a method for stochastic optimization. arXiv preprint arXiv:1412.6980 (2014) arXiv:1412.6980
8. He, K., Zhang, X., Ren, S., Sun, J.: Delving deep into rectifiers: surpassing human-level performance on imagenet classification. In: Proceedings of the IEEE International Conference on Computer Vision, pp. 1026–1034 (2015). https://doi.org/10.1109/ICCV.2015.123
9. Girum, K.B., Lalande, A., Hussain, R., Créhange, G.: A deep learning method for real-time intraoperative us image segmentation in prostate brachytherapy. Int. J. Comput. Assist. Radiol. Surg. **15**(9), 1467–1476 (2020). https://doi.org/10.1007/s11548-020-02231-x
10. Oktay, O., et al.: Anatomically constrained neural networks (acnns): application to cardiac image enhancement and segmentation. IEEE Trans. Med. Imaging **37**(2), 384–395 (2017). https://doi.org/10.1109/TMI.2017.2743464
11. Girum, K.B., Créhange, G., Hussain, R., Walker, P.M., Lalande, A.: Deep generative model-driven multimodal prostate segmentation in radiotherapy. In: Workshop on Artificial Intelligence in Radiation Therapy, pp. 119–127 (2019). https://doi.org/10.1007/978-3-030-32486-5_15
12. Larrazabal, A.J., Martínez, C., Glocker, B., Ferrante, E.: Post-dae: anatomically plausible segmentation via post-processing with denoising autoencoders. IEEE Trans. Med. Imaging **39**(12), 3813–3820 (2020). https://doi.org/10.1109/TMI.2020.3005297
13. Painchaud, N., Skandarani, Y., Judge, T., Bernard, O., Lalande, A., Jodoin, P.M.: Cardiac segmentation with strong anatomical guarantees. IEEE Trans. Med. Imaging **39**(11), 3703–3713 (2020). https://doi.org/10.1109/TMI.2020.3003240

Uncertainty-Based Segmentation of Myocardial Infarction Areas on Cardiac MR Images

Robin Camarasa[1,3]($\boxtimes$), Alexis Faure[1,2,3], Thomas Crozier[1,2,3], Daniel Bos[3,4], and Marleen de Bruijne[1,3,5]

[1] Biomedical Imaging Group Rotterdam, Erasmus MC, Rotterdam, The Netherlands
{r.camarasa,marleen.debruijne}@erasmusmc.nl
[2] Ecole des Mines de Saint-Etienne, Saint-Etienne, France
{t.crozier,a.faure}@etu.emse.fr
[3] Department of Radiology and Nuclear Medicine, Erasmus MC,
Rotterdam, The Netherlands
d.bos@erasmusmc.nl
[4] Department of Epidemiology, Erasmus MC, Rotterdam, The Netherlands
[5] Department of Computer Science, University of Copenhagen,
Copenhagen, Denmark

Abstract. Every segmentation task is uncertain due to image resolution, artefacts, annotation protocol etc. Propagating those uncertainties in a segmentation pipeline can improve the segmentation. This article aims to assess if segmentation can benefit from uncertainty of an auxiliary unsupervised task - the reconstruction of the input image. This auxillary task could help the network focus on rare examples that are otherwise poorly segmented. The method was applied to segmentation of myocardial infarction areas on cardiac magnetic resonance images.

1 Introduction

Medical imaging knows a rising interest in Bayesian deep learning. This rising interest can be observed in the organization of specialized workshops [3] and dedicated tasks in challenges [6]. The popularization of Bayesian techniques to medical imaging can be mostly attributed to the work of Gal et al. [2], in which the authors describe Monte-Carlo dropout, a simple technique to make a neural network Bayesian using dropout at test time.

Mehta et al. [5] showed that propagating uncertainties in cascading networks can lead to improvement of downstream tasks. Their method consists of two cascading U-nets: the first one uses Monte-Carlo dropout to generate the uncertainty map of the segmentation of an Magnetic Resonance (MR) image, and the second one takes the MR image and this uncertainty map as input to output a segmentation. The method proposed in this article is based on the same principle

---

A. Faure and T. Crozier—Equally contributed.

© Springer Nature Switzerland AG 2021
E. Puyol Anton et al. (Eds.): STACOM 2020, LNCS 12592, pp. 385–391, 2021.
https://doi.org/10.1007/978-3-030-68107-4_40

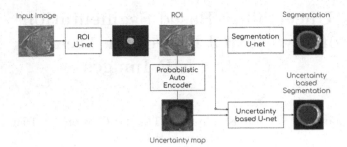

Fig. 1. Method overview

of cascading networks. However, it uses the uncertainty maps of an unsupervised auxiliary task instead of the uncertainty map of a supervised segmentation. This option offers more possibilities as the auxiliary task does not require annotations.

To the best of our knowledge, this is the first work, that investigates the benefit of applying propagation of uncertainties to the segmentation of cardiac MR images.

2 Dataset

The presented methods were developed on the hundred Delayed Enhancement Magnetic Resonance (DEMR) exams of the segmentation dataset of the EMIDEC challenge [4]. From this data only the DEMR exams and the voxel-wise annotations of the five classes (background, cavity, normal myocardium, myocardial infarction and no-reflow) were used to tune the methods.

3 Methods

3.1 Method Overview

Two methods, described in Fig. 1, are proposed for this challenge :

- **Baseline segmentation :** First a region of interest (ROI) U-net selects the ROI, a tile of shape $[n_x, n_y, n_z]$ centered on the four non-background classes. Then, a segmentation U-net segments the four non-background classes from the ROI. The selection of the ROI allows the segmentation U-net to only focus on the segmentation of those four classes.
- **Uncertainty-based segmentation :** The ROI is determined as in the baseline segmentation, using the ROI U-net. Then a probabilistic Auto-Encoder generates an uncertainty map of the reconstruction of the ROI. Finally, an uncertainty-based U-net segments the ROI from the uncertainty map and the ROI.

3.2   U-Net

The different U-nets [8], described in this article, have n_{in} input channels, n_{out} output channels and depends on the number of features n_f as in Fig. 2. Note that due to the low resolution of the DEMR images in the z-axis compared to the other directions, the max-pooling and up-sampling are 2D instead of 3D [1].

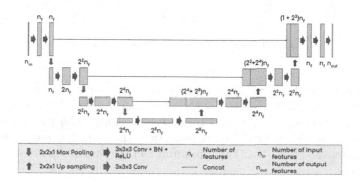

Fig. 2. U-Net structure

3.3   Probabilistic Auto-Encoder

The probabilistic Auto-Encoder, described in Fig. 3, uses Monte-Carlo dropout with a dropout rate p_d to compute the uncertainty map corresponding to ROI reconstructio [2]. Note that, similarly to the U-net, the up-sampling and max-pooling are 2D. To obtain several estimates of the reconstruction at test time, we sample $T = 25$ sets of parameters $(\theta_1, ..., \theta_T)$. From those T set of parameters, T outputs, $(f^{\theta_1}(x^t), ..., f^{\theta_T}(x^t))$, represent a sample of the region of interest reconstruction distribution $q(\hat{x}^t|x^t)$. From this sample, one can derive the variance of the output probabilities at voxel level, which is referred later on as the uncertainty map, as

$$\text{Var}(q(\hat{x}_j^t|x^t)) = \frac{1}{T}\sum_{t=1}^{T} f_j^{\theta_t}(x^t)^2 - \mathbb{E}((q(\hat{x}_j^t|x^t))^2 \qquad (1)$$

where f^θ is the Auto-Encoder with parameters θ, $q(\hat{x}_j^t|x)$ is the reconstructed distribution of the voxel j.

3.4   Data-Augmentation

On the fly, three types of data-augmentation are performed in the following order: rotations, elastic deformations and flips. The rotation applied along the z-axis depends on an angle θ randomly sampled from $\mathcal{U}(\theta_{min}, \theta_{max})$. The random elastic deformation depends on a displacement grid of shape $[g_x, g_y, g_z]$ generated

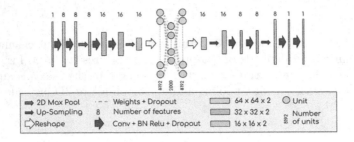

Fig. 3. Probabilistic Auto-Encoder structure

via the module elasticdeform[3]. The values of the component of each displacement vector v_x, v_y and v_z are respectively sampled from the distribution $\mathcal{U}(-d_x, d_x)$, $\mathcal{U}(-d_y, d_y)$, and $\mathcal{U}(-d_z, d_z)$. Finally random flips along x and y axis are applied with a probability p. After data-augmentation, the resulting image intensities are linearly rescaled such that the minimum and the maximum are set to 0 and 1 respectively.

3.5   ROI U-Net

This U-net is trained to segment a binary mask of the myocardial areas. Then the center of mass of the four largest connected components of the prediction determine the predicted center of the ROI. All input DEMR images are rescaled to the network input size $[n_x, n_y, n_z]$, using linear interpolation.

3.6   Segmentation U-Net

This U-net is trained on tiles of shape $[n_x, n_y, n_z]$ that contain the segmentation. If n_z is smaller than the z-dimension of the image is padded with zeros. At train time, a proportion p_c of the training sample have the ground truth segmentation centered.

3.7   Uncertainty Map Based U-Net

This network is comparable to the segmentation U-net but takes a tile of shape $[n_x, n_y, n_z]$ of the input image and of the uncertainty map as input. The padding strategy is also the same.

3.8   Parameters

The parameters and the implementation details of the networks are described in Table 1. Note that baseline U-net uses elastic deformations where uncertainty-based U-net did not.

---

[3] https://github.com/gvtulder/elasticdeform.

Table 1. Implementation details

Parameters	ROI U-Net	Auto-Encoder	Baseline U-Net	Uncertainty-based U-Net
n_{in}	1	1	1	2
n_{out}	1	1	5	5
n_f	8	8	16	16
Batch size	5	5	5	5
Loss	Dice [7]	$\mathcal{L}_1$	Averaged dice [7]	Averaged dice [7]
Optimizer	Adadelta [9]	Adadelta [9]	Adadelta [9]	Adadelta [9]
$[n_x, n_y, n_z]$	[104, 176, 8]	[64, 64, 2]	[64, 64, 8]	[64, 64, 8]
$[g_x, g_y, g_z]$			[20, 20, 3]	
$[d_x, d_y, d_z]$			[1, 1, 0.1]	
$[\theta_{min}, \theta_{max}]$			[-10, 10]	
p	0.5	0.5	0.5	0.5
p_d		0.1		
p_c			0.5	0.5

3.9   Ensemble

The training dataset is randomly split in four subsets. Then, for all networks, four models are trained, each using three of the subsets as training set and the remaining subset as validation set. Segmentations on the test images are then obtained as the voxel-wise mean of the outputs of these four models.

4   Results

Challenge organizers evaluated both the uncertainty-based and the baseline method on an independent test set of fifty patients. The evaluated metrics vary from class to class: Dice index, Haussdorf distance and volume difference assessed the quality of the segmentation of myocardium; Dice index, volume difference and volume difference ratio according to volume of myocardium assessed the quality of the segmentation of infarction and no-reflow areas (Table 2). Both uncertainty-based and baseline method segment the myocardium better than the no-reflow areas and the no-reflow areas better than infarction areas. For every metric, the baseline method out-performed the uncertainty-based method.

Table 2. Performance of the baseline and the uncertainty-based methods. (The best observed results per metrics are in bold).

	Baseline method	Uncertainty-based method
Myocardium		
Dice index (%)	**75.74**	68.86
Haussdorf distance (mm)	**25.44**	35.15
Volume difference (mm^3)	**17108.13**	19646.13
Infarction		
Dice index (%)	**30.79**	21.63
Volume difference (mm^3)	**4868.56**	7476.56
Volume difference ratio according to volume of myocardium (%)	**3.64**	6.09
No-reflow		
Dice index (%)	**60.52**	60.00
Volume difference (mm^3)	**867.86**	944.92
Volume difference ratio according to volume of myocardium (%)	**0.52**	0.57

5   Discussion and Conclusion

We developed and compared two methods to segment myocardial infarction areas using deep learning. Two elements explain partly the poor performances of the uncertainty-based method compared to the baseline method. First, the comparison is not entirely fair as the implementation of the data augmentation of the uncertainty-based method did not include elastic deformations (Table 1). Secondly, fine-tuning the probabilistic Auto-Encoder properly would have required more time (for some images the probabilistic Auto-Encoder reconstructed areas predicting a constant value). A more in depth analysis would be required to draw a definitive conclusion on the comparison of both methods.

Despite the performances, the uncertainty based method gives more information than the baseline method. Indeed, it provides a voxelwise out-of-distribution detection as the probabilistic Auto-Encoder should be uncertain in reconstructing the areas of an image that are different from areas observed at training time.

The challenge organizers ranked the baseline method: seventh out of seven for the segmentation of the myocardium, fifth out of seven for the segmentation of the infarcted area, third out of seven for the segmentation of the no-reflow area and sixth out of seven for the overall segmentation.

In conclusion, this article compares two deep learning methods on a dataset of cardiac MR images: a baseline method and an uncertainty based method. The experiments shows that the baseline method out-performs the uncertainty based one but the uncertainty based method could detect out-of-distribution images.

Acknowledgments. This work was partly funded by Netherlands Organisation for Scientific Research (NWO) VICI project VI.C.182.042.

References

1. Camarasa, R., et al.: Quantitative comparison of monte-carlo dropout uncertainty measures for multi-class segmentation. In: Sudre, C.H., et al. (eds.) UNSURE/GRAIL -2020. LNCS, vol. 12443, pp. 32–41. Springer, Cham (2020). https://doi.org/10.1007/978-3-030-60365-6_4
2. Gal, Y., Ghahramani, Z.: Dropout as a bayesian approximation: representing model uncertainty in deep learning. In: International Conference on Machine Learning, pp. 1050–1059 (2016)
3. Greenspan, H., et al.: CLIP/UNSURE -2019. LNCS, vol. 11840. Springer, Cham (2019). https://doi.org/10.1007/978-3-030-32689-0
4. Lalande, A., et al.: Emidec: a database usable for the automatic evaluation of myocardial infarction from delayed-enhancement cardiac MRI. Data 5(4), 89 (2020)
5. Mehta, R., Christinck, T., Nair, T., Lemaitre, P., Arnold, D., Arbel, T.: Propagating uncertainty across cascaded medical imaging tasks for improved deep learning inference. In: Greenspan, H., et al. (eds.) CLIP/UNSURE -2019. LNCS, vol. 11840, pp. 23–32. Springer, Cham (2019). https://doi.org/10.1007/978-3-030-32689-0_3
6. Mehta, R., Filos, A., Gal, Y., Arbel, T.: Uncertainty Evaluation Metric for Brain Tumour Segmentation. arXiv preprint arXiv:2005.14262 (2020)
7. Milletari, F., Navab, N., Ahmadi, S.A.: V-net: fully convolutional neural networks for volumetric medical image segmentation. In: 2016 Fourth International Conference on 3D Vision (3DV), pp. 565–571. IEEE (2016)
8. Ronneberger, O., Fischer, P., Brox, T.: U-Net: convolutional networks for biomedical image segmentation. In: Navab, N., Hornegger, J., Wells, W.M., Frangi, A.F. (eds.) MICCAI 2015. LNCS, vol. 9351, pp. 234–241. Springer, Cham (2015). https://doi.org/10.1007/978-3-319-24574-4_28
9. Zeiler, M.D.: Adadelta: an adaptive learning rate method. arXiv preprint arXiv:1212.5701 (2012)

Anatomy Prior Based U-net for Pathology Segmentation with Attention

Yuncheng Zhou, Ke Zhang, Xinzhe Luo, Sihan Wang, and Xiahai Zhuang[✉]

School of Data Science, Fudan University, Shanghai, China
{19210980093,20210980130,zxh}@fudan.edu.cn

Abstract. Pathological area segmentation in cardiac magnetic resonance (MR) images plays a vital role in the clinical diagnosis of cardiovascular diseases. Because of the irregular shape and small area, pathological segmentation has always been a challenging task. We propose an anatomy prior based framework, which combines the U-net segmentation network with the attention technique. Leveraging the fact that the pathology is inclusive, we propose a neighborhood penalty strategy to gauge the inclusion relationship between the myocardium and the myocardial infarction and no-reflow areas. This neighborhood penalty strategy can be applied to any two labels with inclusive relationships (such as the whole infarction and myocardium, etc.) to form a neighboring loss. The proposed framework is evaluated on the EMIDEC dataset. Results show that our framework is effective in pathological area segmentation.

Keywords: Pathology segmentation · Attention map · Cardiac MRI

1 Introduction

Cardiovascular diseases are the leading cause of death around the world [3]. Automated segmentation of cardiac magnetic resonance (CMR) image is an important step towards analyzing cardiac pathology on a large scale, and ultimately the development of diagnosis and treatment methods. Many pioneering works on cardiac MRI segmentation have been done, using traditional machine learning methods or deep learning methods. In [5,6,14], prior knowledge of statistical models is employed to achieve cardiac MRI segmentation. In the last few years, with the rapid improvement of convolutional neural networks (CNNs) in the area of computer vision, they have been proven to be the new state-of-the-arts [1,2,12,13].

Late gadolinium enhancement (LGE) MRI is a sophisticated imaging technique for myocardial infarction (MI) enhancement. Accurate segmentation of

This work was funded by the National Natural Science Foundation of China (Grant No. 61971142), and Shanghai Municipal Science and Technology Major Project (Grant No. 2017SHZDZX01).

Y. Zhou and K. Zhang—Contributed equally.

E. Puyol Anton et al. (Eds.): STACOM 2020, LNCS 12592, pp. 392–399, 2021.
https://doi.org/10.1007/978-3-030-68107-4_41

the infarction area in medical images has significant prognostic value for infarction diagnosis. However manual segmentation can be time-consuming and suffer from inter-observer heterogeneity. Hence, automated cardiac myocardial infarction segmentation is still desirable. However, the MI segmentation is a challenging task, for example, the effectiveness of the network could be limited due to low soft-tissue contrast, limited training data, irregularity of pathological targets, heterogeneous intensity distributions. Some deep learning-based segmentation methods have been used in the literature to perform MI segmentation [8–10]. One prominent method is the deep spatiotemporal adversarial network (DST-GAN). The deep DSTGAN leverages a conditional generative model, which conditions the distributions of the objective CMR image to directly optimize the generalized error of the mapping between the input and the output [8].

Our uniqueness is that we use U-net with attention and couple the U-net with neighborhood penalty strategy. Different from the conditional generative model-based segmentation framework, our method includes customized attention block [7]. The main contributions of this work are summarized as follows:

(1) We propose a scheme to combine our segmentation network with the attention blocks.
(2) We introduce a weight generator to automatically balance the gradient with different labels, the weight generator acts like an assistant and can be plugged to any existing framework (such as U-net, Dense-net, etc.).
(3) We propose a neighborhood penalty strategy to gauge the myocardium internal pathology inclusion relationship.

2   Method

In this section, we introduce our proposed anatomy prior based U-net for pathology segmentation with attention. The proposed method is based on a deep-learning framework and uses a U-net variant with attention block as the segmentation network. Firstly, we introduce the pathology mix-up augmentation method, which is applied to different myocardium of image pair. Secondly, we present the architecture of the segmentation network with attention block. Thirdly, we describe the proposed neighborhood penalty strategy and the derived neighboring loss.

Pathology Mixup. We introduce a data augmentation method particularly for segmentation of cardiac pathology. The method is based on the mix-up strategy [11] to interpolate between two images and their segmentation maps. Mix-up strategy assume that the corresponding probabilities of a single label are local linear functions of the intensity for a pixel in the images. Hence, in order to let the model be able to simulate such functions, a linearly interpolated probability map for each label was generated for a pair of images. Specifically, we randomly select two slices of similar z-positions as long as their segmentation maps. By treating one as the fixed image F and the other as the moving image M, we register the cardiac region in the moving image to the fixed image using affine transformation T including only translation and scaling. Denote the foreground center (LV + Myocardium) of the fixed and moving image as $c^F = (c_x^F, c_y^F)$ and $c^M = (c_x^M, c_y^M)$, respectively. The translation offset becomes $\Delta c = (c_x^F - c_x^M, c_y^F - c_y^M)$. Besides, denote the average distance from foreground pixels to the foreground center as d^F and d^M. The affine matrix we use for affine transformation is

$$\begin{pmatrix} s & 0 & c_x^F - s \cdot c_x^M \\ 0 & s & c_y^F - s \cdot c_y^M \\ 0 & 0 & 1 \end{pmatrix} \tag{1}$$

where $s = d^F/d^M$ is the scaling factor. The warped moving image $M \circ T$ and the fixed image F as well as their corresponding transformed segmentation maps are linearly interpolated to create an augmented image-segmentation pair within the foreground area, i.e.

$$(1 - \lambda)M \circ T + \lambda F, \tag{2}$$

where $\lambda \sim Uniform(0,1)$. The background area in the fixed image was used in the augmented image to preserve the outside information of the image. The process is illustrated in Fig. 1.

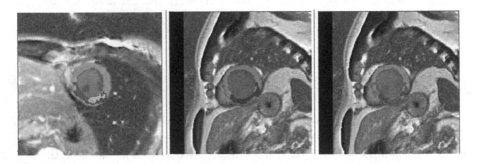

Fig. 1. The pathology mix-up process. The middle image is the fixed image, to which left moving image is registered. The right image indicates the mix-up augmentation result.

Network Architecture. We adopt the classical 2D U-net as the backbone. Motivated by the weighted cross-entropy loss which widely used in segmentation tasks, automated weighted cross-entropy loss was used, i.e.

$$L_{AWCE} = -\frac{1}{N} \sum_x \sum_{i=1}^{s} w_i Y_i(x) \log \hat{Y}_i(x), \tag{3}$$

where N is the number of all pixels, x represents a plixed in the image space, i stands for the segments while $\hat{Y}_i(x)$ represents the predicted probability for label i and $Y_i(x)$ takes 1 if the gold standard label of pixel x is i while it is set to 0 otherwise. To avoid redundant parameter adjustment, we formulate the hyperparameters w_i as an output of the deep neural network. Denote the input augmented image pair as X_A, Y_A, the proposed segmentation framework as S. S takes as input an image from X_A, which is first passed through separate sub-networks, weight generator, $S_{subnet,w}$ and segmentation generator, $S_{subnet,s}$. $S_{subnet,w}$, consists of 6 convolution layers, while $S_{subnet,s}$ consists of U-net encoder and decoder, interleaved with an attention block. The outputs of the two sub-networks are of the same channels. The output of $S_{subnet,w}$ is a weight vector w of dimension 5, which is used to calculate the weighted cross-entropy between the ground truth Y_A and the output segmentation result of $S_{subnet,s}$. Figure 2 presents the pipeline of the proposed anatomy prior-based segmentation network with attention.

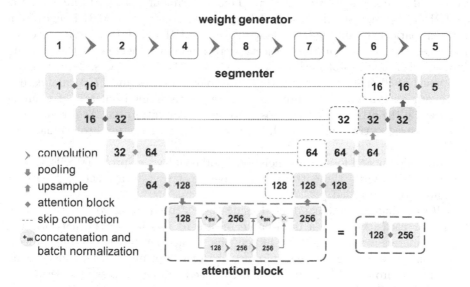

Fig. 2. The framework of anatomy prior based U-net with attention. Each diamond in the graph represents an attention block while the cross in the attention block means point-wise multiplication.

Neighborhood Penalty Strategy. Suppose that "the whole" denotes the entire tissue including normal and pathological areas. Leveraging the fact that the no-reflow is contained in the whole infarction, the whole infarction area is contained in the whole myocardium, We propose a neighboring penalty as a weak constraint strategy. Denote L_{NP} as the neighborhood penalty loss, a, b as the supporting function for the two labels, which maps the inner part of the region to 1 while the outer part 0. If a and b are not close together, that is $a * b < 1 - \varepsilon$ (where ε is a small given number), the neighboring penalty loss is formulated as:

$$L_{NP}(a, b) = mean(\mathbf{1}_{a>0} \times \mathbf{1}_{b>0} \times (1 - a - b)). \tag{4}$$

The penalty encourages two close regions to stick together and thus it prevents cracks. In the pathological segmentation of the cardiac MRI, the whole infarction represents the total area of labels for infarction and no-reflow, and the whole myocardium represents the total area of labels for myocardium and the whole infarction. Denote the derived neighboring loss as L_N, the no-reflow, infarction, myocardium labels as Y_{nf}, Y_{in}, Y_{myo} respectively, the integral neighboring loss is obtained,

$$L_N = L_{NP}(\hat{Y_{nf}}, \hat{Y_{nf}} + \hat{Y_{in}}) + L_{NP}(\hat{Y_{nf}} + \hat{Y_{inf}}, \hat{Y_{nf}} + \hat{Y_{inf}} + \hat{Y_{myo}}). \quad (5)$$

3 Experiment

Materials and Implementation. This experiment was performed on the EMIDEC challenge dataset [4], which comprises 150 cardiac MRI images from different patients including 50 cases with normal MRI and 100 cases with myocardial infarction, of which 100 subjects were randomly selected as the training dataset with 67 cases of pathological cases and 33 normal cases. Test data are composed of 50 other patients with 33 pathological cases and 17 normal cases.

In our experiments, the proposed models were trained on 90% of training slices while the other 10% was chosen to be the validation set. As a result, 639 slices are used as training data while the other 69 slices are used to evaluate the model.

For the proposed loss terms, neighborhood penalty was provided with a coefficient of 10^{-2} while the entropy loss's coefficient was always set to 1 with or without AWCE applied.

All experiments were conducted using the PyTorch framework with Python 3.7.4. For each iteration, a mini-batch of size 16 was fed to the network and the parameters were trained for 400 epochs. To accelerate the training process and avoid convergence to local minima, an Adam optimizer with a decreasing learning rate from 10^{-2} to 10^{-6} was used. The training process was performed on a NVIDIA GTX1080Ti GPU.

Ablation Study. Under a running environment with previously mentioned hard and software, a dense U-net was trained and evaluated. This model was designed after the proposed segmentation generator shown in Fig. 2. In this network, all auxiliary attention branches were removed from the proposed architecture. By using cross-entropy as the energy function, the optimal model was regarded as the baseline model.

A step-by-step increment of the algorithm was performed as an ablation study to prove the effectiveness of the implanted modules and techniques. First, we regard the attention mechanism as a good way to improve the segmentation of small and sensitive regions such as the infarcted areas. Specifically speaking, two convolutions were applied to the input feature maps of the U-net blocks and the output was normalized between 0 and 1 by the sigmoid function. This normalized output was then regarded as used to activate the output feature maps. Secondly, the automated weighted cross-entropy (Eq. 3) was used to automatically balance the gradient caused by different labels. The trained weight generator acts as an assistant to the main network but it does not affect the evaluation process. Thirdly, the neighborhood penalty was also added to the method to regularize the segmentation by shape prior. The coefficient for this penalty was set to 0.01 when the magnitude of the penalty is similar to that of the entropy loss.

Results and Discussion. Table 1 presents quantitative results of our experiments. The reported numbers are the mean Dice score coefficient (DSC) and the standard deviation on the validation set. The improvements can also be visually observed in Fig. 3. In the rest of this section, we discuss the results of the ablation study experiments. It can be observed that the proposed attention block method provides substantial improvements over dense U-net baseline. The proposed generator based weighted cross entropy method improves performance as compared to the U-net with attention method. These results show the benefit of encouraging the automatic label weighting to solve the label imbalance problem.

The no-reflow and infarcted areas are commonly difficult to be segmented due to its blurry boundary and its small area. However, the design of the weighted entropy solved the problem of small area while the attention blocks helped to refine the detailed segmentation boundary by decreasing the propagated gradients of unwanted areas. These techniques, as expected, improve the results.

As an additional experiment, we investigated the effect of including the neighboring loss from neighborhood penalty strategy. While the neighborhood penalty loss resulted in worse performance than the generator based weighted cross-entropy in general, it provides improvements for the most difficult no-reflow label. This shows that although the neighboring loss does not ensure the learning of accurate shape prior, it is still advantageous to enforce soft inclusive constraints on anatomy labels.

Table 1. DSC of segmentation on the validation slices. In this table, LV stands for left ventrical, Myo stands for myocardium while Inf and NoR stand for the infarcted area and no-reflow area respectively.

	LV	Myo	Inf	NoR
Baseline	0.955 ± 0.009	0.871 ± 0.045	0.622 ± 0.080	0.246 ± 0.102
Attention	0.962 ± 0.009	0.900 ± 0.033	0.718 ± 0.067	0.354 ± 0.130
Attention+AWCE+penalty	0.970 ± 0.007	0.916 ± 0.029	0.747 ± 0.082	$\mathbf{0.538 \pm 0.143}$
Attention+AWCE	$\mathbf{0.971 \pm 0.014}$	$\mathbf{0.926 \pm 0.029}$	$\mathbf{0.769 \pm 0.082}$	0.535 ± 0.153

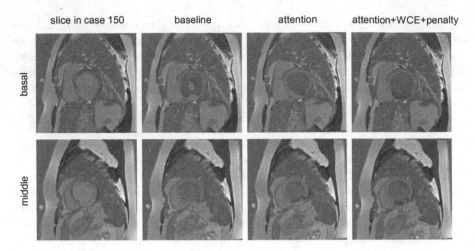

Fig. 3. Qualitative comparison of the ablation study experiments

4   Conclusion

We have proposed an anatomy prior based framework that consists of a weight generator and segmentation generator with attention block. Also, we have presented a neighborhood penalty strategy, which measures the inclusive relationship and acts as a weak constraint. The experiment demonstrates that the proposed framework can segment cardiac MRI images with pathology effectively. Our future research aims to optimize the augmentation methods and investigate our proposed framework in diverse training scenarios.

References

1. Bernard, O., et al.: Deep learning techniques for automatic MRI cardiac multi-structures segmentation and diagnosis: is the problem solved? IEEE Trans. Med. Imaging **37**(11), 2514–2525 (2018)
2. Duan, J., et al.: Automatic 3D bi-ventricular segmentation of cardiac images by a shape-refined multi-task deep learning approach. IEEE Trans. Med. Imaging **38**(9), 2151–2164 (2019)
3. Huertas-Vazquez, A., Leon-Mimila, P., Wang, J.: Relevance of multi-omics studies in cardiovascular diseases. Front. Cardiovasc. Med. **6**, 91 (2019)
4. Lalande, A., et al.: Emidec: a database usable for the automatic evaluation of myocardial infarction from delayed-enhancement cardiac MRI. Data **5**(4), 89 (2020)
5. Qian, X., Lin, Y., Zhao, Y., Wang, J., Liu, J., Zhuang, X.: Segmentation of myocardium from cardiac MR images using a novel dynamic programming based segmentation method. Med. Phys. **42**(3), 1424–1435 (2015)

6. Sun, H., Frangi, A.F., Wang, H., Sukno, F.M., Tobon-Gomez, C., Yushkevich, P.A.: Automatic cardiac MRI segmentation using a biventricular deformable medial model. In: Jiang, T., Navab, N., Pluim, J.P.W., Viergever, M.A. (eds.) MICCAI 2010, Part I. LNCS, vol. 6361, pp. 468–475. Springer, Heidelberg (2010). https://doi.org/10.1007/978-3-642-15705-9_57

7. Woo, S., Park, J., Lee, J.-Y., Kweon, I.S.: CBAM: convolutional block attention module. In: Ferrari, V., Hebert, M., Sminchisescu, C., Weiss, Y. (eds.) ECCV 2018, Part VII. LNCS, vol. 11211, pp. 3–19. Springer, Cham (2018). https://doi.org/10.1007/978-3-030-01234-2_1

8. Xu, C., Howey, J., Ohorodnyk, P., Roth, M., Zhang, H., Li, S.: Segmentation and quantification of infarction without contrast agents via spatiotemporal generative adversarial learning. Med. Image Anal. **59**, 101568 (2020)

9. Xu, C., Xu, L., Brahm, G., Zhang, H., Li, S.: MuTGAN: simultaneous segmentation and quantification of myocardial infarction without contrast agents via joint adversarial learning. In: Frangi, A.F., Schnabel, J.A., Davatzikos, C., Alberola-López, C., Fichtinger, G. (eds.) MICCAI 2018, Part II. LNCS, vol. 11071, pp. 525–534. Springer, Cham (2018). https://doi.org/10.1007/978-3-030-00934-2_59

10. Zhang, D., et al.: A multi-level convolutional LSTM model for the segmentation of left ventricle myocardium in infarcted porcine cine MR images. In: 2018 IEEE 15th International Symposium on Biomedical Imaging (ISBI 2018), pp. 470–473. IEEE (2018)

11. Zhang, H., Cisse, M., Dauphin, Y.N., Lopez-Paz, D.: mixup: beyond empirical risk minimization. arXiv preprint arXiv:1710.09412 (2017)

12. Zhuang, X.: Multivariate mixture model for cardiac segmentation from multi-sequence MRI. In: Ourselin, S., Joskowicz, L., Sabuncu, M.R., Unal, G., Wells, W. (eds.) MICCAI 2016, Part II. LNCS, vol. 9901, pp. 581–588. Springer, Cham (2016). https://doi.org/10.1007/978-3-319-46723-8_67

13. Zhuang, X.: Multivariate mixture model for myocardial segmentation combining multi-source images. IEEE Trans. Pattern Anal. Mach. Intell. **41**(12), 2933–2946 (2019)

14. Zhuang, X., Rhode, K.S., Razavi, R.S., Hawkes, D.J., Ourselin, S.: A registration-based propagation framework for automatic whole heart segmentation of cardiac mri. IEEE Trans. Med. Imaging **29**(9), 1612–1625 (2010)

Automatic Scar Segmentation from DE-MRI Using 2D Dilated UNet with Rotation-Based Augmentation

Xue Feng[1]([✉]) [iD], Christopher M. Kramer[2], Michael Salerno[1,2,3], and Craig H. Meyer[1,3]

[1] Biomedical Engineering, University of Virginia, Charlottesville, VA 22903, USA
xf4j@virginia.edu
[2] Medicine, University of Virginia, Charlottesville, VA 22903, USA
[3] Radiology & Medical Imaging, University of Virginia, Charlottesville, VA 22903, USA

Abstract. Delayed enhancement (DE) MRI is an important tool in diagnosis of cardiovascular disease as it can reveal different characteristics of the myocardium scars including infarction and no-reflow. Automatic segmentation of different regions has the advantage of improved accuracy and reduced inter-observer variability in quantifying key imaging biomarkers such as percentage of scars. In recent years deep learning has led to drastic performance improvement in automatic segmentation tasks using the UNet architecture. Cardiac MRI segmentation is a challenging task due to the high variability in imaging contrast, orientation and signal-to-noise ratio; specifically, for short-axis views, as they are double-oblique slices, due to the different orientations of the heart and operator choice, the overall appearance of the image can vary significantly, which poses a challenge to the neural networks as they cannot learn a consistent global anatomy. In this paper we developed a rotation-based augmentation to address this issue in both training and testing steps by eliminating the variance in orientations and demonstrated its effectiveness. 2D dilated UNet was used as the backbone network structure. On the test dataset, Dice scores of 0.933 for LV blood pool, 0.824 for myocardium, 0.578 for infarction and 0.697 for no re-flow regions were archived using the proposed framework.

Keywords: Scar segmentation · DE-MRI · LGE · Dilated UNet · Rotation-based augmentation

1 Introduction

Delayed enhancement (DE) MRI, also called late gadolinium enhancement (LGE) MRI, is a key MRI technique in cardiac applications. By injecting contrast agent and performing T1-sensitive acquisitions, myocardium regions with different characteristics can be clearly revealed including scars and no-flow regions. Furthermore, by segmenting the whole myocardium, the percentage of scar can be calculated, which is a key imaging biomarker in many diseases for diagnosis and risk stratification [1]. In recent years the

© Springer Nature Switzerland AG 2021
E. Puyol Anton et al. (Eds.): STACOM 2020, LNCS 12592, pp. 400–405, 2021.
https://doi.org/10.1007/978-3-030-68107-4_42

deep learning methods, represented by the UNet architecture [2], have led to drastic performance improvement in many medical image segmentation tasks. Cardiac MRI segmentation is a challenging task due to the high variability of imaging contrast, orientation and signal-to-noise ratio in the images. Specifically, as most sequences acquire the short-axis views of the heart and cover the heart with multiple slices, the overall appearance of the image such as orientations of the heart and locations of the chest wall can vary significantly. For example, some views will have the chest on the left while others will have it on the top. This poses a challenge for neural networks as they cannot learn the global anatomy as efficient as from images acquired on standard axial, coronal or sagittal views. In this study we proposed a rotation-based augmentation in both training and testing stages to address this challenge. The EMIDEC Segmentation Contest [3] provides an open platform with a large and diverse dataset to compare different algorithms. We used the dataset and validated our method in the scope of this contest.

2 Methods

For the DE-MRI segmentation task, the steps in our proposed method include preprocessing of the images, training and deployment of neural networks, and postprocessing. The target regions-of-interests (ROIs) include left ventricle (LV) blood pool, normal myocardium, myocardial infarction area and no-reflow area. Note that not all regions exist in every case as myocardial infraction and no-reflow represent pathological regions. Details are described as follows.

2.1 Image Pre-processing

Substantial image preprocessing was performed by the challenge organizers. Specifically, to address the shifts among different short-axis slices due to different breath holds, all slices were realigned based on the gravity center defined by the epicardial contour. In our pre-processing workflow, we further normalized the pixel spacings to be 1.367 × 1.367 mm^2 and center-cropped or zero-padded the resulting images to have the same matrix size as 224 × 224. All labels were processed using the same workflow to reduce the effect from resampling, each label was converted to binary maps and processed separately. Finally, each slice is normalized to have 0 mean and unit standard deviation to reduce variations in global intensity values.

2.2 Network Structure

Although the pre-processing to align different slices was performed, due to the very large slice thickness and the imperfect alignment, the spatial continuity along the slice direction is assumed to be small. Therefore, even if a 3D network can potentially take advantage of information from neighboring slices, it often leads to inferior performance compared with 2D counterparts. Compared with conventional convolutions, dilated convolutions can increase the receptive field without increasing the number of parameters and have shown advantages in many applications [4]. In cardiac MRI segmentation, we also observed that the dilated convolutions often outperform the conventional convolution when all other parameters are kept the same [5]. Therefore, we decided to use 2D UNet with dilated convolutions as the backbone network structure so that it has an increased robustness against misalignment of different slices as it only takes one slice to make segmentation.

For the final dilated 2D UNet, the number of feature maps at the first layer was set to 48 and was doubled for every encoding block comprising of 2 consecutive convolution layers, batch normalization and rectified linear unit activation layers. The weighted cross entropy plus soft Dice loss averaged over all ROIs without any weighting was used as the loss function to address class imbalance.

2.3 Rotation-Based Augmentation

Augmentation is a key step in training neural networks, especially in medical image applications as the dataset size is often small. As a main variability of the short-axis images is the orientation and in-plane rotation and the neural networks are not designed to be rotation-invariant but instead shift-invariant, we aimed to train a network that can ignore the differences in orientation by providing images that were randomly rotated within the full possible range (−180°–180°). Therefore, during training, we forced the network to ignore the image orientation but instead learn the contrast and anatomical relationships. Specifically, a random affine transformation matrix was generated based on the sampled translation, scaling, shearing and the rotating angle as just described. During testing, in addition to deploying the network to the original image, we also deployed it to images rotated at certain angles as a testing augmentation step. No other augmentations were applied. Practically, to balance the time it takes per image and the expected accuracy, we deployed to images rotated at 9 evenly distributed angles between −180° and 180° and averaged the resulting probabilities after rotating back to the original orientation. This was able to maximumly eliminate the impact from different short-axis orientations and reduce random errors with a single image. In addition to rotation-based augmentation, during training, random shift, scaling and shearing were also applied to emulate more images to further increase the generalizability of the network. In addition, as the myocardial infarction and no-reflow regions are much smaller than normal myocardium in the whole dataset, in each epoch, instead of looping over all images equally, we favored more slices that contain these two regions by looping over them 6 times per epoch to increase the sensitivity of the network to scar tissues.

2.4 Post-processing

As the network is trained end-to-end to directly generate all desired ROIs, no further post-processing was performed except for padding/cropping to the original field-of-view and resampling to the original pixel spacing. Similarly, the label for each ROI was performed separately to reduce any resampling errors.

2.5 Experiments

In model development, we randomly split the 100 training cases into 57 for training and the remaining 43 for internal validation. To evaluate the impact of the rotation-based augmentation, especially during testing, we compared the results with and without the proposed augmentation after the model was trained. Finally, we trained a model using all 100 cases and deployed it to the 50 testing cases with the rotation-based augmentations. All experiments were performed using a Nvidia Ttian Xp GPU with 12 Gb memory using the Tensorflow framework.

3 Results

3.1 Validation

Table 1 shows the calculated Dice scores with (second row) and without (first row) the rotation-based testing augmentation. The rotation-based testing augmentation significantly increased the performance, especially for the infarction and no re-flow regions.

Table 1. Dice scores with and without rotation-based testing augmentation

	LV blood pool	Myocardium	Infarction	No re-flow
No augmentation	0.929	0.813	0.477	0.624
With augmentation	0.933	0.824	0.578	0.697

Figure 1 shows the deployed segmentation results of the apical, mid and basal slices on one case with the ground truth labels (top row), labels without the testing augmentation (middle row) and labels with the testing augmentation (bottom row). The Dice scores increased from 0.891 to 0.903 for LV blood pool and from 0.705 to 0.727 for myocardium with the testing augmentation. However, the infarction and no-reflow regions still have relatively low Dice scores even with rotation augmentation at 0.395 for infarction and 0.140 for no-reflow, indicating that human verification is needed to confirm the automatic results.

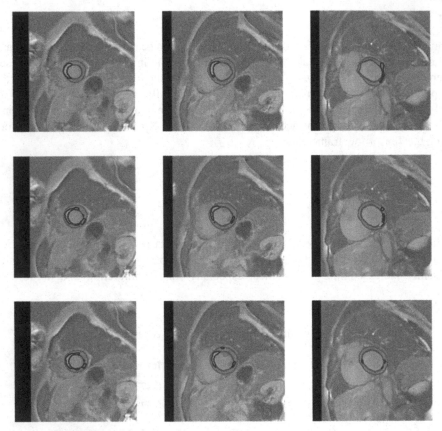

Fig. 1. Automatic segmentation results on one validation case showing the ground truth labels (top row), automatic segmentation without testing augmentation (middle row) and segmentation with testing augmentation (bottom row).

3.2 Testing

The performance on the testing data after submitting to the challenge organizer was given in Table 2.

Table 2. Evaluation metrics on the testing dataset

	Myocardium	Infarction	No re-flow
Dice scores	0.836	0.547	0.722
Volume difference (mm^3)	15187.48	3970.73	883.42
Hausdorff distance (mm)	33.77	N/A	N/A
Ratio difference	N/A	2.89%	0.53%

4 Conclusion

In this paper we developed an automatic DE-MRI segmentation method using dilated 2D UNet with the rotation-based augmentation during training and testing. The effectiveness of rotation in overcoming the varied short-axis orientations was demonstrated in the experiments performed without and with the rotation.

One disadvantage of the rotation augmentation during testing is that it significantly prolongs the deployment time. However, as we adopted the 2D UNet with only 3 encoding blocks and run on a relatively fast GPU, the deployment time is not a major concern.

The final Dice scores for the myocardium, infraction and no-reflow regions were 0.833, 0.547 and 0.722. Although the Dice score for myocardium is high, the pathological regions may not be segmented accurately, especially for infarction region with a mean ratio difference of 2.89%, which may affect the risk stratification if directly using the automatic segmentation results. Therefore, future studies are needed to continue to improve the robustness and accuracy of the model.

References

1. Chan, R.H., et al.: Prognostic value of quantitative contrast-enhanced cardiovascular magnetic resonance for the evaluation of sudden death risk in patients with hypertrophic cardiomyopathy. Circulation **130**(6), 484–495 (2014)
2. Ronneberger, O., et al.: U-Net: Convolutional Networks for Biomedical Image Segmentation. arXiv:1505.04597
3. Lalande, A., et al.: Emidec: a database usable for the automatic evaluation of myocardial infarction from delayed-enhancement cardiac MRI. Data **5**, 89 (2020)
4. Zhou, X.-Y., et al.: ACNN: a Full Resolution DCNN for Medical Image Segmentation. arXiv: 1901.09203
5. Simantiris, G., Tziritas, G.: Cardiac MRI segmentation with a dilated CNN incorporating domain-specific constraints. IEEE J. Sel. Top. Sign. Process. **14**(6), 1235–1243 (2020). https://doi.org/10.1109/JSTSP.2020.3013351

Classification of Pathological Cases of Myocardial Infarction Using Convolutional Neural Network and Random Forest

Jixi Shi[1,2] (✉), Zhihao Chen[1], and Raphaël Couturier[1]

[1] FEMTO-ST Institute, UMR 6174 CNRS, Univ. Bourgogne Franche-Comté, 90000 Belfort, France
[2] ESIGELEC, UNIROUEN, Univ. Normandie, 76000 Rouen, France
jixi.shi@groupe-esigelec.org

Abstract. Myocardial infarction is one of the most common cardiovascular diseases. Clinical information and Delayed Enhancement cardiac MRI (DE-MRI) are crucial to diagnose the myocardial infarction. However, some discrepancies can occur between clinical characteristics and DE-MRI when the disease is diagnosed. In order to deal in an efficient way with the correlation between these data and to be able to automatically classify patients suffering from myocardial infarction, this paper proposes a mixed classification model that takes both the clinical characteristics and DE-MRI into account. In the mixed model, a 3D Convolutional Neural Network (CNN) encodes the MRI as the surface of infarction then the surface is fed with Random Forest and other clinical characteristics to make the final decision.

Keywords: Classification · MRI · Myocardial infarction · CNN · Random Forest

1 Introduction

Myocardial infarction has become one of the most common cardiovascular disease. A clinical diagnosis of physiological data followed by the Delayed Enhancement MRI (DE-MRI) exam [9] has become the clinical routine to identify myocardial infarction. Clinical diagnosis including, for example the troponin level from the blood test and the ST segment from the electrocardiography provides a quick evaluation result for the acute myocardial infarction. For patients likely to suffer from this pathology and whose clinical information reveals the possibility of myocardial infarction, the DE-MRI exam is then organized to get a more robust diagnosis.

This paper responds to the call of EMIDEC, a challenge of the workshop STACOM, co-organized with MICCAI 2020. One of the EMIDEC challenge's contest consists in classifying if a patient has myocardial infarction according to

© Springer Nature Switzerland AG 2021
E. Puyol Anton et al. (Eds.): STACOM 2020, LNCS 12592, pp. 406–413, 2021.
https://doi.org/10.1007/978-3-030-68107-4_43

the clinical physiological data and DE-MRI. To achieve this goal, a two-stage method is proposed which interprets the semantic information from both the physiological data and MRI to strengthen the classification robustness.

In the first stage of the proposal, a 3D Convolutional Neural Network (CNN) predicts the volume of the myocardial infarction from the DE-MRI. The predicted volume is then merged with the physiological features and takes part in the final classification through Random Forest in the second stage.

The paper is organized as follows: introduction of the data including the clinical features and the DE-MRI; presentation of our methods followed by comparative results; discussion and conclusion.

2   Data

The dataset for the EMIDEC challenge [5] consists of 150 exams including DE-MRI associated with clinical physiological data. The training set consisting of 100 exams were publicly available for both the clinical data and the ground truth several weeks before the day of challenge. On the day of the challenge, 50 other cases were provided only with clinical data as the test set. The purpose of the data is to diagnose if a patient has acute myocardial infarction. In the entire dataset, 1/3 of the cases are normal and 2/3 of the cases are pathological. Challenge organizers chose to keep this unbalanced distribution to better reproduce the real clinical conditions of the cardiac emergency department.

The clinical physiological information includes 12 features. Each feature's value is either categorical, Boolean or floating point number. DE-MRI exams include on average 7 slices per case. A DE-MRI study includes the entire left ventricle myocardium from apex to base. The pixel spacing between slices (axis Z) is larger than the one on the X-Y plan. Manual segmentations were provided to challengers for the training set. Each segmentation consists of the normal and the pathological cardiac tissues including the cavity, myocardium, myocardial infarct and microvascular obstruction (MVO, or no-reflow, a subclass of the infarct). A patient that shows signs of infarction in its DE-MRI is considered to be pathologically positive.

3   Method

Each DE-MRI exam has multiple slices of 2D image but the clinical physiological information includes 12 pieces of 1D feature. In order to handle these data that have different dimensions and semantic information for the classification of the myocardial infarction, the proposal contains two stages as showed in Fig. 1: the encoding of DE-MRI then the classification on the fusion of encoded images and their paired clinical physiological features. The image encoding is realized by a 3D CNN and the classification of the myocardial infarction is done by Random Forest [8]. This conception aims at taking the advantage of the correlation between both types of data so that the classification result is more robust than on one single type of data.

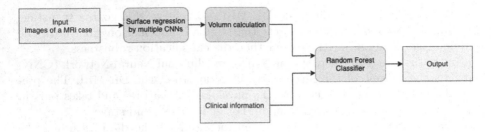

Fig. 1. Overview of our proposed architecture of prediction combining CNNs and Random Forest. A set of three-slice images come from the same MRI case are fed successively to the CNN. The output of the CNN is the regression of the predicted infarct surface of the middle slice in the input. Volume is calculated based on the predicted surfaces and the provided voxel spacing. Predicted volume and the clinical physiological information are concatenated as the input of the Random Forest classifier for the final classification.

3.1   Surface Regression by CNN

First, a 3D CNN is proposed to encode the MRI. The input of the 3D CNN is a three-layer MRI and the output is the predicted surface of the infarct. Since each MRI case can have different numbers of slices, to ensure that each CNN's input has a fixed shape, the image preprocessing was firstly investigated. To match up each three-slice input, an optimized 3D CNN inspired by U-Net [6] was then proposed.

3.2   Image Preprocessing

In order to fully catch semantic information from the DE-MRI, the image preprocessing was executed on each MRI case. To learn the spatial information between adjacent slices and ensure a fixed-size of CNN's inputs, three successive slices were taken as a single 3D input for the CNN. Assuming that an original MRI has N slices, the first and the last slice of the MRI were copied at the top and at the bottom side, hence N new 3D images were obtained and each 3D image was formed by three adjacent slices. Knowing that in the dataset the left ventricle myocardium is centred on the middle of each slice, to reduce the background's size, a centre cropping of size (96, 96) was performed on each slice. Therefore, each CNN's input has the same shape of (3, 96, 96). Figure 2 illustrates the way that the three-slice inputs were created. A 3D input could have had more slices. However, three adjacent slices were sufficient to provide enough spatial information. With more slices, more bottom and top slices should be copied, which was not efficient for the surface regression.

3.3   CNN with 3D Multi-kernel Convolution Block

The 3D convolution was added only at the first layer in our CNN since each input had only three slices. To expand the receptive field, multiple 3D convolution kernels of the size (3, 3, 3), (3, 5, 5) and (3, 7, 7) encoded the input image in parallel. This conception was inspired by Inception structures [7] and the objective was to flexibly extract features for objects of various sizes. For the 3D convolutions, the zero padding was performed only at the width and length dimension in light of the thickness of 3 at the dimension Z. Hence, the feature maps generated by each 3D kernel are in the same size of (96, 96).

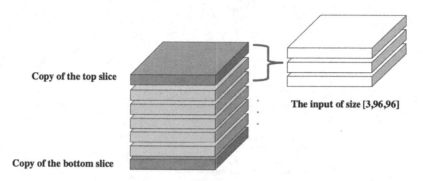

Fig. 2. Preparation of three-slice input for the 3D CNN. For each centre-cropped MRI case, firstly the top and the bottom slice are copied (darkgray). Secondly every three consecutive slices are chosen to form a 3D input for the CNN.

The 2D feature maps obtained by the 3D multi-kernel convolutional layer was then passed to residual modules which is similar to ResNet [3]. Each residual module performed 4 times of convolution + down-sampling + batch normalization [4] + ReLU activation on feature maps. To reinforce the semantic information interpretation, the Dense Atrous Convolution (DAC) block [2] was added at the last layer before the fully connected layers, motivated by the Inception-ResNet-V2 block [7] and atrous convolution. DAC had four cascade branches with the gradual increment of the number of atrous convolution, from 1 to 1, 3, and 5. Therefore, the network could extract high-level semantic information of different scales.

At the end of the CNN, the surface of the pathological tissue was predicted through the 2 fully connected layers as the CNN's final output. Smooth L1 loss was applied to penalize the error between the prediction and the ground truth (Fig. 3).

3.4   Volume Calculation

The output of the CNN is the predicted surface of the infarct in one MRI slice. In order to calculate the predicted volume of the infarct in one MRI case, the sum of the surfaces multiplying the pixel spacing (provided as the MRI meta data) was calculated. The predicted volume was used as an additional feature of the subsequent Random Forest model for the definitive classification of the myocardial infarction (Fig. 4).

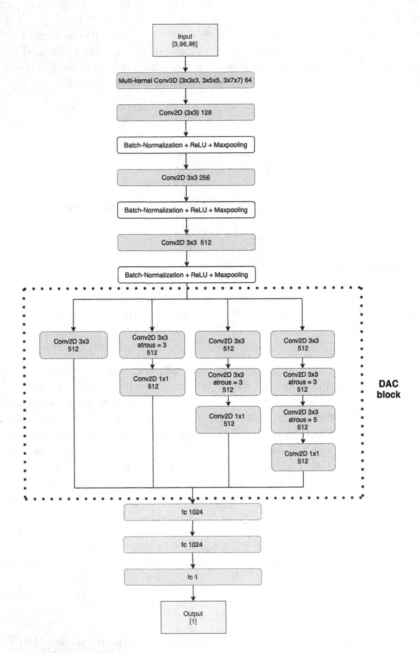

Fig. 3. The structure of the neural network. The 3D convolution is adopted only at the first layer. Before the fully connected layers, the DAC block enhances the cognition for both large and small areas.

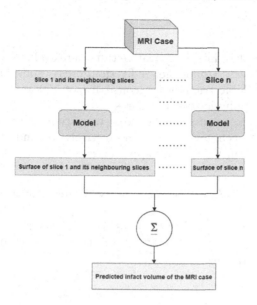

Fig. 4. The algorithm of volume calculation. The CNN predicts the surface of each slice. The predicted volume of a MRI case is calculated based on the predicted surfaces and the voxel spacing provided by the MRI meta data.

3.5   Random Forest Classifier

Random forest, developed by Breiman [1], is a classification algorithm that uses the ensemble of classification trees. Each of the classification trees is built using a bootstrap sample of the data, and at each split the candidate set of variables is a random subset of the variables. Thus, random forest uses both bagging (Bootstrap Aggregation), a successful approach for combining unstable learners, and random variable selection for tree building.

As the predicted volume of infarct from the CNN was obtained during the first stage, at the second stage, the predicted volume was concatenated to the 12 clinical physiological features. The Random Forest was trained on these 13 features and the output was binary that indicated if the case was pathological or not.

4   Implementation Details

The CNN network was implemented in PyTorch. The CNN was trained 500 epochs and the predicted volume is the ensemble of multiple models' prediction of different epochs. To show the advantage of the classification on merged MRI and clinical physiological information, the classifications made only on the CNN and only on Random Forest were also performed. The comparative tests used the same method from stage one or stage two and the data repartition of the cross-validation was identical to that of the above two-stage method, which can be considered as the baseline approaches.

5   Result

The proposed method and its comparative tests were evaluated on the dataset of MICCAI EMIDEC 2020 Challenge. A five-fold cross-validation was performed on the publicly available training set consisting of 100 annotated MRI exams and matched clinical physiological features.

Table 1 details the results of the cross-validation. The two-stage method achieved 95% ± 3% accuracy, which is respectively 4% and 8% superior to the CNN only and Random Forest only approaches.

Table 1. Classification accuracy of three approaches

	Random Forest	CNN	Random Forest + CNN
Accuracy (%)	87	91	95
Standard deviation (%)	3	2	3

6   Conclusion

In this article, a two-stage machine learning framework for the myocardial infarction classification is proposed through the clinical physiological information and DE-MRI exam. A 3D CNN extracts the plane and spatial features from the images, and a Random Forest classifier combines the encoded image feature and physiological information to classify if the case is pathological or not. Our method shows a significant improvement of the classification accuracy on the training dataset compared to the single stage methods. Moreover, our two-stage approach could be applied for other types of disease diagnosis prediction containing significantly different data.

Acknowledgements. This work was partly supported by the ADVANCES project founded by ISITE-BFC project (number ANR-15-IDEX-0003) and by the EIPHI Graduate School (contract ANR-17-EURE-0002). We also thank the Mesocentre of Franche-Comté for the computing facilities.

References

1. Breiman, L.: Random forests. Mach. Learn. **45**(1), 5–32 (2001)
2. Gu, Z., et al.: CE-Net: context encoder network for 2D medical image segmentation. IEEE Trans. Med. Imaging **38**(10), 2281–2292 (2019)
3. He, K., Zhang, X., Ren, S., Sun, J.: Deep residual learning for image recognition. In: Proceedings of the IEEE Conference on Computer Vision and Pattern Recognition, pp. 770–778 (2016)
4. Ioffe, S., Szegedy, C.: Batch normalization: accelerating deep network training by reducing internal covariate shift. Proc. Mach. Learn. Res. **37**, 448–456 (2015)

5. Lalande, A., et al.: Emidec: a database usable for the automatic evaluation of myocardial infarction from delayed-enhancement cardiac MRI. Data 5(4), 89 (2020)
6. Ronneberger, O., Fischer, P., Brox, T.: U-Net: convolutional networks for biomedical image segmentation. In: Navab, N., Hornegger, J., Wells, W.M., Frangi, A.F. (eds.) MICCAI 2015. LNCS, vol. 9351, pp. 234–241. Springer, Cham (2015). https://doi.org/10.1007/978-3-319-24574-4_28
7. Szegedy, C., Ioffe, S., Vanhoucke, V., Alemi, A.A.: Inception-v4, inception-ResNet and the impact of residual connections on learning. In: Proceedings of the Thirty-First AAAI Conference on Artificial Intelligence, pp. 4278–4284 (2017)
8. Than, M.P., et al.: Machine learning to predict the likelihood of acute myocardial infarction. Circulation 140(11), 899–909 (2019)
9. Vogel-Claussen, J., et al.: Delayed enhancement MR imaging: utility in myocardial assessment1. Radiographics 26, 795–810 (2006)

Author Index

Printed in the United States
By Bookmasters